VOLUME ONE HUNDRED AND TWENTY EIGHT

ADVANCES IN
COMPUTERS
Principles of Big Graph:
In-depth Insight

VOLUME ONE HUNDRED AND TWENTY EIGHT

ADVANCES IN
COMPUTERS

Principles of Big Graph:
In-depth Insight

Edited by

RIPON PATGIRI
Department of Computer Science and Engineering,
National Institute of Technology Silchar,
Silchar, Cachar, Assam, India

GANESH CHANDRA DEKA
Ministry of Skill Development and Entrepreneurship,
New Delhi, India

ANUPAM BISWAS
Department of Computer Science and Engineering,
National Institute of Technology Silchar,
Silchar, Assam, India

ACADEMIC PRESS
An imprint of Elsevier

ELSEVIER

Academic Press is an imprint of Elsevier
50 Hampshire Street, 5th Floor, Cambridge, MA 02139, United States
525 B Street, Suite 1650, San Diego, CA 92101, United States
The Boulevard, Langford Lane, Kidlington, Oxford OX5 1GB, United Kingdom
125 London Wall, London, EC2Y 5AS, United Kingdom

First edition 2023

ISBN: 978-0-323-89810-2
ISSN: 0065-2458

For information on all Academic Press publications
visit our website at https://www.elsevier.com/books-and-journals

Publisher: Zoe Kruze
Developmental Editor:
   Cindy Angelita Pe Benito-Gardose
Production Project Manager: James Selvam
Cover Designer: Alan Studholme

Typeset by STRAIVE, India

Working together
to grow libraries in
developing countries

www.elsevier.com • www.bookaid.org

Contents

7. MapReduce based convolutional graph neural networks: A comprehensive review 213

U. Kartheek Chandra Patnaik and Ripon Patgiri

8. Fast exact triangle counting in large graphs using SIMD acceleration 233

Kaushik Ravichandran, Akshara Subramaniasivam, P.S. Aishwarya, and N.S. Kumar

9. A comprehensive investigation on attack graphs 251

M. Franckie Singha and Ripon Patgiri

Contributors

P.S. Aishwarya
Department of Computer Science and Engineering, PES University, Bangalore, India

Mansaf Alam
Jamia Millia Islamia, New Delhi, India

Syed Arshad Ali
Jamia Millia Islamia, New Delhi, India

Chandan Tilak Bhunia
Durgapur Institute of Advanced Technology & Management, Durgapur, India

Anupam Biswas
Department of Computer Science and Engineering, National Institute of Technology Silchar, Silchar, Assam, India

Bhaskar Biswas
Department of Computer Science and Engineering, Indian Institute of Technology (BHU), Varanasi, India

Prasenjit Choudhury
Department of Computer Science and Engineering, National Institute of Technology, Durgapur, India

Soumita Das
Department of Computer Science and Engineering, National Institute of Technology Silchar, Silchar, Assam, India

Ravi Kishore Devarapalli
Department of Computer Science and Engineering, National Institute of Technology Silchar, Silchar, Assam, India

Fitsum Gebreegziabher
Department of Computer Science and Engineering, National Institute of Technology Silchar, Silchar, Cachar, Assam, India

Samiya Khan
Jamia Millia Islamia, New Delhi, India

N.S. Kumar
Department of Computer Science and Engineering, PES University, Bangalore, India

Rahul Chandra Kushwaha
Department of Computer Science and Engineering, Rajiv Gandhi University, Doimukh, Itanagar, India

Xiufeng Liu
Technical University of Denmark, Lyngby, Denmark

V.S. Nageswara Rao Kadiyala
Department of Computer Science and Engineering, National Institute of Technology Silchar, Silchar, Cachar, Assam, India

Sabuzima Nayak
Department of Computer Science and Engineering, National Institute of Technology Silchar, Silchar, Cachar, Assam, India

Joseph L. Pachuau
Department of Computer Science and Engineering, National Institute of Technology Silchar, Silchar, India

Ripon Patgiri
Department of Computer Science and Engineering, National Institute of Technology Silchar, Silchar, Cachar, Assam, India

U. Kartheek Chandra Patnaik
Department of Computer Science and Engineering, National Institute of Technology Silchar, Silchar, Cachar, Assam, India

Kaushik Ravichandran
Department of Computer Science and Engineering, PES University, Bangalore, India

Arnab Roy
Department of Computer Science and Engineering, National Institute of Technology Silchar, Silchar, India

Anish Kumar Saha
Department of Computer Science and Engineering, National Institute of Technology Silchar, Silchar, India

Dhananjay Kumar Singh
Department of Computer Science and Engineering, National Institute of Technology, Durgapur, India

Pawan Singh
Department of CSE, Amity University, Lucknow, Uttar Pradesh, India

Sandeep Kumar Singh
Department of CSE & IT, Jaypee Institute of Information Technology, Noida, India

M. Franckie Singha
Department of Computer Science and Engineering, National Institute of Technology Silchar, Silchar, Cachar, Assam, India

Saurabh Kumar Srivastava
College of Computing Sciences & IT, Teerthanker Mahaveer University, Moradabad, U.P., India

Akshara Subramaniasivam
Department of Computer Science and Engineering, PES University, Bangalore, India

Ankit Vidyarthi
Department of CSE & IT, Jaypee Institute of Information Technology, Noida; Department of CSE&IT, Jaypee Institute of Information Technology, Noida, Uttar Pradesh, India

Lilapati Waikhom
Department of Computer Science and Engineering, National Institute of Technology Silchar, Silchar, Cachar, Assam, India

Preface

Big Graph is an emerging research field that is gaining enormous popularity among academicians, industrialists, and practitioners. The Big Graph is applied in research areas such as bioinformatics, social systems administration, computer networking, complex networks, and data streaming. Big Graph technology is also used for biological networks, scholar article citation networks, protein–protein interaction, and semantic networks.

Big Graph consists of millions of nodes and trillion of edges growing exponentially; hence, Big Graph needs large computing machinery for processing, which is a grand challenge. The Conventional Graph databases and analytics cannot address the versatility of graph information because processing large-scale graph data becomes expensive in terms of computation. Enormous graphs, namely, the WWW, information data sets of web indexes, road maps, atoms, high-energy physics, and science, are growing exponentially these days. Since a graph-structured portrayal in real life is normal, viable and novel strategies are needed to take care of the baffling inexplicable issues inside the diagram. Numerous customary AI approaches have been proposed on top of extricated highlights utilizing different predefined measures from the first information structure. The extricated elements could be pixel insights in picture information or word event measurements in regular language information. Recently, deep learning procedures have acquired huge ubiquity, handling the learning issues effectively, taking in portrayal from crude information, and anticipating utilizing the learned portrayal all the while.

In the current scenario, there are several tools of Big Graph available in the marketplace. However, there are fewer books available in the current marketplace, which emphasize analytics. In addition, the available books are unable to provide rich insight into analytics, visualization, and databases. Hence, this edited book is being planned to bring forth the information regarding Big Graph so that the database and machine learning professionals and academia across the world are supplied with the relevant information. This book comprising 17 chapters presents the state-of-the-art surveys and solutions for various unsolved problems. In addition, the book presents recent findings of diverse research scholars.

DR. RIPON PATGIRI

GANESH CHANDRA DEKA

DR. ANUPAM BISWAS

## CHAPTER ONE

CESDAM: Centered subgraph data matrix for large graph representation

Anupam Biswas[a] and Bhaskar Biswas[b]

[a]Department of Computer Science and Engineering, National Institute of Technology Silchar, Silchar, Assam, India
[b]Department of Computer Science and Engineering, Indian Institute of Technology (BHU), Varanasi, India

Contents

Abstract

In this chapter, a space efficient graph representation scheme is proposed. Unlike the classical approaches where entire graph is represented in single entities such as adjacency matrix or incidence matrix, the proposed approach represents subgraphs of the graph into multiple entities. An algorithm is designed to generate the proposed

Advances in Computers, Volume 128
ISSN 0065-2458
https://doi.org/10.1016/bs.adcom.2021.09.005

representation from the raw data where relationships among objects are identified as edges. The time complexity of the algorithm is bounded by the best case $\Theta(m)$ and worst case $O(mk)$, where m is the number of edges in the graph and k is the number of subgraphs generated for the graph. The methodology is further extended for generating the proposed representation with already available edge list or adjacency matrix. Theoretical, as well as empirical analysis on eight real-world graphs, show efficiency of the proposed approach in terms of memory requirement over existing representations.

1. Introduction

Graphical modeling of relationships among different objects that are present within the data is common to several application domains [1–3], such as social networks in Sociology [4–9], protein-interaction networks in Biology [10, 11], food supply chain networks in Ecology [12, 13], chemical compounds in Chemistry [14, 15], links in the Web [8, 16–21], attribute graphs in image processing [22] and so on. Two distinctive approaches have been evolved for representing graphs: list-based representation and matrix-based representation [23]. The existing matrix-based representations are becoming inefficient with enormously growing sizes of the modern-day real-world graphs. The matrix-based representations of these graphs cannot fit into the memory. The *memory requirement* emerges as the key constraint in the case of matrix-based representation while dealing with the large graphs. Another major disadvantage of existing matrix-based schemes is that even if the graph is sparse, they require very large amount of memory. Various formats are evolved to reduce memory requirement of sparse matrix [24–26]. However, these sparse matrix storage formats are beneficial only when the matrix is significantly sparse, otherwise they cost more for their different data access patterns [27, 28].

In this chapter, an alternative space efficient matrix-based representation scheme is proposed. The existing matrix-based schemes are dependent only on number of nodes and edges, which is the main reason for the high memory requirement. In the proposed approach, connectivity pattern (i.e., the topology of groups of nodes) of the graph is considered along with the nodes and edges. The graph is divided into multiple overlapping subgraphs depending on connectivity pattern. Each subgraph is represented in a special kind of *Matrix* called centered subgraph data matrix (CESDAM). Each CESDAM has an identifier, which is the central node of the subgraph. The primary difference with the existing representation schemes is that the subgraphs of the graph having specific connectivity patterns are represented as multiple entities called CESDAMs in the proposed approach, whereas the existing representation schemes represent the entire graph as a single entity, i.e., as an adjacency matrix or incidence matrix. The empirical results on real-world graphs show that the CESDAM scheme is more efficient than the present matrix-based schemes in terms of memory requirement. The CESDAM scheme requires very small amount of memory even when the graph is sparse in comparison to the existing matrix-based schemes, which is comparable to the list-based schemes.

An algorithm is proposed for generating CESDAMs. It generates CESDAMs in such a way that the graph gets partitioned into smaller subgraphs. The algorithm is designed to generate CESDAMs from the raw data, where relationships among objects can be identified (i.e., the edges of the graph). The algorithm processes each edge of the graph for its possible entries in the CESDAM. The time complexity of the algorithm is bounded by the best case $\Theta(m)$ and worst case $O(mk)$, where m is the number of edges in the graph and k is the number of subgraphs generated for the graph that are represented as CESDAMs. Utilizing this algorithm, a methodology to produce CESDAMs from already available edge list or adjacency matrix of a graph is also proposed.

Rest of the chapter is organized as follows. Section 2 illustrates the background details and the works related to graph representation. Section 3 discusses inter-dependency of graph processing and graph representation with a motivation to utilize the concept of ego network [29] in graph representation. Section 5 performs theoretical analysis of efficiency of proposed CESDAM over adjacency matrix and time complexity of CESDAM generating algorithm. Section 6 presents experimental results and performs comparative analysis of CESDAM with other approaches. Finally concluded in Section 7.

2. Background and related works

An undirected weighted graph $G(V, E)$ is a pair of a finite set V of nodes and E is a function $E : \{V \times V \rightarrow R, R \geq 0\}$ that defines the weights of edges such that $E(u, v) = E(v, u)$ for all $u, v \in V$. Here, $E(u, v)$ means u is source node and v is destination node for the edge. For unweighted graphs, function E is defined as $E : \{V \times V \rightarrow \{0, 1\}\}$, where 0 represents absence of edge and 1 represents presence of edge. Various graphs representation schemes are discussed below.

2.1 List-based approach

2.1.1 Edge list

The list contains generally three columns. First column contains source nodes, second column contains destination nodes, and the third column contains edge weights. If there exists an edge $E(u, v)$ between any pair of nodes $u, v \in V$, there will be an entry for edge $E(u, v)$ in the edge list. For undirected graphs, entry for $E(u, v)$ is enough to represent $E(v, u)$. Absence of an edge for both weighted and unweighted graphs do not have any entry in the edge list. All edges of unweighted graphs are same so it does not require weight column.

2.1.2 Adjacency list and incidence list

It represents same adjacency matrix in efficient way with less storage requirement. An array is maintained with entries of all nodes. An unordered link list is prepared with all adjacent nodes of any node and linked to that node. Similarly, with incidence list an array is maintained with entries of all nodes and unordered link list is prepared with all edges associated with the node.

2.2 Matrix-based approach

2.2.1 Adjacency matrix

Graphs are represented with a matrix $A \in \{0, 1\}^{n \times n}$, where $n = |V|$. Every row $a_u \in A$ is designated to a specific node $u \in V$. For all nodes, $v \in V$ there is an entry in the row a_u. All adjacent nodes v of node u are represented with entry 1 in $a_{u,v}$ for unweighted graph, but for weighted graph it is entry of corresponding edge weight. All nonadjacent nodes v of node u are represented with entry 0. Adjacency matrix is more informative than edge list. Most of the graph processing algorithms require adjacent nodes. Therefore, adjacency matrix is used widely for graph processing.

2.2.2 Incidence matrix

It is same as adjacency matrix representation. Only difference is instead of considering both objects nodes, incidence matrix considers nodes as well as edges for representing graphs. With incidence matrix, graphs are represented with a matrix $I \in \{0, 1\}^{m \times n}$, where $n = |V|$ rows and $m = |E|$ edges. If any node $u \in V$ is associated with edge e then the entry for $i_{u,e}$ is 1 otherwise 0. Therefore, sum of each column is 2 as each edge is associated with only two nodes.

2.3 Other related approaches

In modern-day graph processing, the classical approaches discussed above are utilized several ways to improve performance. The matrix-based scheme such as adjacency matrix is often have large numbers of zero entries, specially when the graph is sparse. Various formats such as compressed sparse row (CSR), compressed sparse column (CSC), and block sparse row (BSR), K^2 − Tree are developed to reduce memory requirement of sparse matrix [24–26, 30]. These formats are applicable only to static graphs. A small change in the original adjacency matrix may require reformatting of entire adjacency matrix into the compressed format.

Unlike static graphs, parts of the graph change with time in dynamic graphs so the graph has to be persistent. The persistent graph means that from any given graph state past states can be accessed. In Ref. [31], a methodology is proposed for representing persistent graphs. The graph processing engine linked-node large multiversioned array (LLAMA) [32] is proposed for efficient processing of dynamic graphs. It partitions adjacency list and stores those in multiple snapshots. The graph processing with LLAMA is fast but it is not space efficient way of utilization of adjacency list. The LLAMA is highly redundant since multiple versions of same parts of adjacency list are stored in different snapshots.

Reformatting of classical representations such as adjacency matrix into CSR, CSC, BSR etc. or adjacency list into LLAMA, both are highly inefficient in terms of memory requirement. Moreover, matrix-based representation such as adjacency matrix is not suitable for extremely large graphs since those cannot fit into the memory. On the contrary, the proposed matrix-based scheme requires very low memory in comparison to the present matrix-based schemes.

3. Graph processing and representation

In this section, dependency of graph processing on graph representation is discussed by incorporating the essence of grouping of nodes in graph representation. A social network-based property called ego network is discussed by extending the notion of grouping.

Graph processing is broadly categorized as sequential and nonsequential processing. In sequential graph processing, graph properties are identified by processing each node one after another. In nonsequential graph processing, multiple nodes are processed simultaneously. During graph processing, either sequential or nonsequential approach is followed to process different properties of graph. Both sequential and nonsequential processing can be used where global properties are processed. However, processing of local properties always follows sequential approach. The nonsequential approach utilizes matrix operation. Thus, graph has to be represented in the form of adjacency matrix for nonsequential processing. On the contrary, sequential processing can be performed in any graph representation. Thus, most of the cases follow conventional sequential processing of graph during identification of a global property. However, efficient graph processing is not possible without an effective graph representation scheme.

The essence of graph representation is to avail information about the graph as much as possible in an efficient way. Nodes and associated edges are the elementary information utilized during graph processing. Thus, representation of graph can be viewed as grouping of nodes and edges. As many nodes cover in the representation of graphs through grouping, it becomes more suitable for processing. For instance, the adjacency matrix is more efficient for processing than edge list. Adjacency matrix covers all neighbors of a node in representation whereas edge list covers only one edges, i.e., one neighbor. Here, a node and all its neighbors are representing nothing but a graph property that is used in adjacency matrix for graph representation.

The notion of grouping of nodes in graph representation can be further broadened from the perceptive of graph processing. Searching of nodes or edges is very important in sequential processing. Breadth-first-search (BFS) and depth-first-search (DFS) are the two widely used approaches for graph processing. Generally, BFS is preferred for sequential processing. Most of such processing require processing any node along with its associated connections, i.e., neighbors, which is simply a personal network of the node. Such personal network is referred as ego network in social network terminology [29]. Ego network is nothing but a graph property that is determined by the central node. Based on the maximum distance from the central node to other nodes, different levels of ego network are defined. Higher level ego network can cover higher numbers of nodes. Various levels of ego network are shown in Fig. 1 and expressed in the following definitions.

Definition 1 (Level 1.0 ego network). In graph $G(V, E)$, the level 1.0 ego network is defined as subgraph $g(u, V_u, E_u)$ with respect to ego node u such that if $\exists v \in V$ then $v \in V_u$ iff $\exists E(u, v) \in E$ and $\forall(x, y) \in E_u$ satisfies following conditions:

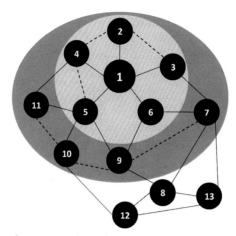

Fig. 1 An example of ego network. Node 1 is the ego or central node. Level 1.0 ego network is the nodes and only *dark edges* inside the *white circle*. Level 1.5 ego network is the nodes and edges (both *dark* and *dash*) inside the *white circle*. Level 2.0 ego network is the nodes and all the edges (excluding *dashed edges* outside white circle) inside the *shaded oval*. Level 2.5 ego network is the nodes and all the edges (both *dark* and *dash*) inside the *shaded oval*.

1. $x = u$
2. $y \in V_u$

Definition 2 (Level 1.5 ego network). In graph $G(V, E)$, the level 1.5 ego network is defined as subgraph $g(u, V_u, E_u)$ with respect to ego node u such that if $\exists v \in V$ then $v \in V_u$ iff $\exists E(u, v) \in E$ and $\forall(x, y) \in E_u$ satisfies following conditions:

1. $(x = u)$ or $(x \in V_u)$
2. $y \in V_u$

Definition 3 (Level 2.0 ego network). In graph $G(V, E)$, the level 2.0 ego network is defined as subgraph $g(u, V_u, E_u)$ with respect to ego node u such that if $\exists v \in V$ then $v \in V_u$ iff $\exists E(u, v) \in E$ and satisfies following conditions:

1. $\forall(x, y) \in E_u \Rightarrow \{(x = u) \text{ or } (x \in V_u)\}$ and $y \in V_u$
2. $V'_u = \bigcup_{\forall y \in V_u}(V_u, V_y)$ and $E'_u = \bigcup_{\forall y \in V_u}(E_u, E_y)$, where $g(y, V_y, E_y)$ is the Level 1.0 ego network
3. $V_u = V'_u$ and $E_u = E'_u$

Definition 4 (Level 2.5 ego network). In graph $G(V, E)$, the level 2.5 ego network is defined as subgraph $g(u, V_u, E_u)$ with respect to ego node u such that if $\exists v \in V$ then $v \in V_u$ iff $\exists E(u, v) \in E$ and satisfies following conditions:

1. $\forall(x, y) \in E_u \Rightarrow \{(x = u) \text{ or } (x \in V_u)\}$ and $y \in V_u$

2. $V'_u = \bigcup_{\forall y \in V_u}(V_u, V_y)$ and $E'_u = \bigcup_{\forall y \in V_u}(E_u, E_y)$, where $g(y, V_y, E_y)$ is the Level 1.5 ego network

3. $V_u = V'_u$ and $E_u = E'_u$

In the above definitions, we refer central node as level 0 node, since the distance from central node is 0. Similarly, nodes having distance from central node 1 is referred as level 1 nodes and so on. Note that both level 1.0 and 1.5 ego network include level 1 nodes. The only difference between level 1.0 ego network and level 1.5 ego network is the edges. Level 1.5 ego network allows edges among level 1 nodes but level 1.0 ego network allows only edges between level 0 and level 1 nodes. Similarly, level 2.5 ego network allows edges among level 2 nodes. Therefore, we refer nodes of both level 1.0 and 1.5 as *level 1 nodes* and nodes of both level 2.0 and 2.5 as *level 2 nodes*. Neighbor nodes of level 1 nodes that have no edge with level 0 node are referred as *level 1 external nodes*. Neighbor nodes of level 2 nodes that have no edge with level 1 nodes are referred as *level 2 external nodes*.

4. Proposed graph representation

In this section, we present the CESDAM scheme for representing graphs and the algorithm to construct CESDAM. We consider the ego network in CESDAM representation to extrapolate the notion of grouping as discussed in Section 3 for covering more nodes. We consider level 2.5 ego network in CESDAM scheme.

4.1 The CESDAM representation scheme

The graph $G(V, E)$ is divided into smaller subgraphs $g(u, V_u, E_u)$, where $u \in V$ is the center of the subgraph. Each subgraph $g(u, V_u, E_u)$ represents a level 2.5 ego network. A generic model of CESDAM for representing a subgraph is shown in Fig. 2. The data structure used in CESDAM representation is a matrix of $A = [\,]^{(m+2) \times d}$, where m is the number of nodes included in subgraph and $d = max(|row\ a_i|)$. The value of d depends on the number of neighbors of nodes included in subgraph, numbers of level 1 external nodes, and numbers of level 2 external nodes. The CESDAM data model contains three parts as follows:

External level 1: All level 1 external nodes are included in this part. It covers the first row of the matrix excluding the first column.

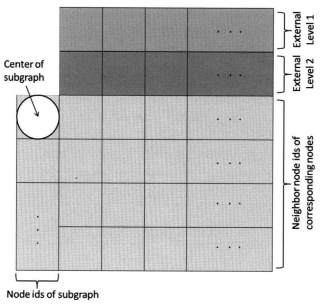

Fig. 2 Generic CESDAM Model to represent subgraphs.

External level 2: All level 2 external nodes are included in this part. It covers the second row of the matrix excluding the first column.

Main subgraph: All nodes associated with the subgraph are included into this part. It covers row 3 onwards rest of the matrix. First element in row 3, i.e., $a_{3,1}$ is the center of the subgraph. With reference to the center node, level 2.5 ego network is constructed. Therefore, the center node is considered as an identifier for the subgraph. Once the center node is known, the corresponding CESDAM for the subgraph can be constructed easily. For all nodes included in the subgraph (excluding level 2 external nodes) there is an entry in the first column (entry in one of $a_{i,1}$ where $i > 2$). If node u is entered in row i (i.e., in $a_{i,1}$) then all neighbor nodes node u are entered in row i in successive columns (i.e., entered in one of $a_{i,j}$ where $j > 1$).

The entire graph is viewed as divisions of the graph into subgraphs, where each subgraph is a level 2.5 ego network. Each subgraph is represented as CESDAM. Thus, the entire graph is represented with multiple CESDAMs. The central node of the ego network is utilized as an identifier for these CESDAMs.

4.2 CESDAM generation

Algorithm 1 presents the procedure for creating new CESDAM. Entry of an edge into the CESDAM is done as follows. One of the two associated nodes of the edge is selected randomly for the entry as center into the new CESDAM. Preparation of CESDAMs requires to ensure that it fulfills necessary conditions for level 2.5 ego network. Therefore, other entries are updated following the generic model as shown in Fig. 2. Depending on the position of edge in the subgraph it attached, corresponding entries for associated nodes into the CESDAM is performed. If CESDAM is already created then there can be five different possibilities for any edge with respect to the subgraph as follows:

ALGORITHM 1 Create_CESDAM(*SGlist*, *SGid*, N_1, N_2).
1: Input: *SGlist*{#*SGid*}[*CESDAM*] // List of subgraphs, here #*SGid* is numeric subgraph id.
    SGids //List of subgraph ids.
    N_1, N_2 //Nodes associated with the edge.
2: Output: *SGids*, *SGlist*
3: *Center* ← select randomly either N_1 or N_2
4: add *Center* to *SGids*
5: *SGlist*{#*Center*}(1, 1) ← (−100) initialize with any dummy value except node ids
6: *SGlist*{#*Center*}(2, 1) ← (−200) initialize with any dummy value except node ids
7: $Node_1$ ← *Center*
8: **if** $Node_1 = N_1$ **then**
9:   $Node_2$ ← N_2
10: **else**
11:   $Node_2$ ← N_1
12: **end if**
13: *SGlist*{#*Center*}(3, 1) ← $Node_1$
14: *SGlist*{#*Center*}(3, 2) ← $Node_2$
15: *SGlist*{#*Center*}(4, 1) ← $Node_2$
16: *SGlist*{#*Center*}(4, 2) ← $Node_1$
17: **return** *SGlist*, *SGids*

Case 1: One of the node is center (level 0)
Case 2: Both nodes are in level 1
Case 3: One of the node is in level 1
Case 4: Both nodes are in level 2
Case 5: One of the node is in level 2

Algorithm 2 determines the possible entries of nodes into the CESDAM that are associated with the edge representing the relationships among objects of raw data. Possible entries in the CESDAM for the edges are explained with examples as follows. An example graph is shown in Fig. 3 with edge positions in the probable subgraph. The CESDAMs of subgraphs prepared for the example graph in Fig. 3 is shown in Fig. 4.

4.2.1 Case 1 edge entry

In the example graph, edge $E(1, 2)$ is an instance of case 1 since the first node 1 is the center (level 0). In edge list or adjacency matrix representation, this edge can be presented as either $E(1, 2)$ or $E(2, 1)$. Therefore, case 1 edge can be found in two forms. Each form needs to be handled separately. In $E(1, 2)$, first node, i.e., node 1 is the center and require following entries of CESDAM to be updated:

- Enter second node at the end of row 3 if not present for the edge representation.
- Enter second node at the end of column 1 (id column)[a] if not present.
- Enter first node at the end of row #second[b] if not present for edge representation.
- Remove entries for second node at row 1 and row 2 if previously node was assumed as level 1 and level 2 external, respectively.

Similarly in $E(2, 1)$, second node, i.e., node 1 is center so same entries of CESDAM has to be updated. Updating of CESDAM for case 1 edges (both type $E(1, 2)$ and $E(2, 1)$) are done with Algorithm 3. For the case like $E(2, 1)$ input to Algorithm 3 are swapped, i.e., first node become second and vice-versa in Algorithm 2.

Case 2 & case 4 edge entry: Edge $E(2, 3)$ is an example of case 2 edge. Both nodes are in level 1. Therefore, edge representation $E(3, 2)$ in edge list or adjacency matrix will mean the same as $E(2, 3)$. There is no special entry in the CESDAM for case 2 other than the entry for edge representation as follows:

---

[a] The column for row entries of specific node.
[b] Here, row #second means the row in the CESDAM that is meant for the #second node and its neighbors.

ALGORITHM 2 Nodes_Entry($SGlist$, $SGid$, CEN_1, CEN_2).

1: Input: $SGlist$ // List of subgraphs.

$SGid$ //subgraph id.

CEN_1, CEN_2 //Nodes of current edge.

2: Output: $SGids$, $SGlist$

3: $flag \leftarrow 0$

4: **for all** $SGid \in SGids$ when $flag = 0$ **do**

5:   **if** $SGlist\{\#SGid\}(3, 1) = CEN_1$ **then**

6:     //Case 1(a): First node is the center (Level 0)

7:     $flag \leftarrow 1$

8:     $SGlist \leftarrow Level01(SGlist, SGid, CEN_1, CEN_2)$

9:   **else if** $SGlist\{\#SGid\}(3, 1) = CEN_2$ **then**

10:     //Case 1(b): Second node is the center (Level 0)

11:     $flag \leftarrow 1$

12:     $SGlist \leftarrow Level01(SGlist, SGid, CEN_2, CEN_1)$

13:   **else if** $CEN_1 \in SGlist\{\#SGid\}(3, j)$ and $CEN_2 \in SGlist\{\#SGid\}(3, j)$ **then**

14:     //Case 2: Both nodes are of Level 1

15:     $flag \leftarrow 1$

16:     $SGlist \leftarrow Levelxx(SGlist, SGid, CEN_1, CEN_2)$

17:   **else if** $CEN_1 \in SGlist\{\#SGid\}(3, j)$ and $CEN_2 \notin SGlist\{\#SGid\}(3, j)$ **then**

18:     //Case 3(a): Only first node is of Level 1

19:     $flag \leftarrow 1$

20:     $SGlist \leftarrow Level12(SGlist, SGid, CEN_1, CEN_2)$

21:   **else if** $CEN_1 \notin SGlist\{\#SGid\}(3, j)$ and $CEN_2 \in SGlist\{\#SGid\}(3, j)$ **then**

22:     //Case 3(b): Only second node is of Level 1

23:     $flag \leftarrow 1$

24:     $SGlist \leftarrow Level12(SGlist, SGid, CEN_2, CEN_1)$

25:   **else if** $CEN_1 \in SGlist\{\#SGid\}(1, j)$ and $CEN_2 \in SGlist\{\#SGid\}(1, j)$ **then**

26:     //Case 4: Both nodes are of Level 2

27:     $flag \leftarrow 1$

28:     $SGlist \leftarrow Levelxx(SGlist, SGid, CEN_1, CEN_2)$

29:   **else if** $CEN_1 \in SGlist\{\#SGid\}(1, j)$ and $CEN_2 \notin SGlist\{\#SGid\}(1, j)$ **then**

30:     //Case 5(a): Only first node is of Level 2

31:     $flag \leftarrow 1$

32:     $SGlist \leftarrow Level23(SGlist, SGid, CEN_1, CEN_2)$

33:   **else if** $CEN_1 \in SGlist\{\#SGid\}(1, j)$ and $CEN_2 \notin SGlist\{\#SGid\}(1, j)$ **then**

34:     //Case 5(b): Only second node is of Level 2

35:     $flag \leftarrow 1$

36:     $SGlist \leftarrow Level23(SGlist, SGid, CEN_2, CEN_1)$

37:   **end if**

38: **end for**

39: **if** $flag = 0$ **then**

40:   $[SGlist, SGids] \leftarrow$ Create_CESDAM($SGlist$, $SGid$, $CEN1$, $CEN2$)

41: **end if**

42: **return** $SGlist$, $SGids$

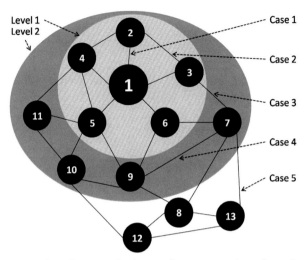

Fig. 3 Example graph with types of edges with respect to the subgraph with central node 1.

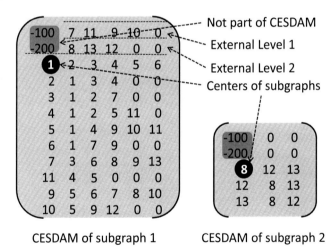

CESDAM of subgraph 1   CESDAM of subgraph 2

Fig. 4 CESDAMs prepared from the example graph.

- Enter first node at the end of row *#second* if not present for edge representation.
- Enter second node at the end of row *#first*[c] if not present for edge representation.

Similarly, case 4 edges also require entries only for edge representation. Edge $E(7, 9)$ is an example of case 2 edge. Updating of CESDAM for both case 2 and case 4 edges are done with Algorithm 4.

---

[c] Here, row *#first* means the row in the CESDAM that is meant for the *#first* node and its neighbors.

ALGORITHM 3 Level01(*SGlist, SGid, N₁, N₂*).

1: **if** $N_2 \notin SGlist\{\#SGid\}(3, j)$, $j = $ all columns **then**
2:    $SGlist\{\#SGid\}(3, Last + 1) = N_2$ //index of Last entry in row 3
3:    $SGlist\{\#SGid\}(Last + 1, 1) = N_2$ //index of Last entry in column 1
4:    Remove N_2 from row 1 and 2 of $SGlist\{\#SGid\}[\ ]$ if present
5: **end if**
6: $Nrow \leftarrow$ index of N_2 entry in column 1
7: $SGlist\{\#SGid\}(Nrow, Last + 1) = N_1$ //index of Last entry in row $Nrow$
8: **return** $SGlist$

ALGORITHM 4 Levelxx(*SGlist, SGid, N₁, N₂*).

1: //Enter N_1 for the edge representation
2: $Nrow \leftarrow$ index of N_2 entry in column 1
3: $SGlist\{\#SGid\}(Nrow, Last + 1) = N_1$ //index of Last entry in row $Nrow$
4:
5: //Enter N_2 for the edge representation
6: $Nrow \leftarrow$ index of N_1 entry in column 1
7: $SGlist\{\#SGid\}(Nrow, Last + 1) = N_2$ //index of Last entry in row $Nrow$
8: **return** $SGlist$

Case 3 edge entry: In case 3 edges, only one node is in level 1. Updating CESDAM for case 3 edges is done with Algorithm 5. Suppose, first node is in level 1. Since, during the time when case 3 is checked already confirmed that second node is not in the level 0 or 1. Thus, definitely the second node of case 3 edges will be in level 2. As explained above for the case 1 edges, there can be two edge representation in adjacency matrix or in edge list. For instance, case 3 edge $E(3, 7)$ can be found as $E(7, 3)$. Though, both represent the same edge, for $E(3, 7)$ first node is in level 1 but for $E(7, 3)$ first nodes is in level 2. Thus, input to Algorithm 5 to is changed by simply swapping the position of first to second for $E(7, 3)$ in Algorithm 2. Following entries in CESDAM are updated for any case 3 edge.

- Enter second node at the end of row #*first* if not present for edge representation.
- Enter second node at the end of row 1 if not present (assumed as level 1 external).
- Enter second node at the end of column 1 (id column) if not present.

ALGORITHM 5 Level12(*SGlist, SGid, N₁, N₂*).

1: //Enter N_2 for the edge representation
2: *Nrow* ← index of N_1 entry in column 1
3: *SGlist*{*#SGid*}(*Nrow, Last* + 1) = N_2 //index of Last entry in row *Nrow*
4: //Enter N_2 in row 1 (assumed Level 1 external)
5: if $N_2 \notin$ *SGlist*{*#SGid*}(1, *j*) then
6:   *SGlist*{*#SGid*}(1, *Last* + 1) = N_2 //index of Last entry in row 1
7: end if
8: //Enter N_2 in column 1
9: if $N_2 \notin$ *SGlist*{*#SGid*}(*i*, 1) then
10:   *SGlist*{*#SGid*}(*Last* + 1, 1) = N_2 //index of Last entry in column 1
11: end if
12: //Enter N_1 for the edge representation
13: *Nrow* ← index of N_2 entry in column 1
14: *SGlist*{*#SGid*}(*Nrow, Last* + 1) = N_1 //index of Last entry in row *Nrow*
15: Remove N_2 from row 2 of *SGlist*{*#SGid*}[ ] if present
16: return *SGlist*

ALGORITHM 6 Level23(*SGlist, SGid, N₁, N₂*).

1: //Enter N_2 for the edge representation
2: *Nrow* ← index of N_1 entry in column 1
3: *SGlist*{*#SGid*}(*Nrow, Last* + 1) = N_2 //index of Last entry in row *Nrow*
4: //Enter N_2 in row 2 (assumed Level 2 external)
5: if $N_2 \notin$ *SGlist*{*#SGid*}(2, *j*) then
6:   *SGlist*{*#SGid*}(2, *Last* + 1) = N_2 //index of Last entry in row 2
7: end if
8: return *SGlist*

- Enter first node at the end of row *#second* if not present for edge representation.
- Remove entry for second node at row 2 if previously node was assumed as level 2 external.

Case 5 edge entry: Case 5 edge means one of the two nodes is in level 2 and other one is sure that it will be level 2 external node. Since, subgraph does not include level 2 external nodes so CESDAM does not contain specific rows for these nodes. Updating of CESDAM for both case 5 edges are done with Algorithm 6. Same swapping mechanism as done for case 1 and case 3 is also followed for case 5 edges in Algorithm 2. If first node (example Edge $E(7, 8)$) is of level 2 then following entries of CESDAM are updated:

- Enter second node at the end of row #*first* if not present for edge representation.
- Enter second node at the end of row 2 if not present (assumed as level 2 external).

As mentioned above if an edge does not fall under any of the five cases for the current CESDAM, it will be checked for another CESDAM. The edge is checked for all CESDAMs that are currently present for possible entries as per the five cases. If the edge does not fall under the five cases on any of the current CESDAMs, it indicates for new CESDAM. After exhausting all the CESDAMs, new CESDAM is created with Algorithm 1.

4.3 CESDAMs from edge list and adjacency matrix

Already available edge list or adjacency matrix of the graph are divided into k parts. Each part is processed for CESDAM construction. Algorithm 7 and Algorithm 8 present the methodology for CESDAM construction from any part of edge list and adjacency matrix, respectively. Each edge representation in edge list or adjacency matrix is processed for the entries into corresponding CESDAM. Processing of each edge determines the possible entries of associated nodes into the CESDAM and it is done with Algorithm 2.

ALGORITHM 7 Create_CESDAM_Edgelist(*ELPart*).
1: Input: *ELPart* // Part of edge list.
2: Output: *SGids, SGlist{#SGid}[CESDAM]*
3: **for all** random $e \in$ *ELPart* **do**
4:    $CEN_1 \leftarrow$ First Node of Current Edge e
5:    $CEN_2 \leftarrow$ Second Node of Current Edge e
6:    [*SGlist, SGid*]$\leftarrow$ Nodes_Entry(*SGlist, SGid, CEN_1, CEN_2*)
7: **end for**
8: **return** *SGlist, SGids*

ALGORITHM 8 Create_CESDAM_Adjacency(*AMPart*).

1: Input: *AMPart* **/** Part of Adjacency Matrix rows.

2: Output: *SGids, SGlist{#SGid}[CESDAM]*

3: **for all** *row* ∈ *AMPart* **do**

4:     CEN_1 ← *#row* **/**First Node of all Current Edges

5:     *N_list* ← list of second Node of all Current Edges (neighbor of node *#row*)

6:     **for all** *CEN2* ∈ *N_list* **do**

7:        [*SGlist, SGid*]← Nodes_Entry(*SGlist, SGid, CEN_1, CEN_2*)

8:     **end for**

9: **end for**

10: **return** *SGlist, SGids*

---

5. Theoretical analysis

In this section, the efficiency of proposed CESDAM representation scheme is analyzed in terms of space complexity. In addition, time complexity of the algorithm to construct CESDAMs for any graph is also analyzed.

5.1 Efficiency of CESDAM representation

In CESDAM representation, the graph with n nodes is partitioned into subgraphs that are presented in the CESDAMs. If there are k CESDAMs that represent the graph then we have k subgraphs each having $m_1, m_2, \ldots, m_k$ number of nodes. The sizes of CESDAMs are not dependent on the number of nodes in the subgraphs, instead it is dependent on connectivity pattern among the nodes. Therefore, we assume that for k matrices of sizes $m_1^r \times m_1^c, m_2^r \times m_2^c, \cdots ., m_k^r \times m_k^c$ that are the CESDAMs for subgraphs. For the simplicity of the analysis, we consider $m_i = m_i^r = m_i^c$. Thus, we have k CESDAMs of sizes $m_1 \times m_1, m_2 \times m_2, \cdots ., m_k \times m_k$. Note that subgraphs that are represented in the CESDAMs are overlapped. Thus, $m_1 + m_2 + \cdots + m_k > n$, if at least one overlapping node exists. In case of adjacency matrix, all the n nodes are represented in a single matrix of size $n \times n$. If k nonoverlapped subgraphs are represented with adjacency matrix then we have $m_1 \times m_1, m_k \times m_2, \cdots ., m_k \times m_k$ sized matrices. Therefore, we have following theorems.

Theorem 1. *Individual adjacency matrix representation of nonoverlapping subgraphs in total requires less space than the adjacency matrix for entire graph.*

Proof. Given a graph $G(V, E)$, the size of adjacency matrix representation of G is $n \times n = n^2$, where $n = |V|$ is number of nodes present in G. Suppose, the graph G is divided into k nonoverlapping subgraphs. Each subgraph $g_i(V_i, E_i)$ contains $m_i = |V_i|$ number of nodes. Hence, size of adjacency matrix representation of each subgraph g_i is $m_i \times m_i$. Since, all k subgraphs are nonoverlapping so $\sum_{i=1}^{k} m_i$, i.e., $m_1 + m_2 + \cdots + m_k = n$.

Now, if G is partitioned into two nonoverlapping subgraphs of sizes 1 and $n - 1$,

$$1 \times 1 + (n-1) \times (n-1) < n \times n$$

Similarly, for sizes 2 and $n - 2$,

$$2 \times 2 + (n-2) \times (n-2) < n \times n...$$

for sizes k and $n - k$,

$$k \times k + (n-k) \times (n-k) < n \times n... \tag{1}$$

Again, if G is partitioned into two nonoverlapping subgraphs of sizes 1 and $n - 1$,

$$1 \times 1 + (n-1) \times (n-1) < n \times n$$

If G is partitioned into three nonoverlapping subgraphs of sizes 1, 1 and $n - 2$,

$$1 \times 1 + 1 \times 1 + (n-2) \times (n-2) < n \times n...$$

for $k - 1$ subgraph of size 1 and one subgraph of size $n - k$,

$$1 \times 1 + 1 \times 1 + \cdots + (k-1)^{th} 1 \times 1 + (n-k) \times (n-k) < n \times n... \tag{2}$$

Therefore, with (1) and (2), for k nonoverlapping subgraphs of sizes $m_1, m_2, \ldots, m_k$,

$$m_1 \times m_1 + m_2 \times m_2 + \cdots + m_k \times m_k < n \times n. \qquad \square$$

Corollary 1. *If k nonoverlapping subgraphs contain equal number of nodes m then for n total number of nodes in the graph $k \times m^2 < n^2$.*
Proof. From Theorem 1, for k nonoverlapping subgraphs containing $m_1, m_2, \ldots, m_k$ nodes,

$$m_1 \times m_1 + m_2 \times m_2 + \ldots + m_k \times m_k < n \times n$$

if $m_1 = m_2, \ldots, m_{k-1} = m_k = m$ then,

$$k \times m \times m < n \times n \Rightarrow k \times m^2 < n^2. \qquad \square$$

Lemma 1. *Level 2 external nodes in CESDAM are the overlapping nodes and shared by only two CESDAMs.*
Proof. Level 2 external nodes are the outermost nodes of the CESDAM representing ego network level 2.5. Thus, these nodes will be shared by other CESDAMs. In proposed approach, a node is allowed to be shared with at most two CESDAMs. However, a CESDAM can have multiple nodes shared with multiple CESDAMs. □

Theorem 2. *CESDAM representation of graph requires less space than adjacency matrix representation.*
Proof. Given a graph $G(V, E)$, $n = |V|$ and k CESDAMs that represent the graph G. With Lemma 1, k CESDAMs can have multiple combinations of overlapping pairs. Suppose, pairing of two CESDAMs follow the pattern as given below:

m_1 with m_2, m_2 with m_3, m_3 with $m_4 \ldots m_{k-2}$ with m_{k-1}, m_{k-1} with m_k, m_k with m_1.

All CESDAMs are paired with two CESDAMs. All CESDAMs pairs have *a* level 2 external nodes, i.e., *a* numbers of overlapped nodes between two CESDAMs. Therefore, all CESDAMs will have $2a$ number of overlapped nodes with their predecessor and successor CESDAMs, each contain *a* number of overlapped nodes. Suppose, we covert subgraphs representing CESDAMs to nonoverlapping subgraphs by assigning overlapped nodes to predecessors as follows:

$$m_1 - a, m_2 - a, m_3 - a \ldots m_{k-1} - a, m_k - a,$$

such that,

$$\left(m_1 - a\right) + \left(m_2 - a\right) + \cdots + \left(m_k - a\right) = n$$
$$m_1 + m_2 + \cdots + m_k = n + ka.$$

Space requirement will be,

$$= \left(m_1 - a\right)^2 + \left(m_2 - a\right)^2 + \cdots + \left(m_k - a\right)^2$$
$$= m_1^2 + m_2^2 + \cdots + m_k^2 - 2a(m_1 + m_2 + \cdots + m_k) + ka^2$$
$$= m_1^2 + m_2^2 + \cdots + m_k^2 - 2a(n + ka) + ka^2$$

With Theorem 1, it is clear that the portion $(-2a(n + ka) + ka^2)$ in above expression is overhead due to overlapped nodes. In order to have less space requirement than adjacency matrix representation of entire graph $- 2a(n + ka) + ka^2 \leq 0$ has to be satisfied. Now,

$$= -2a(n + ka) + ka^2 = -2an - 2ka^2 + ka^2$$
$$= -2an - ka^2 < 0, \forall a, n, k > 0. \qquad \qquad \square$$

Theorem 2 proves that the CESDAM scheme requires less space than the adjacency matrix. It indirectly implies that the CESDAM is also space efficient than the another matrix-based scheme, i.e., the incidence matrix. Incidence matrix requires space $O(e \times n)$ and adjacency matrix requires $O(n^2)$, where e is the total number of edges and n is the total number of nodes. Since most of the connected graphs have $e > n$, adjacency matrix requires less space than incidence matrix. Thus, with Theorem 2 we can show that the CESDAM also requires less space than incidence matrix. On the other hand, sparse matrix format such as CSR requires less space than CESDAM, but CSR has its own limitations [27, 28]. Similarly, the list-based schemes such as adjacency list or incidence list are supposed to require a little less space than CESDAM due to the necessary blank spaces in the matrix, which we have showed in empirically in Section 6.

5.2 Algorithm complexity

Algorithms associated with CESDAM construction from raw data, edge list or adjacency matrix are not straight forward to establish any definite time complexity. The CESDAM construction process is highly dynamic in nature. For the entry of an edge, it is not possible to know the exact positions or even the CESDAM where the entries will be updated. Each entry of edge changes the structure of the appropriate CESDAM. Thus at any stage during execution of associated algorithms, size of the current CESDAM remains unclear. Moreover, for each of the five cases (Section 4.2) and new CESDAM creation, the corresponding algorithms experience different complexities. Overall complexities with respect to Algorithms 1, 3, 4, 5, and 6 are discussed as follows.

Every CESDAM is simply a matrix that contains information about a subgraph. Suppose, c_i and r_i are, respectively, the number of columns and rows present in the ith CESDAM at any stage of in-between execution. If an edge falls under case 1 and it is to be entered into 1st CESDAM then definitely it is the best case for Algorithm 2 with respect to Algorithm 3. If the edge falls under case 1, Algorithm 3 will be executed. Algorithm 3 requires column searching three times and row searching once[d] in order

---

[d] Note that the column searching is done with respect to a particular row and the row searching is done with respect to a particular column of current CESDAM.

to update the entries of the current CESDAM. Thus, time complexity of Algorithm 3 is $O(3c_i + r_i)$. Since, the edge is to be entered in 1st CESDAM, i.e., $i = 1$, the time complexity would be $O(3c_1 + r_1)$. Therefore, best case complexity of Algorithm 3 with respect to Algorithm 3 is $O(3c_1 + r_1)$. If the edge falls under case 2, then Algorithm 4 will execute. In this case, requires row searching twice. Thus, time complexity of Algorithm 4 is $O(2r_i)$. Algorithm 2 has to undergo column searching once to execute Algorithm 4. Therefore, best case complexity of Algorithm 2 with respect to Algorithm 4 is $O(c_1 + 2r_1)$. Similarly, for Algorithms 1, 5, and 6 require searching of columns three times, five times, and six times, respectively, to get executed from Algorithm 2. The time complexities of Algorithms 5 and 6 are $O(c_i + 3r_i)$ and $O(c_i + r_i)$, respectively. Thus, best case complexities of Algorithm 2 with respect to Algorithms 5 and 6 are $O(4c_1 + 3r_1)$ and $O(6c_1 + r_1)$, respectively. Algorithm 1 creates new CESDAMs that incurs constant cost, say x. Therefore, best case time complexity of Algorithm 2 with respect to Algorithm 1 is $O(x + 6c_i) = O(6c_i)$. The analysis shows that the best case time complexity of Algorithm 2 is very much dependent on the number of columns and rows present in the 1st CESDAM.

Suppose, there exist k CESDAMs at any stage in-between execution. If an edge falls under case 1 and it is to be entered into k^{th} CESDAM then it will be the worst case for Algorithm 2 with respect to Algorithm 3. Before reaching k^{th} CESDAM, Algorithm 2 has to dissatisfy all five cases for each of the $k - 1$ CESDAMs. Dissatisfaction of all five cases for any CESDAM will cost $6c_i$ where c_i is the number of columns present in i^{th} CESDAM. This amount of cost will incur of all $k - 1$ CESDAMs with respect to the number of columns present. Therefore, Algorithm has to experience $\sum_{i=1}^{k-1} 6c_i$ cost before reaching k^{th} CESDAM. Thus, worst case time complexity of Algorithm 2 with respect to Algorithm 3 is $O(3c_k + r_k + \sum_{i=1}^{k-1} 6c_i)$. Similarly, time complexity of Algorithm 2 with respect to Algorithms 4, 5, and 6 would be $O(c_k + 2r_k + \sum_{i=1}^{k-1} 6c_i)$, $O(4c_k + 3r_k + \sum_{i=1}^{k-1} 6c_i)$, and $O(6c_k + r_k + \sum_{i=1}^{k-1} 6c_i)$, respectively. For executing Algorithm 1 from Algorithm 2, all k CESDAMs has to be exhausted. Hence, Algorithm 2 has to experience $\sum_{i=1}^{k} 6c_i$ cost. Therefore, worst case time complexity of Algorithm 2 with respect to Algorithm 1 is $O(x + \sum_{i=1}^{k} 6c_i)$.

Processing $m = |E|$ edges of a graph $G(V, E)$ for CESDAM construction from raw data or with Algorithm 7 and Algorithm 8 have to execute Algorithm 2. Therefore, additional cost incurred on Algorithm 2 for each of m edge will be the overall time complexity of CESDAM construction. However, it is not possible to sum costs incurred by Algorithm 2 because of multiple uncertainties as mentioned above. Thus, *overall complexity of CESDAM construction is explained under certain assumption that may not practically possible for real-world graphs.* Assuming that constant k CESDAMs are there and all having equal number of columns and rows c. With these assumptions following theorems are stated.

Theorem 3. *Best case time complexity of CESDAM construction process is $\Theta(m)$.*
Proof. Assumptions:

- Constant k number of CESDAMs.
- All CESDAMs have c columns and c rows.

Under these assumptions, best case time complexities of Algorithm 2 with respect to Algorithms 1, 3, 4, 5, and 6 would be $O(5c)$, $O(3c)$, $O(7c)$, $O(7c)$, and $O(6c)$, respectively, to process any edge. Among these complexities $O(3c)$ is the least. Assume that processing of all edges experience same cost $O(3c)$. There are m edges in the graph which needs to be processed for CESDAM construction. Processing of each edge cost $O(3c)$ for the best case. Thus, total cost incurred for CESDAM construction would be $O(3mc)$. Here, 3 is very small compared to m and c is constant so time complexity can be lower bounded by $\Theta(m)$.   □

Theorem 4. *Worst case time complexity of CESDAM construction process is $O(mk)$.*
Proof. Assumptions:

- Constant k number of CESDAMS.
- All CESDAMs have c columns and c rows.

Under these assumptions, worst case time complexities of Algorithm 2 with respect to Algorithms 1, 3, 4, 5, and 6 would be $O(6kc - 2c)$, $O(6ck - 3c)$, $O(c + 6kc)$, $O(c + 6kc)$, and $O(6kc)$, respectively, to process any edge. Among these complexities $O(c + 6kc)$ is the highest. Assume that processing of all edges experience same cost $O(c + 6kc)$. There are m edges in the graph which needs to be processed for CESDAM construction. Processing of each edge costs $O(c + 6kc)$ for the worst case. Thus, total cost incurred for CESDAM construction would be $O(mc + 6mkc)$, i.e., $O(mkc)$. If $c \ll m$ is constant then the time complexity can be upper bounded by $O(mk)$.   □

6. Empirical analysis

6.1 Evaluation strategy

The primary claim of this work is that the proposed CESDAM scheme is very efficient in terms of memory requirement. The CESDAM scheme also has an advantage over existing matrix-based scheme especially when the graph is sparse since it incorporates connectivity pattern of the graph. Considering both these aspects, we have considered data set comprising sparse and large enough to prove the claims. We have considered following measures to compare the efficiency with existing schemes. If the memory requirement of CESDAM scheme is less than the existing schemes then we compute percentage of memory saved by the CESDAM scheme over existing schemes as follows:

$$\text{Memory Saving Percentage } (\%) = \frac{E - P}{E} \times 100 \qquad (3)$$

where, E and P are the memory required by the existing schemes and the proposed scheme, respectively. If the memory requirement of CESDAM scheme is more than the existing schemes then we compute required memory ratio as follows:

$$\text{Required Memory Ratio } (\text{RMR}) = \frac{E}{P} \qquad (4)$$

where, E and P are the memory required by the existing schemes and the proposed scheme, respectively. We also examine exactly how many times, i.e., $x = \lfloor RMR \rfloor$ more memory is required by the CESDAM scheme in comparison to others.

The experimental analysis is divided into three folds. As a matter of priority, the efficiency of CESDAM scheme in terms of memory requirement is compared with four existing schemes. In the second part, the influence of the number of nodes, the number of edges and graph density on CESDAM scheme is analyzed and discussed advantages over existing schemes. Lastly, showed how the CESDAM scheme overcomes the major disadvantage of existing matrix-based schemes related to sparse graph with the incorporation of connectivity pattern of the graph.

6.2 Experimental setup

Proposed CESDAM representation scheme and CESDAM generating algorithm are analyzed with real-world data sets. We have considered eight data sets of different domains from SNAP database [33] that are publicly available. The data set comprises one Wikipedia voting network, two email networks, two collaboration networks, two citation networks and one Amazon purchasing network. Detail of the data set is summarized in Table 1. The CESDAMs of these graphs are generated with the CESDAM generating algorithm. The CESDAM generating algorithm is implemented on MATLAB version R2010a using MATLAB scripting language. All experiments are done on a 64-bit Computer having Intel (R) Core (TM) i3-3217U CPU@ 1.80 GHz processor and 6 GB memory.

6.3 Efficiency in terms of memory requirement

A comparative analysis of memory required by other representation schemes with CESDAM scheme is performed. Memory saved with the CESDAM is analyzed for matrix-based schemes. While additional memory required with the CESDAM is analyzed for list-based schemes.

Results obtained on the eight graphs are presented in Table 2. For all of the eight graphs in the data set, matrix-based approaches required very high space in comparison to the CESDAM scheme. For instance, com–amazon graph requires $8.56E + 05MB \approx 836GB$ of memory with

Table 1 Details of the real-world data sets.

Networks	Data sets	#Nodes	#Edges
Wikipedia	wiki-Vote	7, 115	1, 03, 689
Email	email-Enron	36, 692	3, 67, 662
	email-EuAll	2, 65, 214	4, 20, 045
Collaboration	ca-AstroPh	18, 772	3, 96, 160
	ca-CondMat	23, 133	1, 86, 936
Citation	cit-HepTh	27, 770	3, 52, 807
	cit-HepPh	34, 546	4, 21, 578
Purchasing	com-amazon	3, 34, 863	9, 25, 872

Table 2 Comparison of space requirement for CESDAM with other representation schemes.

Data sets	Incidence list	Incidence matrix	Adjacency list	Adjacency matrix	CESDAM
		Memory required (MB)			
wiki–Vote	4.8008	5.6286E+003	3.2186	386.2246	27.5614
email–Enron	17.1102	1.0292E+005	11.5001	1.0271E+004	34.9688
email–EuAll	44.9379	8.4993E+005	14.8422	5.3664E+005	82.5742
ca–AstroPh	18.2780	5.6738E+004	12.2331	2.6885E+003	45.6549
ca–CondMat	8.7337	3.2992E+004	5.8813	4.0828E+003	10.2281
cit–HepTh	16.3621	7.4749E+004	10.9787	5.8836E+003	42.6021
cit–HepPh	19.5619	1.1111E+005	13.1291	9.1051E+003	47.6678
com–amazon	21.2516	2.3654E+006	30.8102	8.5551E+005	94.2859

adjacency matrix, which is very high in comparison to CESDAM scheme that requires only $\approx 94 MB$ memory. Thus, with CESDAM we have 99.88% memory saving over adjacency matrix. The com-amazon graph is large and quite sparse so matrix-based schemes that are only dependent on number of nodes and number of edges, they required large amount of memory. On the contrary, CESDAM is dependent on connectivity pattern in addition to the number of nodes and number of edges, which result in significant amount of memory reduction (detailed in coming paragraph). The CESDAM scheme also requires comparatively less memory than existing matrix-based schemes in case dense graph. For instance, CESDAM scheme has saved 99% and 92% of memory required by the incidence matrix and adjacency matrix, respectively, for wiki-vote graph that is comparatively dense. The results in Table 3 show that for most of the graphs, more than 99% of the memory required by the existing matrix-based schemes have been saved with CESDAM.

The list-based schemes are the compact way of graph representation, which does not maintain information other than the presence of nodes and edges. Thus, it is obvious that matrix-based schemes to require more memory than list-based schemes because existing matrix-based schemes maintain additional information about the absence of edges. The results presented in Table 4 show that CESDAM scheme requires almost three

Table 3 Space saving percentage (%) with CESDAM in relative to the memory required by the matrix-based approaches, i.e., incidence matrix and adjacency matrix.

	Space saving (%) over	
Data sets	Incidence matrix	Adjacency matrix
wiki–Vote	99.51	92.86
email–Enron	99.97	99.66
email–EuAll	99.93	99.81
ca–AstroPh	99.92	98.30
ca–CondMat	99.96	99.75
cit–HepTh	99.94	99.28
cit–HepPh	99.96	99.48
com–amazon	99.99	99.88
Average	99.89	98.63

Table 4 Memory required by the proposed CESDAM scheme in reference to the list-based schemes.

	CESDAM requires "x" times memory required by	
Data sets	Incidence list	Adjacency list
wiki–Vote	5	8
email–Enron	2	3
email–EuAll	1	5
ca–AstroPh	2	3
ca–CondMat	1	1
cit–HepTh	2	3
cit–HepPh	2	3
com–amazon	4	3
Average	2.375	3.625

Here, $x = \lfloor RMR \rfloor$.

times more memory than list-based schemes on average. A little higher memory requirement of CESDAM scheme than list-based schemes is completely justifiable because it has to maintain certain blank spaces. However, the blank spaces are not meant for the absence of edges as in the case of existing matrix-based schemes. Overall, the CESDAM scheme is very efficient in comparison to the present matrix-based schemes, and it is also comparable to the list-based approaches since CESDAM took only three times more memory than list-based schemes on average.

6.4 Influence of graph size and density

The size of a graph is defined in terms of both number of nodes and number of edges. Memory saving percentage is considered in the case of matrix-based schemes. The impact of both number of nodes and number of edges on memory saving with CESDAM over matrix-based schemes is presented in Fig. 5. The percentage of memory saving increases with the increment in number of nodes. The percentage of memory saving also

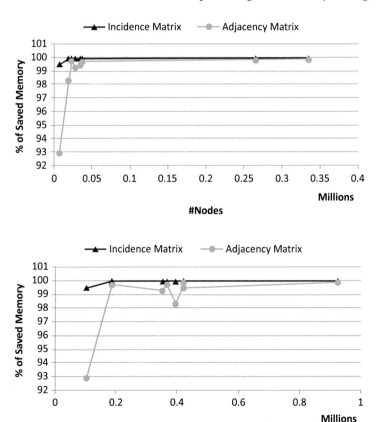

Fig. 5 Impact of #Nodes and #Edges on memory saved % of CESDAM over matrix-based schemes.

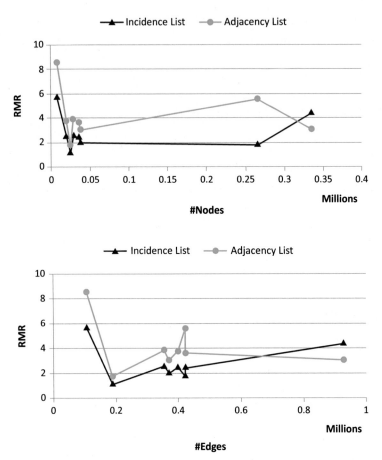

Fig. 6 Impact of #Nodes and #Edges on memory required by CESDAM in reference to list-based schemes.

increases with the increment in number of edges. On the other hand, required memory ratio (RMR) is considered in case of list-based schemes. The impact of both number of nodes and number of edges on RMR of CESDAM scheme with respect to list-based schemes is presented in Fig. 6. The RMR decreases with the increment in number of nodes. The RMR also decreases with the increment in number of edges. Most of the cases, RMR values remain in between two and four. However, for large and sparse graph RMR becomes little bit high. Larger the graph implies higher the memory saving with CESDAM over existing matrix-based schemes, and in this case, if the graph is sparse then CESDAM requires a little more memory than existing list-based schemes.

The major disadvantage of existing matrix-based schemes is that even if the graph is sparse, they require very large amount of memory. Existing matrix-based schemes are designed based on the nodes and edges so both sparse and dense graph of the same number of nodes require the same amount of memory. On the contrary, the designing of CESDAM incorporates the connectivity pattern in addition to both number of nodes and number edges. Therefore, sparse graph requires less memory than the dense graph of the same number of nodes. We have presented the influence of graph density on CESDAM in contrast to existing schemes in Fig. 7. Memory saving percentage of CESDAM over matrix-based schemes decreases with increment in graph density. The rate of decrement

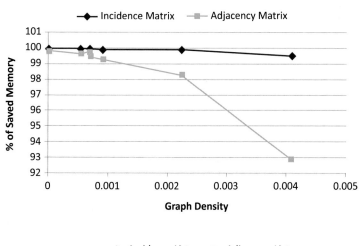

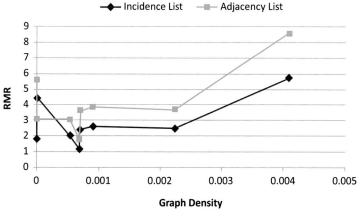

Fig. 7 Impact of graph density on CESDAM scheme.

in memory saving percentage of CESDAM is quite low for incidence matrix in comparison to adjacency matrix. Nevertheless, the memory saving percentage remains above 90% for comparatively dense as well as sparse graphs. The RMR of CESDAM with respect to list-based schemes increases with increment in graph density. The RMR value of CESDAM with respect to incidence matrix is less than that of adjacency matrix. The RMR increases and memory saving decrease for CESDAM over existing list-based and matrix-based schemes, respectively. Thus, the CESDAM scheme would be better for sparse graphs. To ensure effectiveness of CESDAM when the graph is large, one can notice the trend of graph density on real-world networks. The trend of graph density for the data set is presented in Fig. 8. Note that the density reduces with increment of

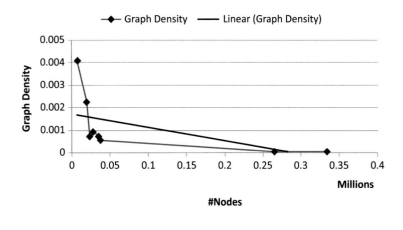

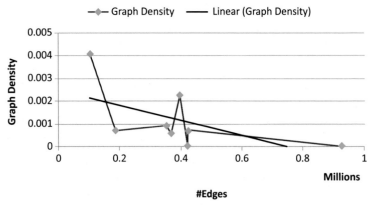

Fig. 8 Graph density trend of the data set. The *black lines* are the linear regression lines, which indicate the trend whether the density is decreasing or increasing.

both number of nodes and number of edges. That is, large real-world graphs have comparatively less density. Therefore, the memory saving percentage of CESDAM over matrix-based will be high, and RMR of CESDAM with respect to list-based schemes will be low.

6.5 Influence of connectivity pattern

Details of CESDAMs obtained for all the graphs are presented in Table 5. The CESDAMs of subgraphs have different number of rows and columns. The size of CESDAM (i.e., rows × columns) varies with the connectivity or topology of the subgraphs. Thus, CESDAM avails the flexibility to represent subgraphs of same number of nodes in different sized CESDAMs. Since the size of a CESDAM is determined by the number of rows and number of columns, we have first analyzed both for the CESDAMs generated for the eight graphs. Fig. 9 presents variation of rows and columns of CESDAMs. The analysis is done with reference to the Matrix with same number of nodes having equal rows and columns, i.e., adjacency matrix of subgraph. (Since incidence matrix mostly have higher size than adjacency matrix, we considered only adjacency matrix of subgraphs.) Clearly, most of the CESDAMs have higher columns than rows. It can also be noted in Table 5 that average number of columns is higher than rows. The unequal number of rows and number of columns results in smaller sizes than the equal sized matrix, which can be noted in Fig. 10. Clearly, the results show almost all CESDAMs have smaller sizes than the corresponding adjacency matrix representation of the subgraphs. Interestingly, if we go vertically through any specific number of nodes one can notice that sizes of CESDAMs are different even though the number of nodes is same in the subgraph. Since the sizes of CESDAMs are influenced by the connectivity pattern of the subgraph, the overall memory required by the CESDAM scheme for the same number of nodes will be different.

7. Conclusion and future work

In this chapter, a matrix-based graph representation scheme called CESDAM is proposed. The proposed scheme represents subgraphs of graph in multiple entities called CESDAM, whereas classical graph representation schemes represent entire graph in single entity such as adjacency matrix or incidence matrix. Moreover, connectivity dependent CESDAM representation allows different rows and columns for the subgraphs with same number of nodes. Empirically shown that most of the CESDAMs have

Table 5 Details of CESDAMs generated.

Data sets	#CESDAMs	#Nodes in CESDAM		#Rows		#Columns		Memory (MB)	
		Average	Maximum	Average	Maximum	Average	Maximum	Average	Maximum
wiki-Vote	510	189.6373	510	43.3941	145	148.249	474	0.054	0.3883
email-Enron	5326	60.8019	249	16.7092	95	45.7918	244	0.0066	0.1018
email-EuAll	17730	26.6950	7640	9.5765	7639	20.0893	7636	0.0329	45.034
ca-AstroPh	3370	89.9861	306	24.5407	84	68.2000	299	0.0135	0.0724
ca-CondMat	4908	30.3739	94	14.034	39	18.0322	64	0.0021	0.0164
cit-HepTh	5116	68.1618	205	21.5295	79	47.6665	197	0.0083	0.0500
cit-HepPh	6534	64.7567	127	23.1411	60	42.3647	119	0.0073	0.0292
com-amazon	18469	52.9539	94	22.9277	41	30.4702	87	0.0051	0.0088

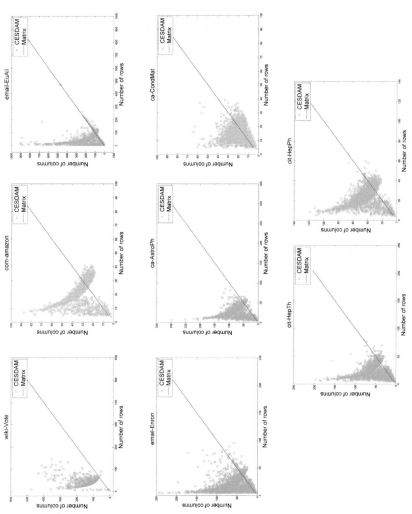

Fig. 9 *Rows vs columns* in CESDAMs. Here, Matrix is the adjacency matrix representations of same subgraphs that are presented in CESDAMs. Since, *row* and *columns* are equal, a line is obtained. Points above the line means *row < column*, points below the line means *row > column* and points on the line means *row = column*.

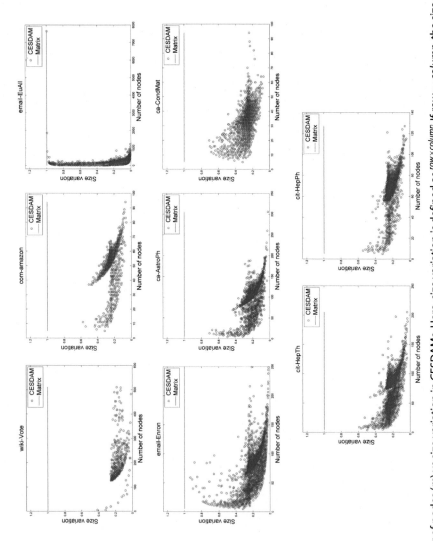

Fig. 10 Number of nodes (m) vs size variation in CESDAMs. Here, size variation is defined as $\frac{row \times column}{m^2}$. If $row = column$, the size variation gives value 1, otherwise it gives value less than 1.

higher columns than rows and sizes of CESDAM (*row × column*) are less than the corresponding matrix-based representation of those subgraphs. In addition, an algorithm for generating CESDAMs is also proposed. The time complexity of the algorithm is bounded by the best case $\Theta(m)$ and worst case $O(mk)$, where m is the number of edges in the graph and k is the number of subgraphs generated for the graph that are represented as CESDAMs. The CESDAMs of all eight graphs are generated with this algorithm.

The CESDAM is very efficient in comparison to the present matrix-based representation schemes in terms of memory requirement. Both theoretical, as well as empirical analysis, are evident for this. Empirical analysis on real-world data sets shows more than 99% memory saving over present matrix-based schemes is possible with CESDAM. Moreover, existing matrix-based schemes require large amount of memory even when the graph is sparse, whereas CESDAM scheme requires less memory because the size of CESDAM is dependent on connectivity pattern. Though CESDAM scheme is space efficient than matrix-based schemes, it requires a little more memory than list-based schemes due to blank spaces in the matrix. With CESDAM scheme, unweighted, undirected and directed graphs can be represented very easily. However, at present weighted graph cannot be represented with CESDAM scheme since only presence and absence of any link is stored in CESDAM. The proposal of CESDAM scheme opens up the doors for the diverse applications of different domains that are confronting with memory requirement. In this regard, an immediate research problem emerges is the designing of generic or application specific methodology for efficient processing of the CESDAMs. The subgraphs of the graph is represented in multiple CESDAMs. Therefore, preparation of meta-graph on top of CESDAMs would be helpful in designing the algorithms for the processing of CESDAMs.

References

[1] X. Wu, X. Zhu, G.-Q. Wu, W. Ding, Data mining with big data, IEEE TKDE 26 (1) (2014) 97–107, https://doi.org/10.1109/TKDE.2013.109.

[2] A. Harth, K. Hose, M. Karnstedt, A. Polleres, K.-U. Sattler, J. Umbrich, Data summaries for on-demand queries over linked data, in: Proceedings of the 19th International Conference on World Wide Web, ACM, New York, NY, USA, 2010, pp. 411–420, ISBN: 978-1-60558-799-8, https://doi.org/10.1145/1772690.1772733.

[3] S. Stadtmüller, S. Speiser, A. Harth, R. Studer, Data-Fu: a language and an interpreter for interaction with read/write linked data, in: Proceedings of the 22nd International Conference on World Wide Web, International World Wide Web Conferences

Steering Committee, Republic and Canton of Geneva, Switzerland, 2013, pp. 1225–1236, ISBN: 978-1-4503-2035-1. http://dl.acm.org/citation.cfm?id=2488388.2488495.

[4] R. Guha, R. Kumar, P. Raghavan, A. Tomkins, Propagation of trust and distrust, in: Proceedings of the 13th International Conference on World Wide Web, ACM, New York, NY, USA, 2004, pp. 403–412, ISBN: 1-58113-844-X, https://doi.org/10.1145/988672.988727.

[5] M. Gupta, Y. Sun, J. Han, Trust analysis with clustering, in: Proceedings of the 20th International Conference Companion on World Wide Web, ACM, New York, NY, USA, 2011, pp. 53–54, ISBN: 978-1-4503-0637-9, https://doi.org/10.1145/1963192.1963220.

[6] R. Balakrishnan, S. Kambhampati, M. Jha, Assessing relevance and trust of the deep web sources and results based on inter-source agreement, ACM Trans. Web 7 (2) (2013) 11:1–11:32, https://doi.org/10.1145/2460383.2460390.

[7] Y. Qian, S. Adali, Foundations of trust and distrust in networks: extended structural balance theory, ACM Trans. Web 8 (3) (2014) 13:1–13:33, https://doi.org/10.1145/2628438.

[8] Z. Su, L. Liu, M. Li, X. Fan, Y. Zhou, Reliable and resilient trust management in distributed service provision networks, ACM Trans. Web 9 (3) (2015) 14:1–14:37, https://doi.org/10.1145/2754934.

[9] J. Zhang, Z. Fang, W. Chen, J. Tang, Diffusion of "following" links in microblogging networks, IEEE Trans. Knowl. Data Eng. 27 (8) (2015) 2093–2106, https://doi.org/10.1109/TKDE.2015.2407351.

[10] K.S.M.T. Hossain, D. Patnaik, S. Laxman, P. Jain, C. Bailey-Kellogg, N. Ramakrishnan, Improved multiple sequence alignments using coupled pattern mining, IEEE/ACM Trans. Comput. Biol. Bioinform. 10 (5) (2013) 1098–1112, https://doi.org/10.1109/TCBB.2013.36.

[11] A. Birlutiu, F. d'Alche Buc, T. Heskes, A Bayesian framework for combining protein and network topology information for predicting protein-protein interactions, IEEE/ACM Trans. Comput. Biol. Bioinform. 12 (3) (2015) 538–550, https://doi.org/10.1109/TCBB.2014.2359441.

[12] C.A. Hill, G.P. Zhang, G.D. Scudder, An empirical investigation of EDI usage and performance improvement in food supply chains, IEEE Trans. Eng. Manag. 56 (1) (2009) 61–75, https://doi.org/10.1109/TEM.2008.922640.

[13] M. Eskandarpour, P. Dejax, J. Miemczyk, O. Péton, Sustainable supply chain network design: an optimization-oriented review, Omega 54 (2015) 11–32.

[14] Y. Zhu, C. Yan, Graph methods for predicting the function of chemical compounds, in: 2014 IEEE International Conference on Granular Computing (GrC), October, 2014, pp. 386–390, https://doi.org/10.1109/GRC.2014.6982869.

[15] W. Zheng, L. Zou, X. Lian, D. Wang, D. Zhao, Efficient graph similarity search over large graph databases, IEEE Trans. Knowl. Data Eng. 27 (4) (2015) 964–978, https://doi.org/10.1109/TKDE.2014.2349924.

[16] V. Fionda, G. Pirrò, C. Gutierrez, NautiLOD: a formal language for the web of data graph, ACM Trans. Web 9 (1) (2015) 5:1–5:43, https://doi.org/10.1145/2697393.

[17] B. Wu, V. Goel, B.D. Davison, Topical TrustRank: using topicality to combat web spam, in: Proceedings of the 15th International Conference on World Wide Web, ACM, New York, NY, USA, 2006, pp. 63–72, ISBN: 1-59593-323-9, https://doi.org/10.1145/1135777.1135792.

[18] L. Becchetti, C. Castillo, D. Donato, R. Baeza-YATES, S. Leonardi, Link analysis for web spam detection, ACM Trans. Web 2 (1) (2008) 2:1–2:42, https://doi.org/10.1145/1326561.1326563.

[19] Q. Jiang, L. Zhang, Y. Zhu, Y. Zhang, Larger is better: seed selection in link-based anti-spamming algorithms, in: Proceedings of the 17th International Conference on World Wide Web, ACM, New York, NY, USA, 2008, pp. 1065–1066, ISBN: 978-1-60558-085-2, https://doi.org/10.1145/1367497.1367658.

[20] X. Zhang, Y. Wang, N. Mou, W. Liang, Propagating both trust and distrust with target differentiation for combating link-based web spam, ACM Trans. Web 8 (3) (2014) 15:1–15:33, https://doi.org/10.1145/2628440.

[21] A. Biswas, B. Biswas, Investigating community structure in perspective of ego network, Expert Syst. Appl. 42 (20) (2015) 6913–6934, https://doi.org/10.1016/j.eswa.2015.05.009.

[22] J. Cai, Z.-J. Zha, M. Wang, S. Zhang, Q. Tian, An attribute-assisted reranking model for web image search, IEEE Trans. Image Process. 24 (1) (2015) 261–272.

[23] D.B. West, Introduction to Graph Theory, vol. 2, Prentice Hall Upper Saddle River, 2001.

[24] Y. Saad, SPARSKIT: A Basic Tool Kit for Sparse Matrix Computations, RIACS, NASA, USA, 1990.

[25] J.E. Gonzalez, R.S. Xin, A. Dave, D. Crankshaw, M.J. Franklin, I. Stoica, Graphx: graph processing in a distributed dataflow framework, in: Eleventh USENIX Symposium on Operating Systems Design and Implementation (OSDI 14), 2014, pp. 599–613.

[26] D. Merrill, M. Garland, Merge-based sparse matrix-vector multiplication (SpMV) using the CSR storage format, in: Proceedings of the 21st ACM SIGPLAN Symposium on Principles and Practice of Parallel Programming, ACM, 2016, p. 43.

[27] T. Oberhuber, A. Suzuki, J. Vacata, New row-grouped CSR format for storing the sparse matrices on GPU with implementation in CUDA, Acta Tech. 56 (2011) 447–466.

[28] F. Khorasani, K. Vora, R. Gupta, L.N. Bhuyan, CuSha: vertex-centric graph processing on GPUs, in: Proceedings of the 23rd International Symposium on High-Performance Parallel and Distributed Computing, ACM, New York, NY, USA, 2014, pp. 239–252, ISBN: 978-1-4503-2749-7, https://doi.org/10.1145/2600212.2600227.

[29] V. Arnaboldi, M. Conti, A. Passarella, F. Pezzoni, Analysis of ego network structure in online social networks, in: 2012 International Conference on Social Computing (SocialCom) Privacy, Security, Risk and Trust (PASSAT), September, 2012, pp. 31–40, https://doi.org/10.1109/SocialCom-PASSAT.2012.41.

[30] S. Álvarez, N.R. Brisaboa, S. Ladra, O. Pedreira, A compact representation of graph databases, in: Proceedings of the Eighth Workshop on Mining and Learning With Graphs, ACM, New York, NY, USA, 2010, pp. 18–25.

[31] S. Kontopoulos, G. Drakopoulos, A space efficient scheme for persistent graph representation, in: 2014 IEEE 26th International Conference on Tools With Artificial Intelligence (ICTAI), November, 2014, pp. 299–303, https://doi.org/10.1109/ICTAI.2014.52.

[32] P. Macko, V.J. Marathe, D.W. Margo, M.I. Seltzer, LLAMA: efficient graph analytics using Large multiversioned arrays, in: 2015 IEEE 31st International Conference on Data Engineering (ICDE), April, 2015, pp. 363–374, https://doi.org/10.1109/ICDE.2015.7113298.

[33] J. Leskovec, A. Krevl, SNAP Datasets: Stanford Large Network Dataset Collection, 2014. http://snap.stanford.edu/data.

About the authors

Anupam Biswas is currently working as an Assistant Professor with the Department of Computer Science and Engineering, National Institute of Technology Silchar, Silchar, India. He has received the BE degree in Computer Science and Engineering from the Jorhat Engineering College, Jorhat, India, in 2011, MTech degree in Computer Science and Engineering from the Nehru National Institute of Technology Allahabad, Prayagraj, India, in 2013, and the PhD degree in Computer Science and Engineering from IIT (BHU) Varanasi, Varanasi, India, in 2017. He has authored or coauthored several research articles in reputed international journals, conference, and book chapters. His research interests include machine learning, deep learning, computational music, information retrieval, social networks, and evolutionary computation. He has served as a Program Chair for the International Conference on Big Data, Machine Learning and Applications (BigDML 2019). He has served as a General Chair for the 25th International Symposium on Frontiers of Research in Speech and Music (FRSM 2020) and coedited proceedings of FRSM 2020 published as book volume in Springer AISC Series. He has edited three books titled Health Informatics: A Computational Perspective in Healthcare, Principles of Social Networking: The New Horizon and Emerging Challenges in different Springer book series.

Bhaskar Biswas is currently working as Associate Professor with the Department of Computer Science and Engineering, Indian Institute of Technology (BHU). He has received the BTech degree in Computer Science and Engineering from Birla Institute of Technology, Mesra, Ranchi, India, in 2000, and the PhD degree in Computer Science and Engineering from the Indian Institute of Technology (BHU), Varanasi, India, in 2010. His research interests include data mining, text analysis, machine learning, fuzzy systems, online social network analysis, and evolutionary computation.

## CHAPTER TWO

Bivariate, cluster, and suitability analysis of NoSQL solutions for big graph applications

Samiya Khan[a], Xiufeng Liu[b], Syed Arshad Ali[a], and Mansaf Alam[a]
[a]Jamia Millia Islamia, New Delhi, India
[b]Technical University of Denmark, Lyngby, Denmark

Contents

Abstract

With the explosion of social media, the Web, Internet of Things, and the proliferation of smart devices, large amounts of data are being generated each day. However, traditional data management technologies are increasingly inadequate to cope with this growth in data. NoSQL has become increasingly popular as this technology can provide consistent, scalable and available solutions for the ever-growing heterogeneous data. Recent years have seen growing applications shifting from traditional data management systems to NoSQL solutions. However, there is limited in-depth literature reporting on NoSQL storage technologies for big graph and their applications in various fields. This chapter fills this gap by conducting a comprehensive study of 80 state-of-the-art NoSQL technologies. In this chapter, we first present a feature analysis of the NoSQL

Advances in Computers, Volume 128
ISSN 0065-2458
https://doi.org/10.1016/bs.adcom.2021.09.006

solutions and then generate a data set of the investigated solutions for further analysis in order to better understand and select the technologies. We perform a clustering analysis to segment the NoSQL solutions, compare the classified solutions based on their storage data models and Brewer's CAP theorem, and examine big graph applications in six specific domains. To help users select appropriate NoSQL solutions, we have developed a decision tree model and a web-based user interface to facilitate this process. In addition, the significance, challenges, applications and categories of storage technologies are discussed as well.

1. Introduction

With the rapid development attributed to widespread digitization in our society, the type, volume and variety of data has undergone dramatic changes in recent years, which is referred to as *big data*. Digitization and the increasing popularity of modern technologies such as the Web, smartphones and the Internet of Things (IoT), have contributed immensely to the "data deluge." In addition, since data are generated from heterogeneous sources, the data formats, types and structures are complex, making it difficult to manage with traditional database technologies [1]. Traditional relational data management systems are designed to manage well-structured data with a strictly defined schema, which has the challenge to meet performance and scaling requirements for big data. In fact, the performance of relational databases tends to decline as the volume of data increases, especially when dealing with semistructured data. In addition, the ability to provide real-time analysis will also decrease as the volume of data grows [2]. Therefore, there is a need for a novel data management system to address the big data problems [3].

Over the last decade, NoSQL databases have been increasingly used. Unlike relational technologies, NoSQL databases provide the support for schemaless data, flexible data models, and horizontal scalability needed for big data use cases. To support scalability, distributed NoSQL databases are emerging, which support parallel querying and data replication across different nodes, different data centers, or even different geographic locations. Data consistency is a challenge for distributed storage systems that keep data copies up-to-date at different nodes and locations. Therefore, the most basic option is to use single-operation, single-row ACID (atomicity, consistency, isolation, durability) and eventual consistency. To date, a wealth of such distributed NoSQL technologies is available, including HBase [4], Cassandra [5], Redis [6], DynamoDB [7], and many others. The provision of highly scalable, reliable, and efficient storage for dramatically growing volumes of data is the main driver for the development of NoSQL technologies, and today NoSQL databases have been used in a variety of operational systems or applications, including big graph applications such as web applications, social media, and IoT systems.

However, there is no in-depth study of the features in the literature and no recommendation as to which NoSQL technologies should be used under different user requirements and specific application domains. There were some attempts at reviewing NoSQL technologies, but they are far from being comprehensive. For example, the work [8, 9] compared the performance of a few NoSQL storage systems. Han et al. examined seven NoSQL databases according to the categories including key-value, document, and column-oriented on an abstract level and classified them based on the CAP (consistency, availability, partition tolerance) theorem [10]. Further qualitative studies on NoSQL database technologies were found in [11–16]. These studies examined a limited number of NoSQL solutions and explained their applications in detail. In addition, NoSQL solutions may have different data models and have different features that can be best suited for specific big graph applications or domains. It becomes difficult to understand the domain and difficult to choose an appropriate solution. However, none of the existing studies provide a statistical analysis of NoSQL solutions or suggest a suitability model for their applications. This motivates us to conduct a comprehensive study for the available state-of-the-art NoSQL solutions.

In this chapter, we will conduct a comparative study of 80 NoSQL solutions and explain how user requirements must be analyzed to determine the best solution for a given application. The qualitative study of the NoSQL solutions is transformed into a data set with nine features: document-oriented, graph, key-value, wide-column, consistent, available, partition-tolerant, free and proprietary. It is important to note that NoSQL solutions can be determined principle features in classification, and for this reason we focus on considering the following four main features, including document-oriented, graph, key-value, and wide-column in our study. Some NoSQL solutions support multiple data models. In addition, we will perform a bivariate analysis to investigate the relationships between the different features, closing the gap of inadequacy in the existing classification schemes. The results of this analysis indicate that there is a dependency between the identified features and therefore, in order to propose a realistic classification scheme, a clustering analysis will be carried out to segment the NoSQL solutions. The application domains supported by the NoSQL solutions will be identified and discussed as well. This analysis will further used to determine relevant features for specific classes of big graph applications or big data applications that can potentially be transformed into big graph applications in the future, which in turn is used to propose a predictive model to determine the suitability of NoSQL solutions for an application. In summary, this chapter aims to: (1) provide a comprehensive investigation of NoSQL solutions; (2) provide analytical models for users to select the appropriate NoSQL solutions to meet their needs; and (3) identify both challenges and opportunities in this area.

The rest of the chapter is structured as follows. Section 2 outlines the study; Section 3 describes the NoSQL technologies and their characteristics. Section 4 describes the classification criteria for NoSQL solutions; Section 5 lists the big graph application domains; Section 6 presents an exploratory analysis for the NoSQL solutions, including bivariate and clustering analysis; Section 7 presents a suitability model for the selection of NoSQL solutions; Section 8 discusses the technical and nontechnical aspects of NoSQL selection; and Section 9 summarizes the chapter and outlines future research directions.

2. Outline of study

Fig. 1 gives an overview of this study, including the classification of NoSQL solutions by features or characteristics and their applications. This study will cover a total of 80 NoSQL solutions.

First, we will evaluate these existing NoSQL solutions from both qualitative and quantitative perspectives. According to our literature review, NoSQL solutions are usually classified according to the data models they support [17], the CAP features [18] and all other features a solution supports. Data models include key-value, wide-column, document-oriented, and graph; and CAP features include consistency, partition tolerance, and availability. Others may include features such as free and proprietary. Based on this classification, a data set of NoSQL solutions and their characteristics is created for further analysis. This includes exploratory analysis to uncover the relationships between the features using Spearman's rank correlation analysis [19] and the Chi-square test [20]. The clustering analysis for segmenting NoSQL solutions will be conducted by using the k-modes clustering method [21].

Next, the big graph applications supported by individual NoSQL solutions will be investigated. The review of the big graph applications will fall

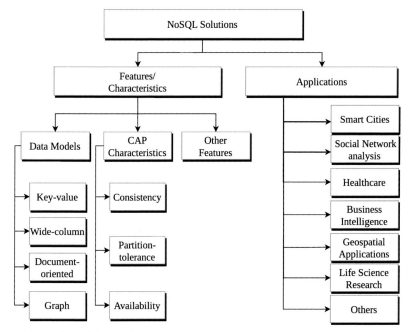

Fig. 1 Overview of the study.

into the following seven categories: Smart Cities, Business Intelligence, Life Science Research, Health Care, Social Network Analysis, Geospatial Applications, and Other. For each category, the most relevant characteristics will be determined and a prediction model will be created to determine whether a NoSQL solution is suitable for that category. The used techniques include the random forest classification [22] and the decision tree classification [23].

3. NoSQL for solving big data storage issues

The idea behind the development of relational databases was to provide a data storage paradigm that supports a structured query language (SQL) [24]. The introduction of relational databases can be dated back to the 1970s, when data schemas were not as complicated as they are today. In addition, storage was expensive and data archiving had a high cost. With the advent of the Web, smartphones and IoT, etc., the amount of data stored about events, objects and people has grown exponentially. The use of data today is not limited to storage or archiving, but extends to frequent data retrieval and processing to serve among many other purposes such as the generation of real-time feeds [25] and customized advertising [26].

Due to the complexity of the data and the need to handle multiple database queries to answer API requests or render webpages, the demands on modern database systems have been growing. Some of the key drivers are about the interactivity, increasing complexity, and ever-evolving networks of users [27]. To meet these growing demands, sophisticated deployment strategies and an improved computing infrastructure [28] are required. For these reasons, single-server deployments are expensive and highly complex, which have led to a tendency to use the Cloud [29]. In addition, the use of agile methods has also shortened development and deployment time [30], enabling faster response to user demands.

However, relational databases are insufficient to meet the agility and scalability requirements of modern systems. Moreover, they are not equipped to work with the cloud and take optimum advantage of its cheaper storage and processing capabilities. These shortcomings can be addressed with two main technical approaches, which are discussed below:

- **Manual sharding.** To use the distributed paradigm, a table has to be segmented into smaller units, which are then stored on different nodes. This process of segmentation is called manual sharding [31]. However, this functionality is not available in traditional data management systems, which requires to be implemented by developers. In addition, the data on each node are stored in an anonymous mode.

    It is the responsibility of an application to segment data, store it in a distributed manner, perform queries, aggregate results and present to users. Additional, the application has to support data rebalancing, performs join operations, handle failover and replication. But it is important to note that manual sharding might compromise some benefits of relational databases, such as transactional integrity.

- **Distributed cache.** Caching is a commonly used technique, primarily for improving the reading performance of a system [32]. It is worth noting that using a cache has no impact on write performance and can significantly increase the complexity of the overall system. Therefore, if the system requirements are read-intensive, the use of a distributed cache should be considered. On the other hand, write or read/write-intensive applications do not require a distributed cache [33].

NoSQL databases are known to alleviate the challenges associated with traditional databases [34]. In addition, they also unleash the true power of the cloud by leveraging commodity hardware, which reduces costs and simplifies deployment, making developers' life much easier as they do not have to manage multiple cache layers. NoSQL is an umbrella term to describe a myriad of technologies, all of which have some common characteristics that will be discussed later in this section. Some of the advantages of NoSQL solutions over traditional databases are presented as follows:

- **Scalability.** NoSQL allows a system to scale horizontally. The Cloud allows this to be accomplished quickly and without affecting the overall performance of the system. However, scaling traditional databases requires manual sharding, which is expensive and complex. In contrast, NoSQL solutions offer automatic sharding, which reduces both complexity and cost of the system [34].
- **Performance.** As already mentioned, NoSQL systems can be scaled out as needed. Increasing the number of systems can also improve the overall performance of the system. Automatic sharding means that the associated overhead can be eliminated, further contributing to the performance.
- **High and global availability.** Relational databases depend on primary and secondary nodes to meet availability requirements. This not only increases the complexity of the system but also makes the system moderately available. In contrast, NoSQL solutions use a masterless architecture and the data are distributed across multiple nodes. Therefore, even if a node fails, the availability of the application for both read and write operations remains unaffected. NoSQL solutions provide data replication across multiple locations. As a result, the user experience is consistent regardless of the user's location. This also significantly reduces latency, with the added benefit of shifting the developer's focus from database administration to business priorities.
- **Flexible data modeling.** It is possible to implement fluid and flexible data models in NoSQL. This allows developers to implement query options and data types that fit the application rather than fit the schema. This simplifies the interaction between the database and the application, making NoSQL a better option for agile development.

4. Classification criteria for NoSQL solutions

NoSQL is the technology developed to address the problems of relational databases and has been implemented in many ways through various models. Common characteristics of NoSQL models include efficient storage, reduced operational costs, high availability, high concurrency, minimal management, high scalability and low latency [35]. NoSQL solutions can be classified according to various criteria, the most commonly used of which are the supported data models (i.e., document-oriented, graph, key-value, and wide-column) [17] and CAP characteristics (i.e., consistency, availability, and partition tolerance) [18]. In the following of this section, NoSQL solutions will be examined using these two broad classification criteria.

4.1 Big data models

Chen et al. [3] presented a detailed classification of NoSQL solutions comprising the following nine categories: wide columnar store, document store, object database, tuple store, data structures server, key-value store, key-value cache, ordered key-value store, and eventually consistent key-value store. Another taxonomic study was made by North [36], which provided a comprehensive classification incorporating cloud-based solutions for the analysis. The NoSQL solutions were divided into six categories: entity-attribute-value data stores, Amazon platform column stores, key-value data stores, and distributed hash-table document stores.

Cattell [37] and Leavitt [38] each proposed a data model-based classification scheme. Cattell [37] divided NoSQL solutions into three categories: Key-value storage, document storage and extensible record storage. Conversely, Leavitt [38] proposed to use three categories: Document-based, key-value stores, and column-based stores. Scofield [17] introduced the scheme for classifying databases into relational, graph, document, column, and key-value stores, which is the most widely accepted to date. This method actually uses the classification based on a four-data model because, by its very nature, it contains all the different categories mentioned in other classification schemes.

4.1.1 Document-oriented data model

The document-oriented data model uses documents for storage and retrieval [39], also known as the document database or document store [40, 41]. It is primarily used for the management of semistructured data because of its flexibility and support for variable schemas. As mentioned above, document databases use documents for their work. These documents can be in PDF or Microsoft Word format. However, JSON and XML are the formats most commonly used by document databases.

A relational database contains columns that are described by their names and data types. In contrast, in a document database, data types, descriptions, and values are within a document itself [39]. The structure of the documents in the same database can be similar or different. Because a document database does not have the schema like a relational database, it does not require to add or update schema when new documents are added to the database. The documents in a document-oriented database are organized in the data structure, called *collection* [39], which is similar to the table in a relational database. A database can consist of many collections. Document-oriented databases support executing queries on collections to retrieve data that meets attribute-specific requirements. This approach has several advantages as follows:

- To date, the data originates from heterogeneous sources, especially many of them from IoT devices and social media. These data are not well structured, so they cannot be easily fitted into the standard relational data model. Document-oriented databases support flexible data models that can meet more data storage requirements than relational databases.
- In general, the writing performance of document-oriented databases is better than that of traditional relational databases. While data consistency can be less weighted if making a system available for faster writing, in this case data replication may take longer.
- Due to the use of inverted indexing structure, the query performance for a document-oriented database is known to be fast and efficient [39].

The NoSQL databases classified on the basis of the four models have their preferred and nonpreferred use cases. Table 1 makes a summary and comparison. In a document-oriented database, the document is its atomic unit for managing data. If the data of a given domain can be organized in documents, they can be managed by a document database; such successful use cases include CMS, blog software, and wiki software [42]. While in some other cases, where structured data are generated, a relational data model may also be a good option.

4.1.2 Graph data model

Graph is often used to model the data as a graph or network with nodes, edges and properties. It is often a challenge to fit large amounts of connected data into a relational database, as it is not optimized to connect tables or NoSQL databases by set of foreign keys. A graph database can circumvent this complexity by presenting data in a graph format—that is, as a collection of objects and their relationships, which can be regarded as nodes and edges in a graph, respectively. This data model is suitable to the use in applications like social network analytics and the Web search engine, which have a

Table 1 Use cases for different big data models.

Big data model	Preferred for use cases	Not preferred for use cases
Document-oriented	1. Content management systems (CMS) 2. E-commerce platforms 3. Blogging platforms 4. Analytics platforms	1. Applications requiring complex search queries. 2. Applications requiring complex transactions with multiple operations.
Key-value	1. Storage of user preferences. 2. Maintenance of user profiles that have no specific schema. 3. Storage of session data for users. 4. Storage of shopping carts' data for multiple users.	1. Specific data value needs to be queried. 2. Multiple unique keys need to be worked upon. 3. Frequent update of a part of the value. 4. Data values have established relationships with each other and the application requires exploitation of the same.
Graph	1. Network and IT operations 2. Graph based searches 3. Social networks 4. Fraud detection	Such a model is inappropriate for any application for which the data cannot be modeled as a graph. Transactional data, which is disconnected and in which relationship between data is not important, is an example.
Wide-column	1. Blogging platforms 2. Content management systems 3. Counter-based systems 4. Applications with write-intensive processing	1. Application requires complex querying. 2. Application has varying patterns of queries. 3. In scenarios where the database requirement is not established, the use of such a store must be avoided.

complex graph structure [43]. This requires the base technology to integrate the data from heterogeneous sources and establish links between the different data sets. The data can best be handled, for example, using a semantic graph database. Semantic graph database is a type of graph database that focuses on the relationships between different elements of the data and generate analytic results.

Graph databases are a good option for storing unstructured graph data generated by diverse sources with high velocity [44]. There is no necessary to define a schema before storing data, which makes the database quite flexible. Furthermore, graph databases are cost-effective and dynamic when it comes to integrating data from diverse sources [44]. In addition, graph databases are better suited for handling, storing, and processing data at high speed compared to relational databases. Graph databases can be used for real-time analysis because they are able to handle large data sets without defining schema in beforehand. The advantages of using graph databases can be summarized as follows:

- The integration of incoming data from different sources is limited if the schema needs to be specified before adding data, because adding a new source may require a change to the schema, which is both time-consuming and difficult. Since there is no such need for graph databases, data integration becomes simple, fast, and cost-effective [45].
- The graph database such as semantic web databases provide additional support for ontology or semantically rich data schemas [45]. Therefore, users can make use of ontology and the semantic data to create reasoning models.
- Semantic graph databases use the standard format (*subject, predicate, object*), called *triple*, to represent data on the Web [46]. The database for storing triple data is called triple store or RDF store [47]. The uniform resource identifier (URI) [48] is one of the standards used for data representation in the database with semantic graphs. URI is a unique ID to distinguish between linked entities and is used to identify the entities for access or search. Linked data are easy to integrate and share through the URI, when comes to open (linked) data. In addition, the challenges such as vendor lock-in can be solved.

Graph databases provide an effective way to manage the data with network-like relationships. A lot of enterprise data are linked data, which can use graph databases as the storage system to simplify the management. In addition, the concept of the networked world has gained increasing attention with the advent of social networks and IoT, where semantic graph databases can be used to provide personalized services. The semantic graph databases can integrate heterogeneous, interlinked data from different sources on the Web and smart devices. Table 1 summarizes the preferred and nonpreferred use cases of graph databases.

4.1.3 Key-value data model

A key-value store uses the policy of being schema-free and making data values opaque, making it be one of the most flexible data models available.

The flexibility of the database is manifested by the fact that applications have a complete control over the values through the keys. The values can be any types, including string, number, image, binary, counter, XML, JSON, HTML, video, among many others [49]. The key benefits of using key-value stores are as follows:

- The database does not impose on applications to structure their data in a specific form. Thus, the application is allowed to freely model its data according to the requirements of its use cases.
- An object can be easily accessed using a key that is assigned to the object. When using a key-value store, it is no longer required to perform operations such as union, join, and lock on objects [49], which makes this data model the most efficient and high performance.
- Key-value stores have a high flexibility, allowing for flexible scaling out as needed. Furthermore, this can be done using commodity hardware without the need of for any redesigning.
- It is easy for a key-value store to provide high availability and scalability. The distributed architecture and masterless configuration for available databases ensure a high resilience [49].
- The design of key-value stores is to facilitate the capacity. Moreover, their design are resilient to network failures and hardware malfunctions [49], lowering the downtime considerably.

Key-value stores are the frequently used data models, preferably for the applications involving the data such as user profiles, emails, blog/article comments, session information, shopping cart data, product reviews, product details and Internet Protocol (IP) forwarding tables [50], among many others. A key-value store can store complete web pages [51]. In this case, the URL can be used as the key and the content of the web page as the value. Table 1 lists the use cases.

4.1.4 Wide-column data model

A wide-column data model is composed of columns and column families, as base entities [52]. Facts or data are grouped together to form columns, which are further organized in the form of column families whose constructs are similar to the tables in a relational database. For example, the data about an individual like name, account name and address are the facts about the individual, which can be grouped together to form a row in a relational database. On the contrary, the same facts can be organized in the form of columns in a wide-column store and each of the columns includes multiple groups. Therefore, a single wide-column can store the data equivalent to the same stored by many rows in a relational database. Other alternative names of this data model include column-oriented DBMS [53], columnar databases [54], and column families [55].

The key advantages of using a wide-column store include:

- Data partitioning and compression can be performed efficiently using a wide-column store.
- Aggression queries like AVG, SUM, and COUNT can be performed efficiently because of the column-oriented structure.
- It is highly scalable and well suited for massively parallel processing (MPP) systems.
- Tables with large amounts of data can be loaded and queried with relatively short response times.

Wide-column stores are regarded as most suitable for distributed systems [56]. In other words, if the available data is large and can be distributed to different nodes, then a wide-column store is suitable. The most outstanding advantage of using a wide-column store is its high performance for some queries, especially for those whose values can be queried by keys, while it might be suboptimal for more complex queries with projections, joins and filter conditions. For these use cases, a conventional RDBMS solution may be better suited. Therefore, the decision to use wide-column store depends on specific use cases which are listed in Table 1.

4.2 CAP theorem

When discussing the applicability of NoSQL solutions to real-world problems, it is important to mention the CAP theorem [18]. This theorem introduces the concept of consistency (C), availability (A), and partition tolerance (P) for distributed systems and states that all these three characteristics cannot be guaranteed by a solution at the same time. In fact, a solution can have at most two characteristics. Consistency is about ensuring that all nodes of a distributed system must read the same data at all times. If a data change is made, the change must be consistent for all the nodes. However, if the change leads to an error, a rollback must be performed to ensure the consistency.

Availability defines the operational requirement of a system, which ensures that when a user makes a request to the system, it must respond despite its status. Partition tolerance refers to the ability of a system to operate despite the failure of a partition and loss of messages. It can also be described as the ability of a system to operate independently of a network failure.

The NoSQL solutions and their CAP status will be described in the follow-ing sections. Rijo [57] discovered that a distributed system can only have two characteristics at the same time. Based on this assertion, A NoSQL system can be CA (consistent-available), AP (available-partition tolerant), or CP (consistent-partition tolerant).

4.3 Other features

In addition, several other features can be considered for NoSQL solution clas-sification, including ownership (free or proprietary), concurrency control, replication model, partitioning scheme, supported programming languages, compression, and indexing methods [58]. In this chapter, we consider own-ership as the feature for classification, while the other features will be consid-ered in the future work, which will improve the proposed classification scheme (Section 6) and the proposed suitability model (Section 7).

5. Big graph applications

Big data technologies have been applied in a variety of areas and domains. In light of this, 152 related resources were examined to determine the possible uses of the various NoSQL solutions (see Tables 4–9). In this context, big graph applications can be roughly divided into the following seven categories. The objective of this categorization is to identify popular application areas in which NoSQL solutions have been used. This catego-rization will be used in Section 7 for further analysis.

- **Smart cities.** The existing literature [59] suggests that smart cities have an ample implementation of IoT applications, aiming at improving the quality of life in urban. Given the wide scope of applications, a large proportion of the research studies are belonging to this category. The applications include smart education [60–67], smart waste management [68], smart agriculture [69], smart governance [70, 71], intelligent natural resources management [72, 73] and intelligent traffic control systems [74–76], among many others.
- **Social networks analysis.** The rise of the Internet has made social net-works popular, and the development of this technology has revolutionized the way people connect and communicate. The social network is a big source of data generation, with a huge amount of data being generated every second, including not only text data but also various forms of multimedia data. Research studies related to social network analysis include storage systems for social network data [77, 78], data management

for social networks [79], processing [80], graph analysis [81], real-time processing [82] and analysis [83], and social intelligence applications [71, 84]. Papers related to microblogging and sentiment analysis have also been included in this category. Applications related to microblogging include data management [85] and text extraction and real-time sentiment analysis [86].

- **Geospatial data analysis.** Geospatial data are another big data type that needs to be analyzed in order to gain valuable insights and make predictions for mission critical applications and projects. The applications associated with this domain span across data representation [87], semantic data management [88], large geospatial raster data management [89], and geospatial/GIS applications [90, 91].
- **Life sciences.** Research in the life sciences is one of the most affected areas by big data technologies, as many applications in bioinformatics depend largely on the analysis of genomics big data. Given that biologists and scientists in this field work with mammoth data sets that cannot be stored and managed by traditional systems, big data technologies can significantly help, e.g., for data administration and analysis. In fact, most of the work in this area focuses on data processing, storage and analysis, e.g., [92–94, 94, 94–98], with only one the application found in comparative genomics [99] by this survey.
- **Healthcare.** The rapid development of emerging technologies such as IoT, Cloud and Big Data has a significant impact on the healthcare industry. The literature examined for the use of NoSQL technologies in healthcare covers a wide range of applications. Data are typically collected by IoT-based sensors in medical applications [100]. Clinical databases [101] were developed with a special focus on the management of electronic medical records (EMR) [102]. Data analysis and monitoring applications include real-time analysis of the electrocardiogram (ECG) [103], medical imaging applications [102], and breast imaging analysis [104], Health information systems [105], biomedical network mining [106], biomedical applications [107], analysis of large clinical data [108], specific applications for radiology [109]. Some problem-specific solutions have also been found, such as the prediction of health hazards [110], the prediction of health problems based on toxicity assessment [111–114], the application of patient safety and the estimation of calorific expenditure [115].
- **Business intelligence.** Big graph applications in this category are designed to improve the decision-making and operational efficiency

for organizations. The applications include data management systems [116–118], problem-specific applications such as financial services [119] and service performance management [120], and the general applications [71, 121–124] for improving operational efficiency. It is important to note that industry-specific applications such as IT enterprise log analysis are not included in this category. They are classified under the category of Other.

- **Others.** Many industry-specific applications, such as the application for proactive semiconductor equipment maintenance [125] and analytical solutions for the construction industry [126] and the insurance industry [127], have been classified in this category. In addition, the IoT applications that do not fall into the above categories have also been put into this category. Last, task-specific systems such as document management systems [128–132], the indexing engine [133] and the biometric system, among many others, were also classified into this category.

6. Analysis of NoSQL solutions

This chapter examines 80 NoSQL solutions and then makes a qualitative assessment based on the supported features. For this study, we identified the following nine features by which we count the number of NoSQL solutions for each feature, i.e. if a feature is supported by a NoSQL solution, the corresponding value increases by 1. The distribution of NoSQL solutions across these nine features is shown in the histogram in Fig. 2. Finally, we use the generated data set to perform a quantitative analysis,

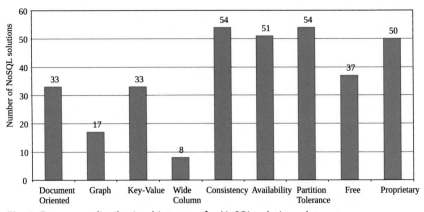

Fig. 2 Frequency distribution histogram for NoSQL solutions data set.

including a bivariate analysis and a cluster analysis, which are described in Section 6.1 and 6.2, respectively.

1. **Document–oriented**: Indicates support for document-oriented data model
2. **Graph**: Indicates support for graph data model
3. **Key–value**: Indicates support for key-value data model
4. **Wide–column**: Indicates support for wide-column data model
5. **Consistent**: Symbolic of consistency from CAP characteristics
6. **Available**: Symbolic of availability from CAP characteristics
7. **Partition–tolerant**: Symbolic of partition tolerance from CAP characteristics
8. **Free**: Represents whether the solution is available free of cost or is open source
9. **Proprietary**: Represents whether the solution is available at a price or subscription

6.1 Bivariate analysis

To investigate the relationship between different features, a bivariate analysis [134] was performed between the pairs of features (or variables). Given the categorical nature of the available data, Spearman's rank correlation [19] and the Chi-square test [20] were chosen for the bivariate analysis. To determine Spearman's rank correlation, the Pearson correlation between the values of two variables is calculated. This measure is used to quantify the statistical dependence between two variables. The value of Spearman's rank correlation coefficient is in $[-1, 1]$. While the negative sign represents a reciprocal association, a positive value indicates a direct association. Values greater than 0.4, whether positive or negative, indicate a moderate to strong association [19]. The heat map for coefficient values between pairs of features is shown to the left of Fig. 3, and four moderate correlations are shown to the right.

To investigate the existence of statistically significant relationships between different features, a Chi-square test was performed. The calculated P-values are compared with the threshold value of 0.05 [20]. If $P < 0.05$, the null hypothesis is rejected and it is concluded that there is a relationship between the two variables. In other words, the value of one variable can help predict the value of the other variable, and they can be called *dependent*. On the other hand, if the P-value is significantly high, the null hypothesis is confirmed and it is concluded that there is no relationship between the variables. The heat map for P-values corresponding to feature pairs is shown to the left of Fig. 4, and the P-values for these nine pairs are shown to the right.

Feature Pair	Value
Document Oriented and Key-Value	-0.4958
Availability and Consistency	-0.5232
Availability and Partition Tolerance	-0.4122
Free and Proprietary	-0.835

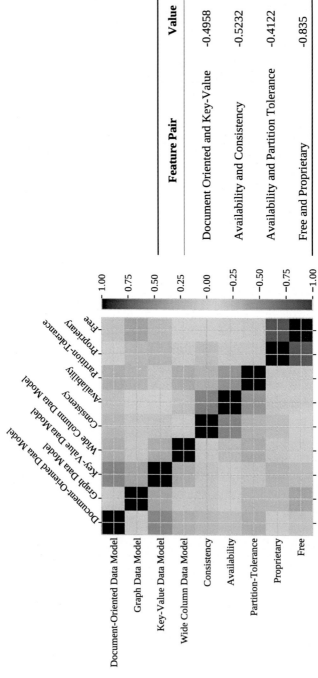

Fig. 3 Heatmap of spearman's rank correlation coefficient.

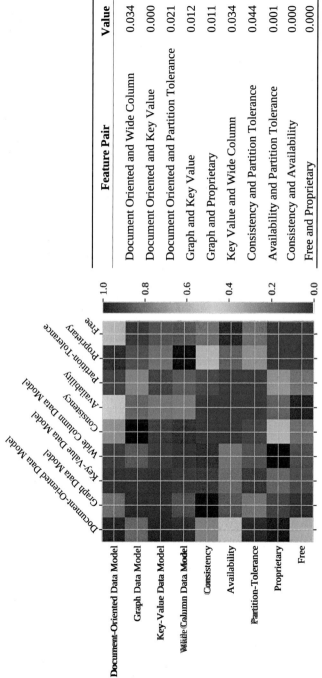

Feature Pair	Value
Document Oriented and Wide Column	0.034
Document Oriented and Key Value	0.000
Document Oriented and Partition Tolerance	0.021
Graph and Key Value	0.012
Graph and Proprietary	0.011
Key Value and Wide Column	0.034
Consistency and Partition Tolerance	0.044
Availability and Partition Tolerance	0.001
Consistency and Availability	0.000
Free and Proprietary	0.000

Fig. 4 Heatmap of chi-square *P*-value.

6.2 Clustering analysis

From the above results in Section 6.1, it is clear that relationships exist between different features and none of the existing classification schemes are independent enough to classify NoSQL solutions. Therefore, a combination of features must be used to create discrete categories for classification. Clustering analysis is performed to form clusters of NoSQL solutions from the available data set of 80 entries based on common features or characteristics. To create categories, the k-modes clustering technique [135] is used, taking into account the fact that the data set contains purely categorical data. Unlike k-means clustering, which clusters data points based on the Euclidean distance between them, the k-modes clustering technique forms clusters based on matching category values for different data points [21]. The implementation of k modes clustering makes use of Cao's initialization scheme [136]. In addition, since there is no established scheme for deciding the number of clusters, data distribution across the clusters is used to determine this value.

Table 2 provides insights into the distribution of data across clusters for different cluster counts. In the table, n is the number of clusters. All represents the execution results when all nine features were considered for clustering. DMCAP represents the execution results when seven features (document-oriented, graph, key-value, wide-column, consistency, availability, and partition tolerance) were considered for clustering. CAPFP represents the execution results when five features (consistency, availability, partition tolerance, free and proprietary) were considered for clustering. DMFP represents the execution results when six features (document-oriented, graph, key-value, wide-column, free and proprietary) were considered for clustering.

The cluster numbers for the four configurations, including All, DMCAP, CAPFP, and DMFP, are shown in Table 2. The results show that the cluster numbers of All are more evenly distributed compared to others. This again confirms that all characteristics should be considered to obtain a good classification scheme. The results show that the number of cluster 0 is largest for All, while all other clusters decrease from $n = 3$ to $n = 6$ and then increase again for $n = 7$. Furthermore, with $n = 7$, the numbers for clusters 3, 4, and 5 remain the same. Therefore, $n = 6$ is the best number of clusters. Therefore, the corresponding six clusters are created and shown in Table 3. The comparative analysis of the cluster from class I to VI is presented in Tables 4–9.

Table 2 Data distribution for different cluster counts.

n	C0				C1				C2				C3				C4			
	All	DMCAP	CAPFP	DMFP	All	DMCAP	CAPFP	DMFP	All	DMCAP	CAPFP	DMFP	All	DMCAP	CAPFP	DMFP	All	DMCAP	CAPFP	DMFP
3	37	37	49	52	28	20	18	25	15	23	13	3	11	10	3	3				
4	37	32	46	39	18	25	18	22	14	13	13	3	13	10	3	16				
5	26	32	45	48	12	20	18	22	20	13	13	3	11	9	3	16	9	5	1	1
6	22	29	43	30	18	15	18	22	15	13	13	3	11	9	3	16	9	5	1	1
7	24	29	41	28	14	15	18	22	11	13	13	3	11	9	3	14	9	3	1	1
8	22	28	26	26	14	15	18	18	11	13	13	13	11	9	3	3	6	3	1	1
9	22	27	15	15	14	15	18	18	11	13	12	12	10	9	3	3	6	3	1	1

n	C5				C6				C7				C8			
	All	DMCAP	CAPFP	DMFP	All	DMCAP	CAPFP	DMFP	All	DMCAP	CAPFP	DMFP	All	DMCAP	CAPFP	DMFP
3																
4																
5																
6	5	9	2	8	6	2	2	4								
7	5	9	2	8	6	2	2	2								
8	5	9	2	2	6	2	2	2	5	1	15	15				
9	5	9	2	2	6	2	2	2	5	1	15	15	1	1	12	12

Table 3 Cluster composition.

Class	NoSQL databases
I	Aerospike [137], Cassandra [5], CDB or Constant Database [138], etcd [139], FoundationDB [140], GT.M [52], Hibari [117], IBM Informix C-ISAM [141], Ignite [142], InfinityDB [143], LevelDB [144], Lightening Memory-Mapped Database (LMDB) [145], Memcached [146], MemcacheDB [147], Project Voldemort [148], Redis [6], Riak [149], RocksDB [150], Scalaris [16], ScyllaDB [151], Tarantool [152], Tokyo Cabinet and Kyoto Cabinet [153]
II	AllegroGraph [154], Cache [155], Cloudant [156], Coherence [15], CosmosDB [157], DocumentDB [158], IBM Informix [159], Lotus Domino [160], Marklogic [161], Microsoft SQL Server [9], MUMP Database [162], ObjectDatabase++ [163], Oracle NoSQL Database [164], Qizx [165], RavenDB [166], RocketU2 [167], SAP HANA [168], SimpleDB [169]
III	Amazon Neptune [170], AnzoGraph [121], Azure Tables [171], DataStax Enterprise Graph [172], Dynamo [173], Hazelcast [174], HyperGraphDB [175], InfiniteGraph [176], JanusGraph [177], KAI [178], Neo4j [179], Oracle Spatial and Graph [180], Sparksee [181], Sqrrl [182], XAP [183]
IV	ArangoDB [72], BaseX [184], CouchDB [185], CrateIO [186], ElasticSearch [187], eXist [188], Jackrabbit [189], OrientDB [190], PostgreSQL [191], Sedna [192], Solr [193]
V	Accumulo [194], Clusterpoint database [195], CouchBase Server [196], HBase [4], HyperTable [197], MongoDB [198], RethinkDB [199], TokuMX [200], TerraStore [201]
VI	BerkeleyDB [8], BigTable [202], GridGain Systems [203], NoSQLz [204], OpenLink Virtuoso [205]

6.2.1 Classification category: Class I

The class I category represents the NoSQL solutions with a key-value or wide-column data model. The comparative analysis includes the supported data model, CAP features, big graph applications, and other related features, which are shown in Table 4.

6.2.2 Classification category: Class II

The class II category represents the NoSQL solutions that do not support a wide-column data model and are free, which are shown in Table 5.

6.2.3 Classification category: Class III

The class III category represents the NoSQL solutions that document-oriented data model and ensure partition tolerance, which are shown in Table 6.

Table 4 Comparison of class I—NoSQL solutions.

S. No.	NoSQL solution	Data model	CAP characteristics	Other features	Big graph applications
1	Aerospike [137]	Key–value	Consistent and partition–tolerant	① Free; ② Highly scalable; ③ Flash-optimized, in-memory; ④ Reliable and consistent; ⑤ Used for applications like dynamic web portals, user profiling and fraud detection	① Data management and query handling for the social networks, LinkedIn and Foursquare for deployment of Voldemort and Riak respectively [79]
2	Cassandra [5]	Wide–column	Available and partition–tolerant	① Free; ② Distributed; ③ Highly available; ④ Master-less replication with robust support for clusters across multiple datacenters	① Analytics for proactive semiconductor equipment maintenance [125]; ② Context-driven analysis in cultural heritage environments [206]; ③ Storage and management of Life Sciences Databases using CumulusRDF, which performs linked data management on nested key-value stores [94]; ④ Management of data and query handling for the social networks, Facebook, Friendfeed, Foursquare and Twitter [79]; ⑤ IoT applications [156]
3	CDB or Constant Database [138]	Key–value	Consistent and available	① Free library; ② On-disk associative array that maps keys to values, allowing a key to have multiple values; ③ It can be used as a shared library	During literature review, no applications were found for this NoSQL solution
4	etcd [170]	Key–value	Available and partition–tolerant	① Free; ② Supports binary data; ③ Allows versioning, validation, collections, triggers, clustering, Lucene full-text search, ACLS and XQuery Update; ④ Uses XML over REST/HTTP	① Telecommunication applications [207]; ② Time critical applications [208]; ③ Traffic forecasting in real time for computing infrastructures [209]
5	FoundationDB [140]	Key–value	Consistent, available, and partition–tolerant	①Free;② Complies with ACID properties; ③ Scalable; ④ Allows replications; ⑤ Bindings for Python, C, PHP and Java, in addition to many other programming languages is available	① IoT applications [210]

Continued

Table 4 Comparison of class I—NoSQL solutions.—cont'd

S. No.	NoSQL solution	Data model	CAP characteristics	Other features	Big graph applications
6	GT.M [52]	Key-value	Available	① Free; ② Developed for transaction processing; ③ Supports ACID transactions; ④ Supports replication and database encryption	① Biomedical applications like electronic health records [107]
7	Hibari [117]	Key-value	Available and partition-tolerant	① Free; ② Distributed big data store; ③ Highly available; ④ Strongly consistent	① Classification and clustering of visual online information [211]; ② Social big data applications [84]; ③ Management of data for digital economy [117] and business intelligence [118]; ④ Data collection and classification from wireless sensor networks [212]
8	IBM Informix C-ISAM [141]	Key-value	Consistent and available	① This API complies with Open Standards [213]; ② Allows management of data files, which have been organized using B+ indexing; ③ It is the file storage used by Informix [159]	① Development of effective handheld solutions [214]
9	Ignite [142]	Key-value	Consistent, available, and partition-tolerant	① Free; ② Distributed, in-memory computing platform; ③ Provides caching and processing platform; ④ Provides support for ACID transactions and MapReduce jobs; ⑤ Allows partitioning, clustering and replication; ⑥ Highly consistent	① Data management of microblogs [85]; ② Data analysis of live traffic for intelligent city traffic management systems [76]
10	InfinityDB [143]	Key-value	Consistent, available, and partition-tolerant	① Proprietary; ② Completely developed in Java and includes DBMS and database engine; ③ Based on B-tree architecture; ④ Provides high performance; ⑤ Reduces risks associated with failures	① On-device database for mobiles [215]
11	LevelDB [144]	Key-value	Consistent and partition-tolerant	① Free; ② Maintains byte arrays for storing key and value pairs; ③ Data compression is supported by means of Snappy; ④ Supports forward/backward iteration and batch writing; ⑤ Used as a library	① Analytical applications for IoT [216]; ② Web-based system to explore tourist network in New Delhi [217]; ③ Mission critical applications like call for fire [218]; ④ Secondary storage for blockchain [219]

#	Name	Type	CAP	Features	Applications
12	Lightening Memor1-Mapped Database (LMDB) [145]	Key-value	Consistent and available	① Free; ② Embedded database; ③ High performance; ④ Provides API bindings for many programming languages; ⑤ Employs multiversion concurrency control is offers high levels of reliability	① Application that provides an estimate of an individual's calorific expenditure [115]; ② Database of violent audio-video content [220]; ③ Aerial sensing applications that process image data in real time [221]; ④ Searching, mapping and visualizing bioinformatics identifiers and keywords [97]
13	Memcached [146]	Key-value	Consistent and partition-tolerant	① Free; ② Memory caching system that is general purpose and distributed; ③ Scalable architecture; ④ Supports sharding	① Filesystem for eScience applications [222]; ② Used on Wikipedia backend to reduce load on database [223]; ③ Used in call centers to alleviate load on database [224]
14	MemcacheDB [147]	Key-value	Consistent and partition-tolerant	① Free; ② A version of memcached that has persistence; ③ It is a memory caching system that is distributed and general purpose; ④ Development has halted on this solution.	① Applications related to IoT [216]
15	Project Voldemort [148]	Key-value	Available and partition-tolerant	① Free; ② Supports horizontal scalability; ③ Availability is high for read/write operations; ④ Fault recovery is transparent; ⑤ Supports automatic partitioning and replication; ⑥ Considered appropriate for applications with read-intensive operations	① Data management and query handling for the social networks, LinkedIn [79]; ② Applications related to IoT [216]
16	Redis [6]	Key-value	Consistent and partition-tolerant	① Free; ② Read/write operations and access to data are efficient; ③ Fault-tolerant; ④ Supports automatic partitioning; ⑤ Appropriate for applications involving structured strings	① Context-driven analysis in cultural heritage environments [206]
17	Riak [225]	Key-value	Available and partition-tolerant	① Free; ② Highly available and fault-tolerant; ③ Highly scalable and easy to operate; ④ Cloud storage and enterprise versions of Riak are also available; ⑤ It supports automatic data distribution and replication for resilience and improved performance.	① Applications based on transactional services like Automatic Vehicle Location System [176]; ② Data management and query handling for the social networks, Foursquare [79]

Continued

Table 4 Comparison of class I—NoSQL solutions.—cont'd

S. No.	NoSQL solution	Data model	CAP characteristics	Other features	Big graph applications
18	RocksDB [150]	Key-value	Consistent and partition-tolerant	① Free; ② Embedded database that assures high performance; ③ Supports all the features of LevelDB [144]. In addition, it also supports geospatial indexing, universal compaction, column families and transactions.	① Facebook's real-time data processing [226]; ② Matric computation for big data applications [149]; ③ Tagging system for blockchain analysis [165]; ④ Blockchain–based Library circulation system [139]
19	Scalaris [16]	Key-value	Consistent and partition-tolerant	① Free; ② Highly available and fault-tolerant; ③ Offers high scalability; ④ Consistent; ⑤ Self-managing; ⑥ Minimal maintenance overhead 7. Considered appropriate for applications that are read/write-intensive	① Applications related to IoT [216]
20	ScyllaDB [151]	Wide-column	Available and partition-tolerant	① Free; ② Distributed; ③ Designed for integration with Cassandra for reducing latency and improving throughput; ④ It supports Thrift and CQL, protocols also supported by Cassandra	① IoT applications [156]; ② Biogas data analytics [72]; ③ Logs analysis for IT establishments [227]
21	Tarantool [152]	Key-value	Consistent and available	① Free; ② Provides crash resistance with the help of maintenance of write ahead logs; ③ Can be integrated with other applications and frameworks written in different programming languages.	① Corpus management system [228]; ② IoT applications [229]
22	Tokyo Cabinet and Kyoto Cabinet [153]	Key-value	Available and partition-tolerant	① Free; ② Provides two libraries for database management; ③ Storage is done via hash tables and B+ trees; ④ Provides limited support for transactions	① Applications related to IoT [216]

Table 5 Comparison of class II—NoSQL solutions.

S. No.	NoSQL solution	Data model	CAP characteristics	Other features	Big graph applications
1	AllegroGraph [154]	Document-oriented and graph	Consistent and partition-tolerant	① Proprietary; ② Supports RDF, JSON and JSON-LD; ③ Provides Multimaster replication, two-phase commit and full-text search	① Biological database creation [92]; ② Semantic indexing engine [133]; ③ Context-driven analysis in cultural heritage environments [206]; ④ Big data analytics for insurance industry [127]; ⑤ Geospatial semantic data management [88]; ⑥ Data management and query handling for the social networks, Friendfeed [79]
2	Cache [155]	Document-oriented	Available	① Proprietary; ② Data are stored in multidimensional arrays. Therefore, structured data that is hierarchical in nature can be stored; ③ Commonly used for business and health-related applications	① Service performance management [120]; ② EMR database [102]
3	Cloudant [170]	Document-oriented	Available and partition-tolerant	① Proprietary; ② Distributed database service; ③ Uses BigCouch [230] and JSON model, at backend	① Environmental sensing applications [231]; ② Application that finds and ranks researchers [232]; ③ Smart campus application [60]
4	Coherence [15]	Key-value	Consistent and available	① Proprietary; ② In-memory data grid and distributed cache; ③ Appropriate for systems requiring high scalability and availability keeping latency at lower levels	① IoT-based analytical applications
5	CosmosDB [157]	Document-oriented	Available and partition-tolerant	① Proprietary; ② Provisioned as Platform-as-a-Service (PaaS); ③ Based on DocumentDB [158]	① Applications like assessment of risk for nuclear power plants [233]; ② Iris-based biometric system; ③ Screening of chemicals for liver toxicity [111], dose toxicity [112, 114] and non-cancer threshold of toxicological concern [113]; ④ Smart city applications like automated waste management [68], traffic monitoring system [74] and automated ticketing system for public transport [234]; ⑤ Prediction of dangers related to human health using IoT-based framework [110]

Continued

Table 5 Comparison of class II—NoSQL solutions.—cont'd

S. No.	NoSQL solution	Data model	CAP characteristics	Other features	Big graph applications
6	DocumentDB [158]	Document-oriented	Consistent and partition-tolerant	① Proprietary; ② Provisioned as a database service; ③ Fully managed version of MongoDB	① Document management system for patent application related documents [129]; ② Paper-digital document management system [130]; ③ Geospatial applications [91]; ④ User-based document management system [131]
7	IBM Informix [159]	Document-oriented	Consistent and available	① Proprietary; ② RDBMS that supports JSON; ③ Complies with ACID rules; ④ Supports sharding and replication	① Enterprise applications for business intelligence [235]
8	Lotus Domino [160]	Document-oriented	Consistent and available	① Proprietary; ② It is a multivalue database [52]	① Business intelligence applications [123]; ② Application to ensure safety of patients by improving supervision [236]; ③ Development of Clinical database for varied purposes [101]; ④ Decision support system for management of graduated student employment [65]
9	Marklogic [161]	Document-oriented and graph	Consistent, available, and partition-tolerant	① Free; ② Supports XML, JSON and RDF triples; ③ Distributed; ④ Provides high availability, full-text search, ACID compliance and security	① Breast Imaging applications [104]; ② Search applications on Bilingual lingual digital libraries [237]
10	Microsoft SQL Server [9]	Graph	Consistent and available	① Proprietary; ② Typically used for modeling many-to-many relationships between data; ③ Integration of relationships are done into Transact-SQL and the foundation DBMS is SQL Server	① Management of scientific database and related applications [238]
11	MUMP Database [162]	Document-oriented	Available	① Proprietary; ② MUMPS is a programming language with inbuilt database; ③ Used for applications related to health sector	① Automated management system for dairy farms [239]; ② Healthcare applications [80, 240]
12	ObjectDatabase+ + [163]	Document-oriented	Consistent and available	① Proprietary; ② Binary; ③ Structure of native C++ class	–

#	Name	Data model	CAP properties	Features	Applications
13	Oracle NoSQL Database [164]	Key-value	Consistent and available	① Proprietary; ② Supports horizontal scalability and transparent load balancing; ③ Supports replication and sharding; ④ It is highly available and fault-tolerant	① Business intelligence applications [124]; ② Analysis of medical imaging data [102]
14	Qizx [143]	Document-oriented	Consistent and available	① Proprietary; ② Distributed XML database; ③ Supports text, JSON and binaries; ④ Provides integrated full-text search	① Applications of the linguistic domain [241]
15	RavenDB [166]	Document-oriented	Consistent, available, and partition-tolerant	① Free; ② Fully transactional and high performance; ③ Highly available; ④ Multiplatform and easy to use; ⑤ Multimodel architecture that allows it to work well with SQL systems	① Storage system for social networks [78]
16	RocketU2 [167]	Document-oriented	Consistent	① Proprietary; ② Provides dynamic support; ③ Scalable; ④ Reliable and efficient; ⑤ Appropriate for business information management	—
17	SAP HANA [168]	Document-oriented and graph	Consistent and available	① Proprietary; ② Supports JSON only; ③ Can be used for ACID transactions	① Web-scale data management for business applications [242]
18	SimpleDB [169]	Document-oriented	Available and partition-tolerant	① Proprietary; ② Distributed database; ③ Used as web service with Amazon EC2 [243] and S3 [244]; ④ Provides availability and partition tolerance	① Large-scale RDF Store development and management for life sciences databases [98]

Table 6 Comparison of class III—NoSQL solutions.

S. No.	NoSQL solution	Data model	CAP characteristics	Other features	Big graph applications
1	Amazon Neptune [170]	Graph	Consistent and partition-tolerant	① Proprietary; ② Fully managed database; ③ Provided as a web service; ④ Supports RDF and property graph models; ⑤ Supports SPARQL [245] and TinkerPop Gremlin [246] query languages	① Metadata repository for smart energy services or intelligent energy management [247]; ② Data Management Systems for BI Applications [116]
2	AnzoGraph [121]	Graph	Consistent and partition-tolerant	① Proprietary; ② Massively parallel; ③ Graph Online Analytics Processing (GOLAP) database; ④ Supports SPARQL [245] and Cypher [248]; ⑤ Originally designed to analyze semantic triple data interactively	① Business intelligence applications [121]
3	Azure tables [171]	Wide-column	Available and partition-tolerant	① Proprietary; ② Provisioned as a service where storage of data is allowed into collections that can be partitioned. Data are accessed by means of primary and partition keys.	① GIS Application [90]; ② Processing of Seismograms [249]; ③ Ground water simulation and related applications [73]; ④ Comparative Genomics applications [99]
4	DataStax Enterprise Graph [172]	Graph	Available and partition-tolerant	① Proprietary; ② Scalable, distributed database; ③ Allows real-time querying; ④ Supports Tinkerpop [246]; ⑤ It is known to integrate well with Cassandra [5]	① Data Management Systems for Business Intelligence (BI) Applications [116]; ② Analytics for center of solar system; ③ Covert network analysis applications [250]; ④ Clinical big data applications [108]
5	Dynamo [173]	Key-value	Available and partition-tolerant	① Proprietary; ② Distributed datastore; ③ Highly available; ④ Supports incremental scalability, symmetry among nodes, decentralization and it exploits the heterogeneity of the infrastructure it works on.	① Data management and query handling for the social networks, MySpace [79]

#	Name	Type	CAP	Features	Applications
6	Hazelcast [174]	Key-value	Available and partition-tolerant	① Free; ② MapStore can be defined by the user; ③ MapStore can be persistent; ④ High consistency and supports sharing in the form of consistent hashing	① Bug prediction system for GitHub projects database [251]; ② Prediction of performance of IoT applications [252]
7	HyperGraphDB [175]	Graph	Available and partition-tolerant	① Free; ② Schemas are dynamic and flexible; ③ Knowledge representation and data modeling are efficient; ④ Non-blocking concurrency; ⑤ Appropriate for semantic web and arbitrary graph use cases	① Document-centric information systems' analysis [253]; ② E-government applications like citizen relationship management [70]; ③ Intelligent manufacturing applications [254]; ④ Mining of biomedical networks [106]; ⑤ Applications in Governance [255]
8	InfiniteGraph [176]	Graph	Consistent and partition-tolerant	① Proprietary; ② Cloud-enabled; ③ Distributed; ④ It is scalable and cross-platform; ⑤ It is capable of handling high throughput	① Smart Grid applications [256]; ② Applications for business, social and government intelligence [71]
9	JanusGraph [177]	Graph	Available and partition-tolerant	① Free; ② Distributed; ③ Scalable and integrates well with backend databases like HBase [4], Cassandra [5], BigTable [202] and BerkleyDB [8]; ④ Integrates well with platforms like Graph [257], Spark [258] and Hadoop [259]; ⑤ Provides support for full-text search by external integration with Solr [193] and Elasticsearch [187]	① Exploration of scholarly networks [63, 64]; ② Smart City applications [23]; ③ Analysis of social graph data [81]; ④ Applications like network security analytics [260]
10	KAI [178]	Key-value	Available and partition-tolerant	① Free; ② Scalable; ③ Highly fault-tolerant; ④ Provides low latency; ⑤ Used for social networks and web repositories	① Analytical applications for IoT

Continued

Table 6 Comparison of class III—NoSQL solutions.—cont'd

S. No.	NoSQL solution	Data model	CAP characteristics	Other features	Big graph applications
11	Neo4j [179]	Graph	Available and partition-tolerant	① Free; ② Can be used for ACID transactions; ③ Supports clustering and high availability; ④ Provides complete administrative support; ⑤ Provides inbuilt REST API for interface with other programming languages	① Analytics for proactive semiconductor equipment maintenance [125]; ② Storage model for healthcare information systems [105]; ③ Storage of social network data [77]; ④ Data management and query handling for the social networks, Facebook, Twitter, Flickr and MySpace [79]; ⑤ Metadata repository for smart energy services or intelligent energy management [247]; ⑥ Representation and enrichment of Geodata [87]; ⑦ Bug prediction system for GitHub projects database [251]; ⑧ Application in supply chain management [261]
12	Oracle Spatial and Graph [180]	Graph	Available and partition-tolerant	① Proprietary; ② Capable for handling RDF and property graphs	① Big data analytics for insurance industry [127]; ② Data Management Systems for BI Applications [116]
13	Sparksee [181]	Graph	Consistent and partition-tolerant	① Proprietary; ② Scalable; ③ High performance; ④ First graph database for mobile devices; ⑤ Bindings available for C++, C#, Objective C, Python and Java	① Social networking applications [262]; ② Smart City applications [263]; ③ IoT applications
14	Sqrrl [182]	Graph	Consistent and partition-tolerant	① Proprietary; ② Distributed; ③ Mass-scalable; ④ Real-time database; ⑤ Provides cell-level security	① Progressive web application that maintains a database of coins and historical figures [182]
15	XAP [183]	Key-value	Available and partition-tolerant	① Proprietary; ② Software platform for in-memory computing; ③ Appropriate use cases for this solution include real-time analytics and transaction processing requiring low latency and high-performance levels.	① BI applications [124]

6.2.4 Classification category: Class IV

The class IV category represents the NoSQL solutions that support document-oriented data model, ensure availability and are proprietary, which are shown in Table 7.

6.2.5 Classification category: Class V

The class V category represents the NoSQL solutions that do not support graph data model, ensure consistency and partition tolerance and are proprietary, which are shown in Table 8.

6.2.6 Classification category: Class VI

The class VI category represents the NoSQL solutions that can ensure consistency and are free, which are shown in Table 9.

7. Suitability study for NoSQL solutions

As mentioned above, for the analysis purposes, the big graph applications were divided into seven categories, six of which are specific fields of application including smart cities, social network analysis, geospatial applications, life sciences research, healthcare and business intelligence; while the last category includes all unclassified works. In all reviewed application papers, one or more of the NoSQL solutions listed were used. Accordingly, we count the number of NoSQL solutions supported by each application (or field). If an application supports a specific NoSQL solution, the corresponding number of supported NoSQL solutions is increased by 1, otherwise by 0.

The random forest classification [22] is carried out to identify the feature importance from nine features for each field. For all experiments we divide the data generated above into 75% and 25% for training and testing. Gini index [290] is used as the metric to determine the feature importance in the classification algorithm. This metric calculates the ratio of tree partitions that contain the feature and the number of samples for the partition. The sum of these values is calculated for all trees created in Random Forest to determine the final value of the feature importance. Fig. 5 shows the relative

Table 7 Comparison of class IV—NoSQL solutions.

S. No.	NoSQL solution	Data model	CAP characteristics	Other features	Big graph applications
1	ArangoDB [154]	Document-oriented, key-value, and graph	Consistent and available	① Free; ② Supports multiple database models with a single core; ③ Possesses unified query language, called ArangoDB Query Language (AQL), which enables a single query for accessing multiple data stores.	① Android application for controlling traffic [75]; ② Progressive web application that maintains a database of coins and historical figures [182]; ③ Predictive analytics involving collaboration platform between academia and industry [61]; ④ Representation and enrichment of Geodata [87]
2	BaseX [184]	Document-oriented	Consistent and available	① Free; ② Provides support for JSON, XML and binary formats; ③ Implements master-slave architecture; ④ Provides support for concurrent structural and full-text update/search	① Annotation and retrieval in digital humanities research [264]
3	CouchDB [185]	Document-oriented	Available and partition-tolerant	① Free; ② Supports JSON over HTTP/REST; ③ Provides limited support for ACID transactions; ④ Supports multiversion concurrency control	① Data management and query handling for the social networks, Foursquare [79]; ② Analytical solutions for radiology [109]; ③ Document management system for software projects [128]; ④ Querying system for graphical music documents [265]; ⑤ Web-based application for monitoring and visualizing energy consumption of a house for improving energy efficiency [266]
4	CrateIO [186]	Document-oriented	Available and partition-tolerant	① Free; ② It is based on the Elasticsearch/Lucene ecosystem; ③ Supports objects that are binary or are also called BLOBs; ④ Makes use of SQL syntax for distributed querying of the system in real time	① Industrial IoT applications [267]
5	ElasticSearch [187]	Document-oriented	Consistent and available	① Free; ② Supports JSON; ③ Basically a search engine	① External Indexing Engine for Smart City Applications [268]; ② Querying system for graphical music documents [265]

#	Name	Type	CAP	Features	Applications
6	eXist [188]	Document-oriented	Consistent and available	① Free; ② Supports text, JSON, HTML and XML formats, in addition to binary formats; ③ XQuery is the provided querying language while XSLT is the corresponding programming language	① Document management system for linguistic applications [132]; ② Search applications on Bilingual lingual digital libraries [237]
7	Jackrabbit [189]	Document-oriented	Consistent and available	① Free; ② Implementation of Java Content Repository	① Data management of biological investigations of systems biology [96]; ② Database of work-related accidents in construction industry [126]
8	OrientDB [190]	Document-oriented and graph	Consistent, available, and partition-tolerant	① Free; ② Supports JSON over HTTP; ③ Supports SQL-type language use; ④ Can be used for ACID transactions; ⑤ Supports sharding, multimaster replication, security features and schemaless modes	① Community detection and related applications [269]; ② Smart education application [66]; ③ Bug prediction system for GitHub projects database [251]; ④ Real-time social networking applications [82]
9	PostgreSQL [191]	Document-oriented	Consistent and available	① Free; ② Supports JSONB, JSON function and JSON store; ③ Supports HStore 2 and HStore [270]	① Storing configuration parameters and catalogs for Context-driven analysis in cultural heritage environments [206]; ② IoT Application [252]
10	Sedna [192]	Document-oriented	Consistent and available	① Free; ② XML database	① IoT applications
11	Solr [193]	Document-oriented	Available and partition-tolerant	① Free; ② Search engine written in Java; ③ Supports real-time indexing, full-text search, database integration, dynamic clustering and rich document handling; ④ Provides index replication and distributed search; ⑤ Scalable and fault-tolerant	① Creation of digital libraries with the help of web crawl [271]

Table 8 Comparison of class V – NoSQL solutions

S. No.	NoSQL solution	Data model	CAP characteristics	Other features	Big graph applications
1	Accumulo [194]	Wide-column	Consistent and partition-tolerant	① Free; ② Scalable and distributed; ③ It is built over and above Hadoop [258], Thrift [272] and Zookeeper [273]; ④ Provides cell-level security and mechanisms for server-side programming	① Backend storage for life sciences databases [93]
2	Clusterpoint database [195]	Document-oriented and key-value	Consistent and partition-tolerant	①Proprietary; ② Distributed JSON/XML database platform; ③ Transactions are compliant with ACID properties; ④ Highly available; ⑤ Provides sharding and replication; ⑥ Uses SQL or JS as query language	① Medical Database Analysis [274]; ② Enterprise content management [275]
3	CouchBase Server [196]	Document-oriented and key-value	Consistent and partition-tolerant	① Free; ② Distributed database; ③ Uses SQL as querying language; ④ Uses JSON model	① Traffic forecasting application that works in real time for environments, which implement the fog computing concept [209]; ② Social network analytics [83]; ③ Collection of sensor data for medical applications [100]; ④ Decentralized social networking application [276]
4	HBase [4]	Wide-column	Consistent and partition-tolerant	① Free; ② Distributed database; ③ It runs on top of Hadoop and provides capabilities similar to that of BigTable; ④ It is fault-tolerant for scenarios where a large amount of sparse data is being dealt with.	① Analytics for proactive semiconductor equipment maintenance [125]; ② SPARQL query engine, Jena HBase, for life sciences databases [95]; ③ Data management and query handling for the social networks, Facebook, Twitter, Friendfeed and LinkedIn [79]; ④ Real-time sentiment analysis and text extraction for microblogging applications [86]
5	HyperTable [197]	Wide-column	Consistent and partition-tolerant	①Proprietary; ② It is based on BigTable; ③ Massively scalable	① BI applications [122]; ② Applications in digital forensics [277]; ③ Smart applications like online vehicle tracking system [278]

#	Database	Type	CAP	Features	Applications
6	MongoDB [198]	Document-oriented	Consistent and partition-tolerant	① Free; ② Supports BSON or binary JSON [279]; ③ Allows replication and sharding	① Biological Database Creation [92]; ② Analytics for proactive semiconductor equipment maintenance [125]; ③ Storage of real-time data from sensors [280]; ④ Context-driven analysis in cultural heritage environments [206]; ⑤ Data Storage for Smart City Applications [268]; ⑥ Data management and query handling for the social networks, LinkedIn, Flickr, Foursquare and MySpace [79]; ⑦ Handling of Big Geospatial Raster Data [89]; ⑧ Secondary storage for blockchain [219]
7	RethinkDB [199]	Document-oriented	Consistent and partition-tolerant	① Free; ② Distributed database; ③ Supports JSON; ④ Provides sharding and replication	① Data analytical applications for IoT-based green house system [281]; ② Application that implements an agent-based Platform for Autonomous Sailing [282]; ③ Real-time data management of aquarium data [283]; ④ Database for learning environment [67]; ⑤ IoT-based real-time ECG monitoring for healthcare applications [103]; ⑥ Agriculture IoT application [69]
8	TokuMX [200]	Document-oriented	Consistent and partition-tolerant	① Free; ② Version of MongoDB [198]; ③ Supports fractal tree indexing [284]	① Server-based monitoring of energy storage systems like lithium–ion battery packs for mobile as well as stationary applications [285]
9	TerraStore [201]	Document-oriented	Consistent and partition-tolerant	① Free; ② In-memory storage; ③ Dynamic cluster configuration; ④ Persistent; ⑤ Supports load balancing and automatic data redistribution; ⑥ Used for structured big data	① IoT applications; ② Storage system for social networks [78]

Table 9 Comparison of class VI—NoSQL solutions.

S. No.	NoSQL solution	Data model	CAP characteristics	Other features	Big graph applications
1	BerkeleyDB [8]	Key–value	Consistent and partition-tolerant	① Free (with commercial versions also available); ② High performing and scalable; ③ Supports complex data management; ④ Most appropriate for applications requiring embeddable database.	① E-learning system [62]; ② Collaborative web search [286]
2	BigTable [202]	Wide–column	Consistent and partition-tolerant	① Proprietary; ② High performance; ③ Provides data compression	① Management of sensor network data [287]; ② Financial services application like audit trail [119]
3	GridGain Systems [203]	Key–value	Consistent and partition-tolerant	① Proprietary; ② Services and software solutions are provided for systems dealing with big data; ③ Supports in-memory computing; ④ Provides improved throughput and reduced latency	① Text documents' classification [288] and clustering [289]
4	NoSQLz [204]	Key–value	Consistent	① Proprietary; ② Complies with ACID properties; ③ Allows CRUD (create, read, update, delete) operations; ④ Easy to implement	–
5	OpenLink Virtuoso [205]	Document-oriented, graph and key–value	Consistent and partition-tolerant	① Proprietary; ② Hybrid of database engine and middleware; ③ High performance and secure; ④ Supports SQL [24] and SPARQL [245] for performing operations on SQL tables and RDF; ⑤ JSON, XML and CSV document types are supported	① Big data analytics for insurance industry [127]; ② Geospatial semantic data management [88]; ③ Metadata repository for Smart energy services or intelligent energy management [247]

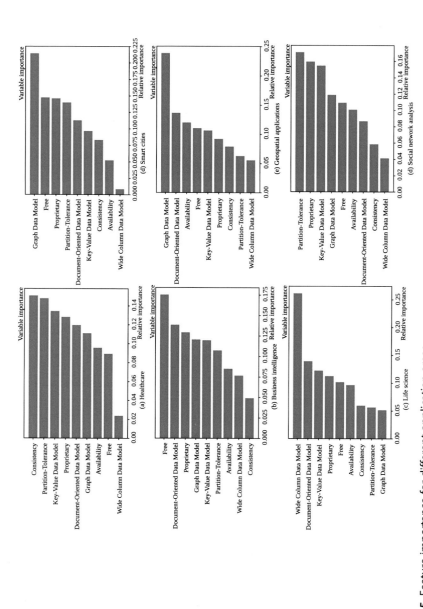

Fig. 5 Feature importance for different application areas.

feature importance for the nine features in each field. It is important to note that the values for the importance of features are comparable for all features for Business Intelligence (BI) applications, except Free, which can be explained by the fact that budget is one of the most important requirements in a BI application. In addition, all features have comparable values for Smart Cities. In theory, comparable values for the importance of features can be linked to the fact that Business Intelligence includes both organizational and company-specific solutions. Similarly, Smart Cities includes a range of applications from waste management to intelligent manufacturing. Therefore, the solution design can be of different types and interests. In addition, consistency, wide column data model, partition tolerance and document-oriented data model are considered the most relevant features for healthcare, life sciences, social network analysis and geospatial applications, respectively.

Next, the decision tree [23] is used as a classification model for predicting the suitability of NoSQL solutions in each application field. Fig. 6A–F illustrates the decision trees of the six application areas. The blocks are symbolic for checking whether a feature is present or not. The labels present and not present indicate whether the solution being tested has the feature specified in the previous block. Suitable and Not Suitable are used as classes that indicate whether the NoSQL solution under test is suitable or unsuitable for that particular application field. The accuracy achieved in the decision tree classification is better than the Random Forest classification, which means that the trees created for the Random Forest had a lower correlation between them. Therefore, it is concluded that the Decision Tree is an appropriate predictive model for determining the suitability of a NoSQL solution for a particular application. The comparison of the accuracy values is shown in Table 10 and illustrated in Fig. 7.

8. Discussion

A big data application is a viable solution for solving real-world problems that depend on the basic technologies including storage systems and a computing environment. Given the specific requirements of big data, such as scalability, availability, integration and security, traditional relational database systems have proven incapable of meeting these challenges, while NoSQL solutions can bridge this gap. NoSQL has demonstrated its feasibility for managing big data in many applications, especially those with the

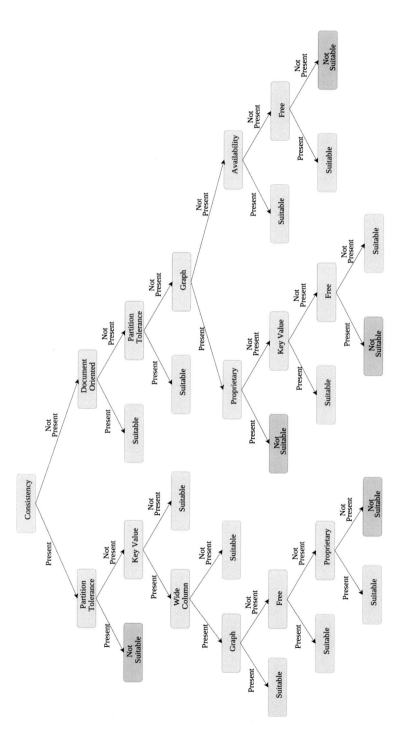

Fig. 6 Decision trees for different application areas (A) healthcare; (B) business intelligence; (C) life sciences; (D) smart cities; (E) geospatial applications; (F) social network analysis.

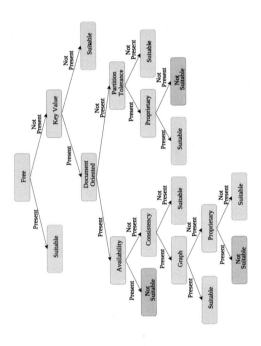

Fig. 6—Cont'd

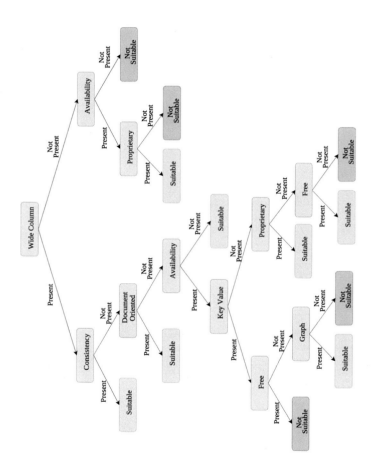

Fig. 6—Cont'd

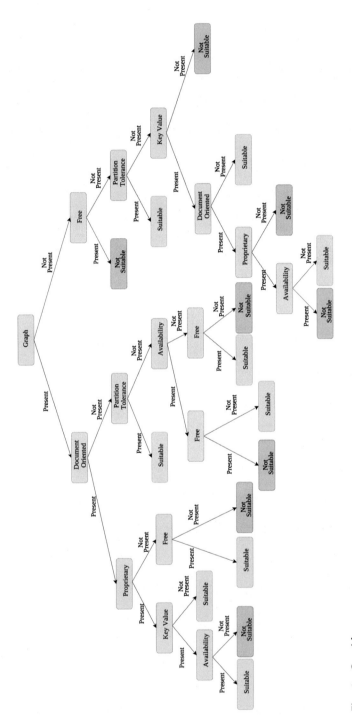

Fig. 6—Cont'd

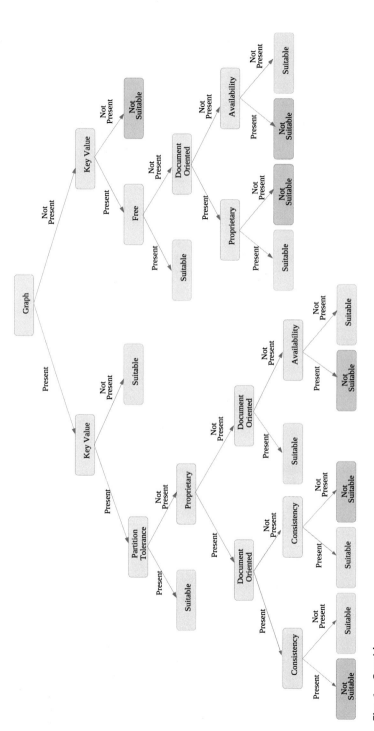

Fig. 6—Cont'd

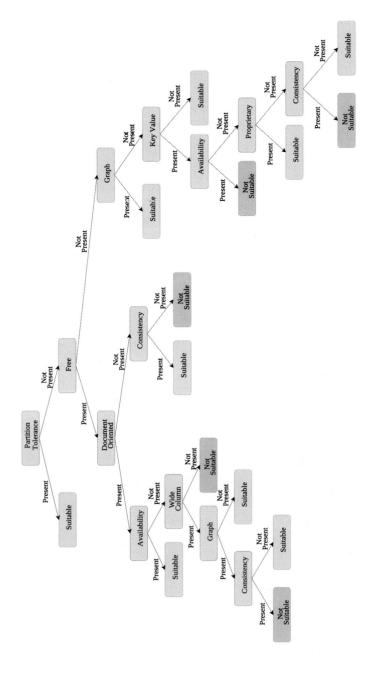

Fig. 6—Cont'd

Table 10 Decision tree classification vs random forest classification.

Application	Random forest	Decision tree
Business intelligence	81.66%	86.67%
Geospatial applications	88.33%	93.33%
Healthcare	75.00%	88.33%
Life sciences	85.00%	86.67%
Smart cities	65.00%	78.33%
Social network analysis	65.00%	80.00%

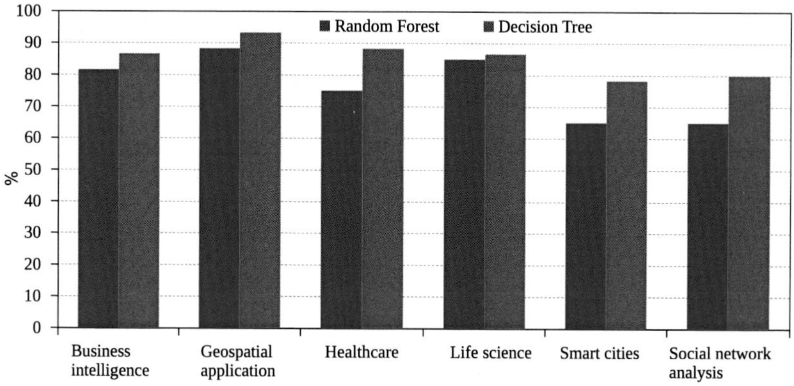

Fig. 7 Comparison of accuracy for decision tree and random forest classifications.

need for high scalability, flexible data modeling, high availability, and pro-visioning required performance. Furthermore, NoSQL solutions are not only capable of handling batch static data, but also real-time data, due to their inherent features including dynamic schema, autosharding, automatic replication and built-in caching. These are usually not available for most traditional relational database management systems.

Data management is the core for applications, and the selection of an appropriate database management system is crucial for the success of an application project. Database selection and database design require knowledge of database management systems and the specific domain of the application. In fact, a majority of system failures are due to inappropriate database selection. In this chapter, we have not only exhaustively reviewed the existing NoSQL solutions, but also examine their characteristics and create a suitability model for selecting NoSQL solutions. But, be aware that a NoSQL solution often supports multiple data models and that it can be suitable for a number of applications. Therefore, it is not realistic to create a sole model for selecting NoSQL solutions for all application domains.

It is noteworthy that the CAP theorem suffers from several shortcomings, due to an overly simplified model. For example, transactions involving multiple objects are not treated for consistency [291]. In addition, only partition tolerance is considered, while other types of system failures can also occur and latency is not taken into account. Tables 4–9 shows that there are some exceptions to the CAP theorem. For example, the solutions like Solr only provide availability, but RavenDB, MarkLogic, FoundationDB, and Ignite can provide all the three characteristics in the CAP theorem. There are also some solutions like CosmosDB that provide variable and configurable consistency. This leads to the conclusion that neither the classification based on data models nor the CAP theorem can be sufficiently used for NoSQL solution classification. Therefore, this chapter proposes a novel classification scheme that covers different characteristics of NoSQL solutions from the two approaches.

To date, many BI vendors offer complete BI solutions that provide customizable storage, data capture, processing and visualization [28]. The decision to choose such a solution depends on many factors, including technical and nontechnical ones. Once the decision is made, the solution should be able to provide the desired functionality for the big data problem. For example, an application may have specific scalability and security requirements. In this regard, a distribution data model will be superior to others because scalability is the main requirement.

The scaling of the read operations is supported by a master-slave architecture. However, if scaling of both read and write operations is desired, a peer-to-peer architecture is the better option. Some NoSQL solutions, such as Cassandra, scale well, while others, such as memory-based databases, may not scale across different machines. In addition, the use of NoSQL databases also has some security issues that must be considered and mitigated before developing and deploying a solution. Therefore, adding more functionality to the proposed classification scheme and prediction model can be useful to make this work more general. In this chapter, a decision-tree-based prediction model is proposed to determine the suitability of NoSQL solutions for an application area based on their features.

This work has some limitations. First, the classification for the NoSQL solutions reviewed was on a very coarse level. In fact there are many different requirements for considering within each application category. For example, Smart Cities and Healthcare have very detailed requirements for data management, including privacy issues, inclusion of streaming data, support for analytical queries, etc. In addition, big graph applications in some other areas such as IoT require real-time and/or search analysis capabilities, which need to be incorporated as well. Therefore, to meet these requirements, further comparative analysis is needed, especially to investigate the underlying details of data storage systems, such as read/write efficiency, storage structure, sharding and caching. These low-level features will contribute to better classification and comparative analysis. This will be our future work. Second, since the data set analyzed in this study is taken from the literature review, data quality such as integrity may not be good enough. For example, hybrid data storage schemes and their applications in different application areas were not considered. This can be improved by a better data collection process, e.g. based on field studies for some industrial applications, to improve data integrity. In addition, data quality can be assessed using quantitative analysis, which examines the impact of data attributes, such as the effect of the presence or absence of a characteristic on the results of data analysis.

As NoSQL database systems and their applications evolve rapidly, the results of this chapter were published at http://www.p-nasa.com, and the data set and prediction model will be regularly updated on the website. The website is a web application prototype that takes NoSQL solution's features as input and predicts the class of applications for which an NoSQL solution shall be appropriate. In the future work, we will make a further improvement for the predictive model, as well as the modules for listing appropriate NoSQL solutions for a specific application.

9. Conclusion and future work

Data model-based or CAP theorem-based classification is not sufficient to provide discrete classification criteria for NoSQL solutions. In this chapter, over 80 NoSQL solutions and their nine application areas were examined. A bivariate analysis was performed to identify the principle features for NoSQL solution classification. A hybrid features was proposed for the classification scheme. A clustering analysis was performed based on the nine characteristics identified in this chapter, including document-oriented, graphic, key-value, wide column, consistency, availability, partition tolerance, free and proprietary. The clustering analysis was used as a basis for creating NoSQL categories. Then a random forest and decision tree classification was used to investigate the suitability of NoSQL solutions in different application areas. In addition, a web application was created to facilitate the use of the results of this study.

There are several directions for future work. First, we will improve the prediction model for recommending NoSQL solutions according to user requirements. Second, we plan to perform a thorough benchmarking for mainstream NoSQL solutions and make quantitative comparisons. Finally, we will identify additional features that can help improve classification and prediction performance.

References

[1] M.D. Assunção, R.N. Calheiros, S. Bianchi, M.A. Netto, R. Buyya, Big data computing and clouds: trends and future directions, J. Parallel Distrib. Comput. 79 (2015) 3–15.

[2] A. Oussous, F.Z. Benjelloun, A.A. Lahcen, S. Belfkih, Comparison and classification of NoSQL databases for big data, in: Proceedings of International Conference on Big Data, Cloud and Applications (vol. 2), May, 2015.

[3] C.P. Chen, C.Y. Zhang, Data-intensive applications, challenges, techniques and technologies: a survey on big data, Inf. Sci. 275 (2014) 314–347.

[4] L. George, HBase: The Definitive Guide: Random Access to Your Planet-Size Data, "O'Reilly Media, Inc", 2011.

[5] S. Gupta, G. Narsimha, Miscegenation of scalable and DEP3K performance evaluation of NoSQL-cassandra for bigdata applications deployed in cloud, Int. J. Bus. Process Integr. Manag. 9 (1) (2018) 12–21.

[6] M. Diogo, B. Cabral, J. Bernardino, Consistency models of NoSQL databases, Future Internet 11 (2) (2019) 43.

[7] S. Sivasubramanian, Amazon dynamoDB: a seamlessly scalable non-relational database service, in: Proceedings of the 2012 ACM SIGMOD International Conference on Management of Data, 2012, pp. 729–730.

[8] A. Ul-Haque, T. Mahmood, N. Ikram, Performance comparison of state of art NoSql technologies using apache spark, in: Proceedings of SAI Intelligent Systems Conference, Springer, Cham, 2018, pp. 563–576.

[9] A. Flores, S. Ramírez, R. Toasa, J. Vargas, R. Urvina-Barrionuevo, J.M. Lavin, Performance Evaluation of NoSQL and SQL Queries in Response Time for the E-government, in: 2018 International Conference on eDemocracy & eGovernment (ICEDEG), IEEE, 2018, pp. 257–262.

[10] J. Han, E. Haihong, G. Le, J. Du, Survey on NoSQL database, in: 2011 Sixth International Conference on Pervasive Computing and Applications, IEEE, 2011, pp. 363–366.

[11] A.B.M. Moniruzzaman, S.A. Hossain, NoSQL database: new era of databases for big data analytics-classification, characteristics and comparison, Int. J. Database Theory Appl. 6 (4) (2013). In this issue.

[12] B. Namdeo, N. Nagar, V. Shrivastava, Survey on RDBMS and NoSQL databases, Int. J. Innov. Knowl. Concepts 6 (6) (2018) 261–264.

[13] S.S. Nicolaescu, H.C. Palade, C.V. Kifor, A. Florea, Collaborative platform for transferring knowledge from university to industry—a bridge grant case study, in: Proceedings of the 4th IETEC Conference, Hanoi, Vietnam, 2018.

[14] G. Bathla, R. Rani, H. Aggarwal, Comparative study of NoSQL databases for big data storage, Int. J. Eng. Technol. 7 (26) (2018) 83.

[15] A.K.N.S.P. Lathar, K.G. Srinivasa, Comparison study of different NoSQL and cloud paradigm for better data storage technology, in: Handbook of Research on Cloud and Fog Computing Infrastructures for Data Science, IGI Global, 2018, pp. 312–343.

[16] A. Davoudian, L. Chen, M. Liu, A survey on NoSQL stores, ACM Comput. Surv. (CSUR) 51 (2) (2018) 40.

[17] B. Scofield, NoSQL-death to relational databases, CodeMash Presentation (2010) 1–14.

[18] M. noz Escoí F.D., R. de Juan-Marín, J.R. González de Mendívil, J.M. Bernabéu-Aubán, G.-E.J. R., CAP theorem: revision of its related consistency models, Comput. J. 62 (6) (2019) 943–960.

[19] Barcelona Field Studies Center, Spearman's Rank Correlation Coefficient Rs and Probability (p) Value Calculator, 2019. Retrieved from: https://geographyfieldwork.com/SpearmansRankCalculator.html. (accessed 01 May, 2020).

[20] S. by Jim, Chi-Square Test of Independence and an Example, 2019. Retrieved from: https://statisticsbyjim.com/hypothesis-testing/chi-square-test-independence-example/. (accessed 01 May, 2020).

[21] Z. Huang, Extensions to the k-means algorithm for clustering large data sets with categorical values, Data Min. Knowl. Discov. 2 (3) (1998) 283–304.

[22] A. Liaw, M. Wiener, Classification and regression by randomForest, R News 2 (3) (2002) 18–22.

[23] S.R. Safavian, D. Landgrebe, A survey of decision tree classifier methodology, IEEE Trans. Syst. Man Cybernetics 21 (3) (1991) 660–674.

[24] H. Köhler, S. Link, SQL schema design: foundations, normal forms, and normalization, Inf. Syst. 76 (2018) 88–113.

[25] P. Rathore, A.S. Rao, S. Rajasegarar, E. Vanz, J. Gubbi, M. Palaniswami, Real-time urban microclimate analysis using Internet of Things, IEEE Internet Things J. 5 (2) (2018) 500–511.

[26] N. Albayrak, A. Özdemir, E. Zeydan, An artificial intelligence enabled data analytics platform for digital advertisement, in: 2019 22nd Conference on Innovation in Clouds, Internet and Networks and Workshops (ICIN), IEEE, 2019, pp. 239–241.

[27] T.R. Rao, P. Mitra, R. Bhatt, A. Goswami, The big data system, components, tools, and technologies: a survey, Knowl. Inf. Syst. 60 (2018) 1–81.

[28] S. Khan, K.A. Shakil, M. Alam, Big data computing using cloud-based technologies: challenges and future perspectives, in: Networks of the Future: Architectures, Technologies, and Implementations, Chapman and Hall/CRC, 2017.

[29] C. Stergiou, K.E. Psannis, B.G. Kim, B. Gupta, Secure integration of IoT and cloud computing, Future Generation Comput. Syst. 78 (2018) 964–975.

[30] P. Lous, P. Tell, C.B. Michelsen, Y. Dittrich, A. Ebdrup, From scrum to agile: a journey to tackle the challenges of distributed development in an agile team, in: Proceedings of the 2018 International Conference on Software and System Process, ACM, 2018, pp. 11–20.

[31] J.R. Cordeiro, O. Postolache, Big data storage for a health predictive system, in: 2018 International Symposium in Sensing and Instrumentation in IoT Era (ISSI), IEEE, 2018, pp. 1–6.

[32] J. Zhao, M. Lai, H. Tian, Y. Chang, U.S. Patent Application No. 10/205,673, 2019.

[33] K.S. Reddy, S. Mohair, N. Karamchandani, Effects of storage heterogeneity in distributed cache systems, in: 2018 16th International Symposium on Modeling and Optimization in Mobile, Ad Hoc, and Wireless Networks (WiOpt), IEEE, 2018, pp. 1–8.

[34] C. Strauch, U.L.S. Sites, W. Kriha, NoSQL Databases, Stuttgart Media University, 20, 2011.

[35] X. Zheng, Database as a service-current issues and its future, arXiv preprint arXiv:1804.00465 (2018).

[36] K. North, Databases in the Cloud: Elysian Fields or Briar Patch?, 2020. https:// www.drdobbs.com/database/databases-in-the-cloud-elysian-fields-or/218900502. (accessed 01 May 2020).

[37] R. Cattell, Scalable SQL and NoSQL data stores, Acm Sigmod Record 39 (4) (2011) 12–27.

[38] N. Leavitt, Will NoSQL databases live up to their promise? Computer 43 (2) (2010) 12–14.

[39] R.A.S.N. Soransso, M.C. Cavalcanti, Data modeling for analytical queries on document-oriented DBMS, in: Proceedings of the 33rd Annual ACM Symposium on Applied Computing, ACM, 2018, pp. 541–548.

[40] J. Sun, C.V.B. Milani, U.S. Patent Application No. 15/406,643, 2018.

[41] S. Geissinger, U.S. Patent Application No. 15/354,921, 2018.

[42] P. Raj, G.C. Deka, A Deep Dive into NoSQL Databases: The Use Cases and Applications, vol. 109, Academic Press, 2018.

[43] S. Khan, X. Liu, K.A. Shakil, M. Alam, A survey on scholarly data: from big data perspective, Inf. Process. Manag. 53 (4) (2017) 923–944.

[44] N.T.W. Khin, N.N. Yee, Query classification based information retrieval system, in: International Conference on Intelligent Informatics and Biomedical Sciences (ICIIBMS), vol. 3, IEEE, 2018, pp. 151–156.

[45] R. Angles, C. Gutierrez, An introduction to graph data management, in: Graph Data Management, Springer, Cham, 2018, pp. 1–32.

[46] X. Liu, C. Thomsen, T.B. Pedersen, 3XL: supporting efficient operations on very large OWL Lite triple-stores, Inf. Syst. 36 (4) (2011) 765–781.

[47] X. Liu, C. Thomsen, T.B. Pedersen, 3XL: an efficient DBMS-based triple-store, in: Proc. of the 23rd International Workshop on Database and Expert Systems Applications, IEEE, 2012, pp. 284–288.

[48] Q. Liu, K. Cao, F.U. Bang, U.S. Patent Application No. 10/110,640, 2018.

[49] R.T.P. Atzeni, L. Cabibbo, Data modeling across the evolution of database technology, in: A Comprehensive Guide Through the Italian Database Research Over the Last 25 Years, Springer, Cham, 2018, pp. 221–234.

[50] X. Ding, L. Chen, Y. Gao, C.S. Jensen, H. Bao, UlTraMan: a unified platform for big trajectory data management and analytics, Proc. VLDB Endowment 11 (7) (2018) 787–799.

[51] A.H. Miller, A.J. Fisch, J.D. Dodge, A.H. Karimi, A. Bordes, J.E. Weston, U.S. Patent Application No. 16/002,463, 2018.

[52] J.K. Chen, W.Z. Lee, A study of NoSQL database for enterprises, in: 2018 International Symposium on Computer, Consumer and Control (IS3C), IEEE, 2018, pp. 436–440.

[53] A. Celesti, M. Fazio, A. Romano, A. Bramanti, P. Bramanti, M. Villari, An oais-based hospital information system on the cloud: analysis of a NoSQL column-oriented approach, IEEE J. Biomed. Health Inf. 22 (3) (2018) 912–918.

[54] S.J. Kamath, K. Kanagaratnam, J.D. Keenleyside, S.S. Meraji, U.S. Patent No. 9,971,808, U.S. Patent and Trademark Office, Washington, DC, 2018.

[55] R. Rudnicki, A.P. Cox, B. Donohue, M. Jensen, Towards a methodology for lossless data exchange between NoSQL data structures, in: Ground/Air Multisensor Interoperability, Integration, and Networking for Persistent ISR IX, vol. 10635, International Society for Optics and Photonics, 2018, p. 106350R.

[56] P.A.P. Nguyen, D. Kryze, T. Vassilakis, A. Lerios, U.S. Patent Application No. 10/176,236, 2019.

[57] A.D. R.M. Rijo, Building Tunable CRDTs (Ph.D. thesis). 2018.

[58] NoSQL, NoSQL databases. Retrieved from: http://nosql-database.org/.

[59] B.N. Silva, M. Khan, K. Han, Towards sustainable smart cities: a review of trends, architectures, components, and open challenges in smart cities, Sustain. Cities Soc. 38 (2018) 697–713.

[60] E. Husni, Front-end and back-end application development for uBeacon smart campus system, Adv. Sci. Lett. 23 (4) (2017) 3786–3791.

[61] S.S. Nicolaescu, H.C. Palade, C.V. Kifor, A. Florea, Collaborative platform for transferring knowledge from university to industry—a bridge grant case study, in: Proceedings of the 4th IETEC Conference, Hanoi, Vietnam, 2018.

[62] B. Song, M. Li, An e-Learning system based on GWT and Berkeley DB, in: International Conference in Swarm Intelligence, Springer, Berlin, Heidelberg, 2012, pp. 26–32.

[63] G.C. Durand, A. Janardhana, M. Pinnecke, Y. Shakeel, J. Krüger, T. Leich, G. Saake, Exploring large scholarly networks with hermes, in: EDBT, 2018, pp. 650–653.

[64] Y. Nedumov, A. Babichev, I. Mashonsky, N. Semina, Scinoon: exploratory search system for scientific groups, in: IUI Workshops, 2019.

[65] H.P. Wu, J.Q. Ma, Research on application of graduated student employment management based on decision support system, Heilongjiang Res. Higher Educ. 6 (2008).

[66] J.S. Hwang, S. Lee, Y. Lee, S. Park, A selection method of database system in bigdata environment: a case study from smart education service in Korea, Int. J. Adv. Soft Comput. Appl. 7 (1) (2015) 9–21.

[67] I.A. Kautsar, S.I. Kubota, Y. Musashi, K. Sugitani, The Use of RethinkDB and Lecturer Based Supportive Tool (LBST) as Database Learning Environment, 2016.

[68] C. Popa, G. Carutasu, C. Cotet, N. Carutasu, T. Dobrescu, Smart city platform development for an automated waste collection system, Sustainability 9 (11) (2017) 2064.

[69] H.Y. Chien, Y.M. Tseng, R.W. Hung, Some study of applying infra-red in agriculture IoT, in: 2018 Ninth International Conference on Awareness Science and Technology (iCAST), IEEE, 2018, pp. 1–5.

[70] B. Iordanov, A. Alexandrova, S. Abbas, T. Hilpold, P. Upadrasta, The semantic web as a software modeling tool: an application to citizen relationship management, in: International Conference on Model Driven Engineering Languages and Systems, Springer, Berlin, Heidelberg, 2013, pp. 589–603.

[71] R.F. van der Lans, InfiniteGraph: Extending Business, Social and Government Intelligence with Graph Analytics, 2010, Tech. rep.

[72] R. Pieper, D. Griebler, A. Lovato, Towards a software as a service for biodigestor analytics, Rev. Eletr. Argentina-Brasil Tecnol. Inf. Comun. 1 (5) (2016).

[73] B. Behzad, A. Padmanabhan, Y. Liu, Y. Liu, S. Wang, Integrating CyberGIS gateway with Windows Azure: a case study on MODFLOW groundwater simulation, in: Proceedings of the ACM SIGSPATIAL Second International Workshop on High Performance and Distributed Geographic Information Systems, ACM, 2011, pp. 26–29.

[74] B. Pawłowicz, M. Salach, B. Trybus, Smart city traffic monitoring system based on 5G cellular network, RFID and machine learning, in: KKIO Software Engineering Conference, Springer, Cham, 2018, pp. 151–165.

[75] R.B. Yeshani, Ridesharing Android Application for Traffic Control, 2018.

[76] M. Weibach, Live traffic data analysis using stream processing, in: 2018 IEEE/ACM International Conference on Utility and Cloud Computing Companion (UCC Companion), IEEE, 2018, pp. 65–70.

[77] F. Holzschuher, R. Peinl, Performance optimization for querying social network data, in: EDBT/ICDT Workshops, 2014, pp. 232–239.

[78] D. Gašpar, M. M, NoSQL databases as social networks storage systems, in: 2017 ENTRENOVA Conference Proceedings, 2017.

[79] A.B. Mathew, S.M. Kumar, Analysis of data management and query handling in social networks using NoSQL databases, in: 2015 International Conference on Advances in Computing, Communications and Informatics (ICACCI), IEEE, 2015, pp. 800–806.

[80] G. Wiederhold, Database technology in health care, J. Med. Syst. 5 (3) (1981) 175–196.

[81] C.F. Reilly, Storing and Querying Social Graph Data on a Variety of Distributed Systems, CIDR, 2019.

[82] A. Pacaci, A. Zhou, J. Lin, M.T. Özsu, Do we need specialized graph databases? Benchmarking real-time social networking applications, in: Proceedings of the Fifth International Workshop on Graph Data-management Experiences & Systems, ACM, 2017, p. 12.

[83] D. Cea, J. Nin, R. Tous, J. Torres, A.E., Towards the cloudification of the social networks analytics, in: International Conference on Modeling Decisions for Artificial Intelligence, Springer, Cham, 2014, pp. 192–203.

[84] S. Matilda, Big data in social media environment: a business perspective, in: Decision Management Concepts, Methodologies, Tools, and Applications, IGI Global, 2017, pp. 1876–1899.

[85] A. Magdy, M.F. Mokbel, Demonstration of kite: a scalable system for microblogs data management, in: 2017 IEEE 33rd International Conference on Data Engineering (ICDE), IEEE, 2017, pp. 1383–1384.

[86] R. Martínez-Castaño, J.C. Pichel, P. Gamallo, Polypus: a big data self-deployable architecture for microblogging text extraction and real-time sentiment analysis, arXiv preprint arXiv:1801.03710 (2018).

[87] A. Agoub, F. Kunde, M. Kada, Potential of graph databases in representing and enriching standardized Geodata, Tagungsband der 36 (2016) 208–216.

[88] K. Patroumpas, G. Giannopoulos, S. Athanasiou, Towards GeoSpatial semantic data management: strengths, weaknesses, and challenges ahead, in: Proceedings of the 22nd ACM SIGSPATIAL International Conference on Advances in Geographic Information Systems, ACM, 2014, p. 301.

[89] F. Hu, M. Xu, J. Yang, Y. Liang, K. Cui, M. Little, C. Yang, Evaluating the open source data containers for handling big geospatial raster data, ISPRS Int. J. Geo-Inf. 7 (4) (2018) 144.

[90] D. Agarwal, S.K. Prasad, Lessons learnt from the development of GIS application on azure cloud platform, in: 2012 IEEE Fifth International Conference on Cloud Computing, IEEE, 2012, pp. 352–359.

[91] E. Baralis, A. Dalla Valle, P. Garza, C. Rossi, F. Scullino, SQL versus NoSQL databases for geospatial applications, in: 2017 IEEE International Conference on Big Data (Big Data), IEEE, 2017, pp. 3388–3397.

[92] N.K. Gundla, Z. Chen, Creating NoSQL biological databases with ontologies for query relaxation, Procedia Comput. Sci. 91 (2016) 460–469.

[93] R. Punnoose, A. Crainiceanu, D. Rapp, Rya: a scalable RDF triple store for the clouds, in: Proceedings of the First International Workshop on Cloud Intelligence, ACM, 2012, p. 4.

[94] G. Ladwig, A. Harth, CumulusRDF: linked data management on nested key-value stores, in: The Seventh International Workshop on Scalable Semantic Web Knowledge Base Systems (SSWS 2011), vol. 30, 2011.

[95] V. Khadilkar, M. Kantarcioglu, B. Thuraisingham, P. Castagna, Jena-HBase: a distributed, scalable and efficient RDF triple store, 2012, pp. 85–88.

[96] J. Boyle, H. Rovira, C. Cavnor, D. Burdick, S. Killcoyne, I. Shmulevich, Adaptable data management for systems biology investigations, BMC bioinformatics 10 (1) (2009) 79.

[97] T. Gur, Biobtree: a tool to search, map and visualize bioinformatics identifiers and special keywords, in: F1000Research, 2019, p. 8.

[98] R. Stein, V. Zacharias, RDF on cloud number nine, in: Fourth Workshop on New Forms of Reasoning for the Semantic Web: Scalable and Dynamic, 2010, pp. 11–23.

[99] I. Kim, J.Y. Jung, T.F. DeLuca, T.H. Nelson, D.P. Wall, Cloud computing for comparative genomics with windows azure platform, Evol. Bioinform. 8 (2012) 527–534.

[100] R. Wadhwa, P. Singh, M. Singh, S. Kumar, An EMR-enabled medical sensor data collection framework, in: 2015 Seventh International Conference on Communication Systems and Networks (COMSNETS), IEEE, 2015, pp. 1–6.

[101] D. Marcelli, J. Kirchgessner, C. Amato, H. Steil, A. Mitteregger, M.V., E. Gatti, EuCliD (European Clinical Database): a database comparing different realities, J. Nephrol. 14 (2001) S94–S100.

[102] L. Byczkowska-Lipińska, A. Wosiak, Multimedia NoSQL database solutions in the medical imaging data analysis, Przegl 12 (2013) 234–237.

[103] J. Granados, T. Westerlund, L. Zheng, Z. Zou, IoT platform for real-time multichannel ECG monitoring and classification with neural networks, in: International Conference on Research and Practical Issues of Enterprise Information Systems, Springer, Cham, 2017, pp. 181–191.

[104] L.R. Margolies, G. Pandey, E.R. Horowitz, D.S. Mendelson, Breast imaging in the era of big data: structured reporting and data mining, Am. J. Roentgenol. 206 (2) (2016) 259–264.

[105] M. Singh, K. Kaur, Sql2neo: moving health-care data from relational to graph databases, in: 2015 IEEE International Advance Computing Conference (IACC), IEEE, 2015, pp. 721–725.

[106] S. Tagore, N. Chowdhury, R.K. De, Analyzing methods for path mining with applications in metabolomics, Gene 534 (2) (2014) 125–138.

[107] S. Meystre, H. Müller, Open source software in the biomedical domain: electronic health records and other useful applications, Swiss Med. Inf. 55 (3) (2005) 1–25.

[108] E.A. Mohammed, B.H. Far, C. Naugler, Applications of the MapReduce programming framework to clinical big data analysis: current landscape and future trends, BioData Mining 7 (1) (2014) 22.

[109] S.J. Rascovsky, J.A. Delgado, A. Sanz, V.D. Calvo, G. Castrillón, Informatics in radiology: use of CouchDB for document-based storage of DICOM objects, Radiographics 32 (3) (2012) 913–927.

[110] M. Gołosz, D. Mrozek, Detection of dangers in human health with IoT devices in the cloud and on the edge, in: International Conference: Beyond Databases, Architectures and Structures, Springer, Cham, 2019, pp. 40–53.

[111] I. Tsakovska, M. Al Sharif, E. Fioravanzo, A. Bassan, S. Kovarich, V. Vitcheva, M. Cronin, In silico approaches to support liver toxicity screening of chemicals: case study on molecular modelling of ligands-nuclear receptors interactions to predict potential steatogenic effects, Toxicol. Lett. 2 (238) (2015) S173.

[112] A.N. Richarz, M.N. Berthold, E. Fioravanzo, D. Neagu, A. Pery, A.P. Worth, M. Cronin, New computational approaches for repeated dose toxicity prediction in view of the safety assessment of cosmetic ingredients, in: 16. International Workshop on Quantitative Structure-Activity Relationship in Environmental and Health Sciences (QSAR 2014), 2014, p. 68.

[113] H.M. Hollnagel, K. Arvidson, S. Barlow, A. Boobis, M.T. Cronin, S.P. Felter, C. Yang, Final report on the development of a non-cancer threshold of toxicological concern (TTC) database to support alternative assessment methods for cosmetics-related chemicals, Toxicologist 150 (1) (2016) 349.

[114] M. Cronin, A.N. Richarz, D. Neagu, C. Yang, M. Pavan, J.M. Zaldívar-Comenges, T. Meinl, COSMOS: Integrated in Silico Models for the Prediction of Human Repeated Dose Toxicity of COSMetics to Optimise Safety, No. 2, Toxicology in the 21st Century: Mechanism-Driven Toxicology Defines the Toxic Dose (2016) 140–173.

[115] B. Wang, L. Tao, T. Burghardt, M. Mirmehdi, Calorific expenditure estimation using deep convolutional network features, in: IEEE Winter Applications of Computer Vision Workshops (WACVW), IEEE, 2018, pp. 69–76.

[116] A. Ghrab, O. Romero, S. Jouili, S. Skhiri, Graph BI & analytics: current state and future challenges, in: International Conference on Big Data Analytics and Knowledge Discovery, Springer, Cham, 2018, pp. 3–18.

[117] V. Jayagopal, K.K. Basser, Data management and big data analytics: data management in digital economy, IGI Global, 2019, pp. 1–23.

[118] J. Lukić, The impact of big data technologies on competitive advantage of companies, FU Econ. Org. 14 (3) (2017) 255–264.

[119] N. Palmer, M. Sherman, Y. Wang, S. Just, Scaling to Build the Consolidated Audit Trail: A Financial Services Application of Google Cloud Bigtable, Fidelity National Information Services Inc., 2015.

[120] M. Chalkiadaki, K. Magoutis, Managing service performance in NoSQL distributed storage systems, in: Proceedings of the Seventh Workshop on Middleware for Next Generation Internet Computing, ACM, 2012, p. 5.

[121] Cambridge Semantics Inc, Anzograph, 2007. Retrieved from: https://dbdb.io/db/anzograph. (accessed 01 May 2020).

[122] L.J. Sandoval, Design of business intelligence applications using big data technology, in: IEEE Thirty Fifth Central American and Panama Convention (CONCAPAN XXXV), IEEE, 2015, pp. 1–6.

[123] C. Mohan, R. Barber, S. Watts, A. Somani, M. Zaharioudakis, Evolution of Groupware for Business Applications: A Database Perspective on Lotus Domino/Notes, VLDB, 2000, pp. 684–687.

[124] J. Duda, Business intelligence and NoSQL databases, Inf. Syst. Manag. 1 (1) (2012) 25–37.

[125] S. Munirathinam, B. Ramadoss, Big data predictive analytics for proactive semiconductor equipment maintenance, in: 2014 IEEE International Conference on Big Data (Big Data), IEEE, 2014, pp. 893–902.

[126] B. Hoła, M. Szóstak, A computer knowledge database of accidents at work in the construction industry, IOP Conf. Ser. Mater. Sci. Eng. 251 (1) (2017) 012049. iOP Publishing.

[127] D.A. Koutsomitropoulos, A.K. Kalou, A standards-based ontology and support for big data analytics in the insurance industry, ICT Express 3 (2) (2017) 57–61.

[128] S. Zhang, Application of document-oriented NoSQL database technology in web-based software project documents management system, in: 2013 IEEE Third International Conference on Information Science and Technology (ICIST), IEEE, 2013, pp. 504–507.

[129] H. Uchio, T. Kaneda, S. Motohasi, U.S. Patent Application No. 09/852,563, 2002.

[130] K. Konishi, N.F. Ikeda, Data model and architecture of a paper-digital document management system, in: Proceedings of the 2007 ACM symposium on Document engineering, ACM, 2007, pp. 29–31.

[131] G. Shi, M. Su, F. Li, J. Lou, Q. Huang, A user-based document management mechanism in cloud, in: 2014 Tenth International Conference on Computational Intelligence and Security, IEEE, 2014, pp. 377–381.

[132] E. Siegel, A. Retter, eXist: A NoSQL Document Database and Application Platform, O'Reilly Media, Inc., 2014.

[133] J. Aasman, M.C. Hadfield, P. Mirhaji, U.S. Patent No. 9,679,041, U.S. Patent and Trademark Office, Washington, DC, 2017.

[134] R.K. Prematunga, Correlational analysis, Australian Critical Care 25 (3) (2012) 195–199.

[135] A. Chaturvedi, P.E. Green, J.D. Caroll, K-modes clustering, J. Classification 18 (1) (2001) 35–55.

[136] F. Cao, J. Liang, L. Bai, A new initialization method for categorical data clustering, Expert Syst. Appl. 36 (7) (2009) 10223–10228.

[137] M. Lee, S. Jeon, M. Song, Understanding user's interests in NoSQL databases in stack overflow, in: Proceedings of the 7th International Conference on Emerging Databases, Springer, Singapore, 2018, pp. 128–137.

[138] H. Hung, K. Rajamani, J. Lee, S. Jain, G. Ravipati, J. Mchugh, J.C. Huang, U.S. Patent Application No. 16/135,769, 2019.

[139] Y. Wu, Research and implementation of library circulation system based on block chain, in: Proceedings of the Second International Conference on Computer Science and Application Engineering, ACM, 2018, p. 36.

[140] C. Chrysafis, B. Collins, S. Dugas, J. Dunkelberger, M. Ehsan, S. Gray, M. McMahon, FoundationDB record layer: a multi-tenant structured datastore, arXiv preprint arXiv:1901.04452 (2019).

[141] IBM, IBM Informix C-ISAM, Retrieved from: https://www.ibm.com/in-en/ marketplace/ibm-informix-cisam (accessed 01 May 2020).

[142] S. Acharya, Apache Ignite Quick Start Guide: Distributed Data Caching and Processing Made Easy, Packt Publishing Ltd, 2018.

[143] M. Płaza, S. Deniziak, M. Płaza, R. Belka, P. Pięta, Analysis of parallel computational models for clustering, in: Photonics Applications in Astronomy, Communications, Industry, and High-Energy Physics Experiments 2018, International Society for Optics and Photonics, vol. 10808, 2018, p. 108081O.

[144] M.A. Qader, S. Cheng, V. Hristidis, A comparative study of secondary indexing techniques in LSM-based NoSQL databases, in: Proceedings of the 2018 International Conference on Management of Data, ACM, 2018, pp. 551–566.

[145] N. Johnsirani Venkatesan, C. Nam, D.R. Shin, Deep learning frameworks on apache spark: a review, IETE Tech. Rev. 36 (2) (2019) 164–177.

[146] J. Li, J. Li, Research on NoSQL Database Technology, Atlantis Press, 2018.

[147] Y. Patel, M. Verma, A.C. Arpaci-Dusseau, R.H. Arpaci-Dusseau, Revisiting concurrency in high-performance NoSQL databases, in: 10th USENIX Workshop on Hot Topics in Storage and File Systems (HotStorage 18), 2018.

[148] M. Banane, A. Belangour, A survey on RDF data store based on NoSQL Systems for the semantic web applications, in: International Conference on Advanced Intelligent Systems for Sustainable Development, Springer, Cham, 2018, pp. 444–451.

[149] Y. Huangfu, J. Cao, H. Lu, G. Liang, Matrixmap: programming abstraction and implementation of matrix computation for big data applications, in: 2015 IEEE 21st International Conference on Parallel and Distributed Systems (ICPADS), IEEE, 2015, pp. 19–28.

[150] A. Papaioannou, K. Magoutis, Replica-group leadership change as a performance enhancing mechanism in NoSQL data stores, in: 2018 IEEE 38th International Conference on Distributed Computing Systems (ICDCS), IEEE, 2018, pp. 1448–1453.

[151] K. Segeljakt, A Scala DSL for Rust Code Generation (Dissertation), Retrieved from http://urn.kb.se/resolve?urn=urn:nbn:se:kth:diva-235358Q20 (accessed 14 January 2022).

[152] O. Kalyonova, N. Akparaliev, I. Perl, Design Of specialized storage for heterogeneous project data, in: Proceedings of the 23rd Conference of Open Innovations Association FRUCT, FRUCT Oy, 2018, p. 21.

[153] X. Yin, Q. Luo, Research and application of large data query technology based on NoSQL database, in: 2018 Third International Workshop on Materials Engineering and Computer Sciences (IWMECS 2018), Atlantis Press, 2018.

[154] D. Fernandes, J. Bernardino, Graph databases comparison: AllegroGraph, ArangoDB, InfiniteGraph, Neo4J, and OrientDB, in: Proceedings of the Seventh International Conference on Data Science, Technology and Applications, 2018, pp. 373–380.

[155] G. Daniel, J. Cabot, S.G., Advanced prefetching and caching of models with PrefetchML, Softw. Syst. Model. 18 (3) (2018) 1–22.

[156] A. Mahgoub, S. Ganesh, F. Meyer, A. Grama, S. Chaterji, Suitability of NoSQL systems—Cassandra and ScyllaDB—For IoT workloads, in: 2017 Ninth International Conference on Communication Systems and Networks (COMSNETS), IEEE, 2017, pp. 476–479.

[157] R. Reagan, Cosmos DB, in: Web Applications on Azure, Apress, Berkeley, CA, 2018, pp. 187–255.

[158] S.A. Noghabi, J. Kolb, P. Bodik, E. Cuervo, Steel: simplified development and deployment of edge-cloud applications, in: 10th { USENIX} Workshop on Hot Topics in Cloud Computing (HotCloud 18), 2018.

[159] Y. Kondratenko, G. Kondratenko, I. Sidenko, Multi-criteria decision making for selecting a rational IoT platform, in: 2018 IEEE Ninth International Conference on Dependable Systems, Services and Technologies (DESSERT), IEEE, 2018, pp. 147–152.

[160] C.I. DelGaudio, S.D. Hicks, W.M. Houston, R.S. Kurtz, V.A. Hanrahan, J.A. Martin Jr, D.C. Rauch, U.S. Patent No. 9,928,480, U.S. Patent and Trademark Office, Washington, DC, 2018.

[161] M. Eshtay, A. Sleit, M. Aldwairi, Implementing Bi-Temporal Properties into Various NoSQL Database Categories, Int. J. Comput. 18 (1) (2019) 45–52.

[162] C.G. Chutea, S.M. Huffb, The pluripotent rendering of clinical data for precision medicine, in: MEDINFO 2017: Precision Healthcare Through Informatics: Proceedings of the 16th World Congress on Medical and Health Informatics, vol. 245, IOS Press, 2018, p. 337.

[163] EkkySoftware, ObjectDatabase++, 2012. Retrieved from: https://web.archive.org/web/20120926131201/http://www.ekkysoftware.com/ODBPP.

[164] F. Bugiotti, Modeling strategies for storing data in distributed heterogeneous NoSQL databases, in: Conceptual Modeling: 37th International Conference, ER 2018, Xi'an, China, October 22–25, 2018, Proceedings, Springer, 11157, 2018, p. 488.

[165] Y. Boshmaf, H.A. Jawaheri, M.A. Sabah, BlockTag: design and applications of a tagging system for blockchain analysis, in: SEC 2019, 2018, pp. 299–313.

[166] O. Eini, Inside RavenDB 4.0, 2018. https://ravendb.net/articles/inside-ravendb-Q23 book.

[167] Rocket, RocketU2. Retrieved from: https://www.rocketsoftware.com/products/rocket-u2/documentation (accessed 14 January 2022).

[168] M. Hwang, Graph processing using SAP HANA: a teaching case, J. Bus. Educ. Scholar. Teach. 12 (2) (2018) 155–165.

[169] P. Martins, M. Abbasi, F. Sá, A study over NoSQL performance, in: World Conference on Information Systems and Technologies, Springer, Cham, 2019, pp. 603–611.

[170] B. Iancu, T.M. Georgescu, Saving large semantic data in cloud: a survey of the main DBaaS solutions, Inf. Econ. 22 (1) (2018) 5–16.

[171] R. Reagan, Azure data storage overview, in: Web Applications on Azure, Apress, Berkeley, CA, 2018, pp. 61–76.

[172] R. Angles, M. Arenas, B.P., P. Boncz, G. Fletcher, C. Gutierrez, O. van Rest, G-CORE: a core for future graph query languages, in: Proceedings of the 2018 International Conference on Management of Data, ACM, 2018, pp. 1421–1432.

[173] R. Montella, D. Di Luccio, S. Kosta, G. Giunta, I. Foster, Performance, resilience, and security in moving data from the fog to the cloud: the DYNAMO transfer framework approach, in: International Conference on Internet and Distributed Computing Systems, Springer, Cham, 2018, pp. 197–208.

[174] S. Yuzuk, M.G. Aktas, M.S. Aktas, On the performance analysis of map-reduce programming model on in-memory NoSQL storage platforms: a case study, in: 2018 International Congress on Big Data, Deep Learning and Fighting Cyber Terrorism (IBIGDELFT), 2018.

[175] G. Vonitsanos, A. Kanavos, P. Mylonas, S. Sioutas, A NoSQL database approach for modeling heterogeneous and semi-structured information, in: 2018 Ninth International Conference on Information, Intelligence, Systems and Applications (IISA), IEEE, 2018, pp. 1–8.

[176] M.T. González-Aparicio, M. Younas, J. Tuya, R. Casado, Testing of transactional services in NoSQL key-value databases, Future Generation Comput. Syst. 80 (2018) 384–399.

[177] T. Kovács, G. Simon, G. Mezei, Benchmarking graph database backends—what works well with Wikidata? in: The 11th Conference of PhD Students in Computer Science, 2018, p. 154.

[178] K. Ahmad, M.S. Alam, N.I. Udzir, Security of NoSQL database against intruders, Recent Patents Eng. 13 (1) (2019) 5–12.

[179] J. Webber, I. Robinson, A Programmatic Introduction to Neo4j, Addison-Wesley Professional, 2018.

[180] D. Agrawal, R.K. Ganti, K. Lee, M. Srivatsa, U.S. Patent No. 9,886,785, U.S. Patent and Trademark Office, Washington, DC, 2018.

[181] N. Franciscus, X. Ren, B. Stantic, Precomputing architecture for flexible and efficient big data analytics, Vietnam J. Comput. Sci. 5 (2) (2018) 133–142.

[182] K. Hu, J. Zhu, A Progressive Web Application on Ancient Roman Empire Coins and Relevant Historical Figures with Graph Database, in: Euro-Mediterranean Conference, Springer, Cham, 2018, pp. 235–241.

[183] H.U. Rahman, R.U. Khan, A. Ali, Programming and pre-processing systems for Big data storage and visualization, in: Handbook of Research on Big Data Storage and Visualization Techniques, IGI Global, 2018, pp. 228–253.

[184] H. Gujral, A. Sharma, P. Kaur, Empirical investigation of trends in NoSQL-based big-data solutions in the last decade, in: 2018 Eleventh International Conference on Contemporary Computing (IC3), IEEE, 2018, pp. 1–3.

[185] Z. Smith, Joining and aggregating data sets using CouchDB (Ph.D. thesis), University of Cape Town 2018.

[186] J. Kepner, V. Gadepally, L. Milechin, S. Samsi, W. Arcand, D. Bestor, M. Jones, A billion updates per second using 30,000 in-memory D4M databases, arXiv preprint arXiv:1902.00846 (2019).

[187] P. Seda, J. Hosek, P. Masek, J. Pokorny, Performance testing of NoSQL and RDBMS for storing big data in e-applications, in: 2018 Third International Conference on Intelligent Green Building and Smart Grid (IGBSG), IEEE, 2018, pp. 1–4.

[188] R. Sánchez-de Madariaga, A. Muñoz, A.L. Castro, O. Moreno, M. Pascual, Executing complexity-increasing queries in relational (MySQL) and NoSQL (MongoDB and EXist) size-growing ISO/EN 13606 standardized EHR databases, J. Vis. Exp. 133 (2018) 57439.

[189] G.M.S.S.M. Agarwal, "Big" data management in cloud computing environment, in: Harmony Search and Nature Inspired Optimization Algorithms, Springer, Singapore, 2019, pp. 707–716.

[190] T. Yamaguchi, M. Brain, C. Ryder, Y. Imai, Y. Kawamura, Application of abstract interpretation to the automotive electronic control system, in: International Conference on Verification, Model Checking, and Abstract Interpretation, Springer, Cham, 2019, pp. 425–445.

[191] A. Makris, K. Tserpes, G. Spiliopoulos, D. Anagnostopoulos, Performance evaluation of MongoDB and PostgreSQL for spatio-temporal data, in: EDBT/ICDT Workshops, 2019.

[192] D. Swami, B. Sahoo, Storage size estimation for schemaless big data applications: a JSON-based overview, in: Intelligent Communication and Computational Technologies, Springer, Singapore, 2018, pp. 315–323.

[193] G.C. Deka, NoSQL web crawler application, in: Advances in Computers, vol. 109, Elsevier, 2018, pp. 77–100.

[194] G.V. Demirci, C. Aykanat, Scaling sparse matrix-matrix multiplication in the accumulo database, Distrib. Parallel Databases 38 (1) (2019) 1–32.

[195] A.A. Mahmood, Automated algorithm for data migration from relational to NoSQL databases, Alnahrain J. Eng. Sci. 21 (1) (2018) 60–65.

[196] J. Chang, O. Gutsche, I. Mandrichenko, J. Pivarski, Striped Data Server for Scalable Parallel Data Analysis, Publishing, IOP, 2018.

[197] M. Marinov, G. Georgiev, E. Popova, NoSQL approach for sensor data storage and retrieval, in: 2018 41st International Convention on Information and Communication Technology, Electronics and Microelectronics (MIPRO), IEEE, 2018, pp. 1427–1432.

[198] K. Chodorow, MongoDB: The Definitive Guide: Powerful and Scalable Data Storage, "O'Reilly Media, Inc"., 2013.

[199] W. Wingerath, F. Gessert, E. Witt, S. Friedrich, N. Ritter, Real-time data management for big data, in: EDBT, 2018, pp. 524–527.

[200] M.E.L. Malki, H.B. Hamadou, N. El Malki, A. Kopliku, MPT: suite tools to support performance tuning in NoSQL systems, in: Conference on Enterprise Information Systems (ICEIS 2018), 2018.

[201] R. Rai, P. Chettri, NoSQL hands on, Adv. Comput. 109 (2018) 157–277.

[202] B. Maity, A. Acharya, T. Goto, S. Sen, A framework to convert NoSQL to relational model, in: Proceedings of the Sixth ACM/ACIS International Conference on Applied Computing and Information Technology, ACM, 2018, pp. 1–6.

[203] M. Imran, M.V. Ahamad, M. Haque, M. Shoaib, Big data analytics tools and platform in big data landscape, in: Handbook of Research on Pattern Engineering System Development for Big Data Analytics, IGI Global, 2018, pp. 80–89.

[204] BigOpenData.eu, NoSQLz, 2017. Retrieved from: https://www.bigopendata.eu/tag/nosqlz-developers/. (accessed 01 May 2010).

[205] J. Yao, A Comparison of Different Graph Database Types (Dissertation), 2018. Retrieved from: https://static.epcc.ed.ac.uk/dissertations/hpc-msc/2017-2018/Jieru_Yao-dissertation_s1702971.pdf (Accessed 14 January 2022).

[206] A. Castiglione, F. Colace, V. Moscato, F. Palmieri, CHIS: a big data infrastructure to manage digital cultural items, Future Gen. Comput. Syst. 86 (2018) 1134–1145.

[207] C. Rotter, L. Farkas, G. Nyíri, G. Csatári, L. Jánosi, R. Springer, Using Linux containers in telecom applications, in: Proc. ICIN, 2016, pp. 234–241.

[208] K. Evans, A. Jones, A. Preece, F. Quevedo, D. Rogers, S.I., other, G. Suciu, Dynamically reconfigurable workflows for time-critical applications, in: Proceedings of the 10th Workshop on Workflows in Support of Large-Scale Science, ACM, 2015, p. 7.

[209] J.L. Pérez, A. Gutierrez-Torre, J.L. Berral, D. Carrera, A resilient and distributed near real-time traffic forecasting application for Fog computing environments, Fut. Gen. Comput. Syst. 87 (2018) 198–212.

[210] X. Yuan, X. Yuan, B. Li, C. Wang, Toward secure and scalable computation in Internet of Things data applications, IEEE Internet Things J. 6 (2) (2019) 3753–3763.

[211] N.A. Shiftan, A.S. Gupta, B.W. Lowry, L. Kratz III, U.S. Patent Application No. 14/773,703, 2016.

[212] M. Giri, S. Jyothi, Big data collection and correlation analysis of wireless sensor networks yielding to target detection and classification, in: Proceedings of International Conference on Computational Intelligence and Data Engineering, Springer, Singapore, 2018, pp. 201–213.

[213] Open Source Initiative, Open Standards Compliance, Retrieved from: https://opensource.org/osr-compliance (accessed 14 January 2022).

[214] S. Coughlan, J.G. Breslin, The application of modern PDA technology for effective handheld solutions in the retail industry, in: IEEE International Conference on Industrial Technology, 2003, 1. IEEE, 2003, pp. 411–415.

[215] H. Wong, U.S. Patent No. 8,478,772, U.S. Patent and Trademark Office, Washington, DC, 2013.

[216] A. Copie, F.T. F., V.I. Munteanu, Benchmarking cloud databases for the requirements of the Internet of Things, in: Proceedings of the ITI 2013 35th International Conference on Information Technology Interfaces. IEEE, 2013, pp. 77–82.

[217] A. Maccioni, M. Collina, Graph databases in the browser: using levelgraph to explore New Delhi, Proc. VLDB Endowment 9 (13) (2016) 1469–1472.

[218] M.H. Diallo, M. August, R. Hallman, M. Kline, H. Au, V. Beach, Nomad: a framework for developing mission-critical cloud-based applications, in: 2015 10th International Conference on Availability, Reliability and Security, IEEE, 2015, pp. 660–669.

[219] M.K. Shrivas, T. Yeboah, The disruptive blockchain: types, platforms and applications, in: Fifth Texila World Conference for Scholars (TWCS) on Transformation: The Creative Potential of Interdisciplinary, 2018.

[220] J.L.R. Martínez, M.H.M. Cruz, M.A.R. Vázqu, L.R. Espejo, A.M. Obeso, M.S.G. Vázquez, A.A.R. Acosta, BDVC (bimodal database of violent content): a database of violent audio and video, in: Applications of Digital Image Processing XL. International Society for Optics and Photonics, vol. 10396, 2017, p. 103961O.

[221] C. Padgett, Real-Time Image Processing in Support of Aerial Sensing Applications, 2012. Retrieved from: https://trs.jpl.nasa.gov/bitstream/handle/2014/43193/12-5419_A1b.pdf?sequence=1 (Accessed 14 January 2022).

[222] A. Uta, A. Sandu, S. Costache, T. Kielmann, MemEFS: an elastic in-memory runtime file system for escience applications, in: 2015 IEEE 11th International Conference on e-Science, IEEE, 2015, pp. 465–474.

[223] S.R. Chalamalasetti, K. Lim, M. Wright, A. AuYoung, P. Ranganathan, M. Margala, An FPGA memcached appliance, in: Proceedings of the ACM/SIGDA International Symposium on Field Programmable Gate Arrays, ACM, 2013, pp. 245–254.

[224] F. Li, S. Zhan, L. Li, Research on using memcached in call center, in: Proceedings of 2011 International Conference on Computer Science and Network Technology. IEEE, 3, 2011, pp. 1721–1723.

[225] M.S. Kumar, Comparison of NoSQL database and traditional database—an emphatic analysis, JOIV: Int. J. Inf. Vis. 2 (2) (2018) 51–55.

[226] G.J. Chen, J.L. Wiener, S. Iyer, A. Jaiswal, R. Lei, N. Simha, S. Yilmaz, Realtime data processing at Facebook, in: Proceedings of the 2016 International Conference on Management of Data, ACM, 2016, pp. 1087–1098.

[227] C. Vega, P. Roquero, R. Leira, I. Gonzalez, J. Aracil, Loginson: a transform and load system for very large-scale log analysis in large IT infrastructures, J. Supercomput. 73 (9) (2017) 3879–3900.

[228] O. Nevzorova, D. Mukhamedshin, R. Gataullin, Developing corpus management system: architecture of system and database, in: Proceedings of the International Conference on Information and Knowledge Engineering (IKE). The Steering Committee of The World Congress in Computer Science, Computer Engineering and Applied Computing (WorldComp), 2017, pp. 108–112.

[229] I.A. Perl, Efficient storage mechanisms for Internet of Things solutions in ESB, in: Exploring Enterprise Service Bus in the Service-Oriented Architecture Paradigm, IGI Global, 2017, pp. 206–215.

[230] T. Wilcox, N. Jin, P. Flach, J. Thumim, A big data platform for smart meter data analytics, Comput. Indus. 105 (2019) 250–259.

[231] S. Andari, M. Caruso, M. Ganis, C.B. Robbins, C. Whit, A.J. Zada, Web Application for Environmental Sensing, Seidenberg School of Computer Science & Information Systems, Pace University, Pleasantville, NY, 2016.

[232] H. Häggander, R. Letterkrantz, F. Rahn, E. Sievers, EIRA—An Application for Finding and Ranking Researchers (Student Essay), 2016. Retrieved from: https://gupea.ub.gu.se/handle/2077/49470 (Accessed 14 January 2022).

[233] T. Kuramoto, Risk monitoring for nuclear power plant applications using probabilistic risk assessment, in: Progress of Nuclear Safety for Symbiosis and Sustainability, Springer, Tokyo, 2014, pp. 145–151.

[234] S. Sulander, Microservices Architecture in Open Retail Interface for Public Transport Tickets (Dissertation), 2018. Retrieved from: https://www.theseus.fi/bitstream/handle/10024/160848/sulander_santtu.pdf?sequence=1 (Accessed 14 January 2022).

[235] C.J. White, IBM enterprise analytics for the intelligent e-business, 2001. Retrieved at: ftp://129.35.224.112/software/data/informix/pubs/papers/bisolution/bisolution. PDF. (accessed 01 May 2020).

[236] V. Chacko, N. Varvarelis, D.G. Kemp, eHand-offs: an IBM® lotus® domino ® application for ensuring patient safety and enhancing resident supervision in hand-off communications, in: AMIA Annual Symposium Proceedings, American Medical Informatics Association, 2006, p. 874. 874.

[237] S.R., C. Krstev, D. Vitas, V.N., K. O, Keyword-based search on bilingual digital libraries, in: Semantic Keyword-Based Search on Structured Data Sources, Springer, Cham, 2016, pp. 112–123.

[238] M. Ivanova, N. Nes, R. Goncalves, M. Kersten, MonetDB/NoSQL meets skyserver: the challenges of a scientific database, in: 19th International Conference on Scientific and Statistical Database Management (SSDBM 2007), IEEE, 2007, p. 13.

[239] J.P.T.M. Noordhuizen, J. Buurman, VAMPP: a veterinary automated management and production control programme for dairy farms (the application of MUMPS for data processing), Vet. Q. 6 (2) (1984) 66–72.

[240] L. Mes, EMR database upgrade from MUMPS to CACHE: lessons learned, Integr. Inform. Technol. Manage. Qual. Care 202 (2014) 142.

[241] M. Stührenberg, D. Goecke, SGF—an integrated model for multiple annotations and its application in a linguistic domain, in: Proceedings of Balisage: The Markup Conference, 2008, p. 1.

[242] F. Faerber, J. Dees, M. Weidner, S. Baeuerle, W. Lehner, Towards a web-scale data management ecosystem demonstrated by SAP HANA, in: 2015 IEEE 31st International Conference on Data Engineering, IEEE, 2015, pp. 1259–1267.

[243] G. Portella, G.N. Rodrigues, E. Nakano, A.C. Melo, Statistical analysis of Amazon EC2 cloud pricing models, Concurr. Comput. Pract. Exp. 31 (18) (2018) e4451.

[244] Z. Daher, H. Hajjdiab, Cloud storage comparative analysis Amazon simple storage vs. Microsoft Azure Blob storage, Int. J. Mach. Learn. Comput. 8 (1) (2018) 85–89.

[245] A. Bielefeldt, J. Gonsior, M. Krötzsch, Practical linked data access via SPARQL: the case of wikidata, in: Proc. WWW2018 Workshop on Linked Data on the Web (LDOW-18). CEUR Workshop Proceedings, CEUR-WS.org, 2018.

[246] H. Thakkar, D. Punjani, Y. Keswani, J. Lehmann, S. Auer, A stitch in time saves nine-SPARQL querying of property graphs using gremlin traversals, arXiv preprint arXiv:1801.02911 (2018).

[247] V. Marinakis, H. Doukas, J. Tsapelas, S. Mouzakitis, A. Sicilia, L. Madrazo, S. Sgouridis, From big data to smart energy services: an application for intelligent energy management, Future Generation Comput. Syst. 110 (2020) 572–586.

[248] N. Francis, A. Green, P. Guagliardo, L. Libkin, T. Lindaaker, V. Marsault, A. Taylor, Cypher: an evolving query language for property graphs, in: Proceedings of the 2018 International Conference on Management of Data, ACM, 2018, pp. 1433–1445.

[249] V. Subramanian, L. Wang, E.J. Lee, P. Chen, Rapid processing of synthetic seismograms using windows azure cloud, in: 2010 IEEE Second International Conference on Cloud Computing Technology and Science, IEEE, 2010, pp. 193–200.

[250] A. Gutfraind, M. Genkin, A graph database framework for covert network analysis: an application to the Islamic State network in Europe, Soc. Netw. 51 (2017) 178–188.

[251] Z. Tóth, P. Gyimesi, R. Ferenc, A public bug database of github projects and its application in bug prediction, in: International Conference on Computational Science and Its Applications, Springer, Cham, 2016, pp. 625–638.

[252] S. Duttagupta, M. Kumar, R. Ranjan, M. Nambiar, Performance prediction of IoT application: an experimental analysis, in: Proceedings of the Sixth International Conference on the Internet of Things, ACM, 2016, pp. 43–51.

[253] A. Béleczki, B. Molnár, Modeling framework for designing and analyzing document-centric information systems based on HypergraphDB, in: CEUR Workshop Proceedings (ISSN: 1613-0073), vol. 2046, 2016, pp. 17–22.

[254] C.H. Zhou, K. Yao, Z.Y. Jiang, W.X. Bai, Research on the Application of NoSQL Database in Intelligent Manufacturing, in: Wearable Sensors and Robots, Springer, Singapore, 2017, pp. 423–434.

[255] B. Molnár, A. Béleczki, A. Benczúr, Application of legal ontologies based approaches for procedural side of public administration, in: International Conference on Electronic Government and the Information Systems Perspective, Springer, Cham, 2016, pp. 135–149.

[256] Z. Aung, Database systems for the smart grid, in: Smart Grids, Springer, London, 2013, pp. 151–168.

[257] K. Fernandes, R. Melhem, M. Hammoud, Dynamic elasticity for distributed graph analytics, in: 2018 IEEE International Conference on Cloud Computing Technology and Science (CloudCom), IEEE, 2018, pp. 145–148.

[258] M. Zaharia, M. Chowdhury, M.J. Franklin, S. Shenker, I. Stoica, Spark: cluster computing with working sets, HotCloud 10 (10) (2010) 95.

[259] T. White, Hadoop: The Definitive Guide, "O'Reilly Media, Inc"., 2012.

[260] S. Noel, A review of graph approaches to network security analytics, in: From Database to Cyber Security, Springer, Cham, 2018, pp. 300–323.

[261] S. Gajendra, Product Recall in Supply Chain Management Using Neo4j graph Database, 2016.

[262] O. Erling, A. Averbuch, J. Larriba-Pey, H. Chafi, A. Gubichev, A. Prat, P. Boncz, The LDBC social network benchmark: interactive workload, in: Proceedings of the 2015 ACM SIGMOD International Conference on Management of Data, ACM, 2015, pp. 619–630.

[263] M. Brugnara, M. Lissandrini, Y. Velegrakis, Graph Databases for Smart Cities, IEEE Smart Cities Initiative, 2016.

[264] C. Mahlow, C. Grün, A. Holupirek, M.H. Scholl, A framework for retrieval and annotation in digital humanities using XQuery full text and update in BaseX, in: Proceedings of the 2012 ACM Symposium on Document Engineering, ACM, 2012, pp. 195–204.

[265] J. Thompson, A. Hankinson, I. Fujinaga, Searching the Liber Usualis: Using COUCHDB and ELASTICSEARCH to query graphical music documents, in: Proceedings of the 12th International Society for Music Information Retrieval Conference, International Society for Music Information Retrieval, Canada, 2011.

[266] G. Ghidini, S.K. Das, Improving home energy efficiency with E 2 Home: a web-based application for integrated electricity consumption and contextual information visualization, in: 2012 IEEE Third International Conference on Smart Grid Communications (SmartGridComm), IEEE, 2012, pp. 471–475.

[267] M. Martinviita, Time Series Database in Industrial IoT and Its Testing Tool, vol. 1, Pentti Kaiteran katu, 2018, p. 90570.

[268] A. Medvedev, A. Zaslavsky, M. Indrawan-Santiago, P.D. Haghighi, A. Hassani, Storing and indexing IoT context for smart city applications, in: Internet of Things, Smart Spaces, and Next Generation Networks and Systems, Springer, Cham, 2016, pp. 115–128.

[269] S. Beis, S. Papadopoulos, Y. Kompatsiaris, Benchmarking graph databases on the problem of community detection, in: New Trends in Database and Information Systems II, Springer, Cham, 2015, pp. 3–14.

[270] D. Pal, T. Triyason, P. Padungweang, Big data in smart-cities: current research and challenges, Indo. J. Elect. Eng. Inf. (IJEEI) 6 (4) (2018) 351–360.

[271] P.B. Teregowda, I.G. Councill, J.P.F. Ramírez, M. Khabsa, S. Zheng, C.L. Giles, SeerSuite: developing a scalable and reliable application framework for building digital libraries by crawling the web, WebApps 10 (2010) 14.

[272] M. Anvari, M.D. Takht, B. Sefid-Dashti, Thrift service composition: toward extending BPEL, in: Proceedings of the International Conference on Smart Cities and Internet of Things, ACM, 2018, p. 13.

[273] A. Erraissi, A. Belangour, Meta-modeling of Zookeeper and MapReduce processing, in: 2018 International Conference on Electronics, Control, Optimization and Computer Science (ICECOCS), IEEE, 2018, pp. 1–5.

[274] S.C. Suh, S. Saffer, N.K. Adla, Extraction of meaningful rules in a medical database, in: 2008 Seventh International Conference on Machine Learning and Applications, IEEE, 2008, pp. 450–456.

[275] J. Rats, G. Ernestsons, Using of cloud computing, clustering and document-oriented database for enterprise content management, in: 2013 Second International Conference on Informatics & Applications (ICIA), IEEE, 2013, pp. 72–76.

[276] R. Lai, Y. Shinjo, Sweets: a decentralized social networking service application using data synchronization on mobile devices, in: International Conference on Collaborative Computing: Networking, Applications and Worksharing, Springer, Cham, 2016, pp. 188–198.

[277] M. Ugen, Scalable performance for a forensic database application (Ph.D. thesis), University of Twente, 2013.

[278] B. Hnich, F.R. Al-Osaimi, A. Sasmaz, O. Sayın, A. Lamine, M. Alotaibi, Smart online vehicle tracking system for security applications, in: 2016 IEEE International Conference on Big Data (Big Data), IEEE, 2016, December, pp. 1724–1733.

[279] J. Sutter, K. Sons, P. Slusallek, Blast: a binary large structured transmission format for the web, in: Proceedings of the 19th International ACM Conference on 3D Web Technologies, ACM, 2014, pp. 45–52.

[280] N.Q. Mehmood, R. Culmone, L. Mostarda, Modeling temporal aspects of sensor data for MongoDB NoSQL database, J. Big Data 4 (1) (2017) 8.

[281] Y.J. Chen, H.Y. Chien, IoT-based green house system with splunk data analysis, in: 2017 IEEE 8th International Conference on Awareness Science and Technology (iCAST), IEEE, 2017, pp. 260–263.

[282] B. Alves, B. Veloso, B. Malheiro, APASail—an agent-based platform for autonomous sailing research and competition, in: Robotic Sailing, Springer, Cham, 2018, pp. 31–38.

[283] D. Sink, A real-time database system for managing aquarium data, Appalachian State University, 2017 (Doctoral dissertation).

[284] S.V. Khobragade, S.L. Nalbalwar, A.B. Nandgaonkar, Fusion execution of NaCl on tree-shaped MSA, Int. J. Antennas Propag. 2018 (2018).
[285] M. Sachenbacher, M. Blankenburg, M. Leucker, CeLiM: Centralized Runtime Monitoring of Lithium-Ion Battery Packs, 2014.
[286] N. Lele, L.S. Wu, R. Akavipat, F. Menczer, Sixearch.org 2.0 peer application for collaborative web search, in: Proceedings of the 20th ACM Conference on Hypertext and Hypermedia, ACM, 2009, pp. 333–334.
[287] B. Yu, A. Cuzzocrea, D. Jeong, S. Maydebura, On managing very large sensor-network data using bigtable, in: Proceedings of the 2012 12th IEEE/ACM International Symposium on Cluster, Cloud and Grid Computing (ccgrid 2012), IEEE Computer Society, 2012, pp. 918–922.
[288] M. Samovsky, T. Kacur, Cloud-based classification of text documents using the Gridgain platform, in: 2012 Seventh IEEE International Symposium on Applied Computational Intelligence and Informatics (SACI), IEEE, 2012, pp. 241–245.
[289] M. Sarnovsky, Z. Ulbrik, Cloud-based clustering of text documents using the GHSOM algorithm on the GridGain platform, in: 2013 IEEE Eighth International Symposium on Applied Computational Intelligence and Informatics (SACI), IEEE, 2013, pp. 309–313.
[290] C. Strobl, A.L. Boulesteix, T. Augustin, Unbiased split selection for classification trees based on the Gini index, Comput. Stat. Data Anal. 52 (1) (2007) 483–501.
[291] M. Kleppmann, A critique of the CAP theorem, arXiv preprint arXiv:1509.05393 (2015).

About the authors

Samiya Khan is an alumna of University of Delhi, India and did her PhD in Computer Science from Jamia Millia Islamia, India. She is currently serving as research fellow at University of Wolverhampton, United Kingdom. She has contributed several research papers and her publications extend across journal articles, conference papers, book chapters, magazine articles and edited books in high impact publications of international repute. Samiya's research interests include the multi-faceted use of Artificial Intelligence and Edge Computing for IoT applications.

Dr. Xiufeng Liu is a senior researcher at the Department of Technology, Economics and Management, Technical University of Denmark (DTU). He received his PhD in Computer Science from Aalborg University, Denmark in 2012. Prior to joining DTU, he worked as a Postdoctoral researcher at Waterloo University, Canada, in 2013–14. His research includes time-series data analytics, data warehousing, the Internet of Things, and big data.

Syed Arshad Ali received his MCA degree from Jamia Hamdard, New Delhi, India. Currently, he is working as a Senior research fellow in the Department of Computer Science, Jamia Millia Islamia, New Delhi, India. His research areas are Dynamic Resource allocation in Cloud Computing, Big data and IoT.

Mansaf Alam received his doctoral degree in Computer Science from Jamia Millia Islamia New Delhi in the year 2009. He is currently working as an Associate Professor in the Department of Computer Science, Jamia Millia Islamia. He is also the Editor-in-Chief, Journal of Applied Information Science. He is an Editorial Board of some reputed International Journals in Computer Sciences and has published several high impact research papers. He also has two books entitled "Concepts of Multimedia, Book" and "Digital Logic Design, PHI" to his credit. His areas of research include Cloud Computing, Big data analytics, Genetic Programming, Bioinformatics, Image Processing, Information Retrieval and Data Mining.

CHAPTER THREE

An empirical investigation on BigGraph using deep learning

Lilapati Waikhom and Ripon Patgiri
Department of Computer Science and Engineering, National Institute of Technology Silchar, Silchar, Cachar, Assam, India

Contents

Advances in Computers, Volume 128
ISSN 0065-2458
https://doi.org/10.1016/bs.adcom.2021.09.007

Abstract

The sparse nature of BigGraphs makes tasks like Graph Classification, Node Classification, and Link Prediction very challenging. Furthermore, homogeneous or heterogeneous characteristics of data in BigGraph make it challenging to do real-time operations. This paper examines some approaches for the mentioned tasks on BigGraph that outperform the previous state-of-the-art approaches. We also empirically investigate these algorithms on various datasets. These comparisons of the various Graph-based deep neural network architectures such as Graph Convolutional Network (GCN), Deep Graph Convolutional Neural Network (DGCNN), etc., on datasets like Mutag, Proteins, etc., shows that deep architecture learns enough patterns. We also show the experimental results that reveal which architecture performs best on a dataset for a given graph-based task.

1. Introduction

A graph is a data structure that consists of a set of nodes and edges representing their relationships. The graph data structure indicates numerous systems in various fields such as knowledge graphs, social networks, natural science, and many other research areas. When the data becomes too big to fit in the traditional way of storing data and grows continuously, it is called BigGraph, which has many nodes and edges. It is a nontrivial task to handle the updates of the addition and deletion of new nodes and edges in the BigGraph taxonomy. Computation on BigGraph becomes even more difficult as they might have heterogeneous nodes and edges. The heterogeneity comes as they are made up by combining the various sources and categories, making it more complex.

Graph data is being generated at an unprecedented pace. With billions of nodes and edges, they are currently stacking in terabytes and moving toward petabytes. An example of a BigGraph is social networks. Facebook is the worldwide largest social network [1], with over 2.7 billion active users monthly in the second quarter of 2020. The number of daily Facebook active users reached 1 billion in the third quarter of 2012, making it the first social network platform (Users signed in to Facebook in the last 30 days are considered active). According to the firm, throughout the most recent quarters [1], 3.14 trillion people utilize at least one of the corporate key products such as Facebook, Messenger, Instagram, and WhatsApp every month. With 2 billion subscribers, YouTube comes in second, followed by WhatsApp, Facebook Messenger, WeChat, and Instagram, all of which have over a billion users. TikTok, which launched in 2016, now has 800 million subscribers and seems to be the one to follow. These large graphs enable one to discover patterns and anomalies, leading to various exciting applications such as computer protection, fraud detection, information flow, and social network analysis.

Traditional graph algorithms presume that the input graph fits in a single machine's memory or discs. However, the recent increase in graph size defies this theory. We automatically shift to distributed algorithms because single machine algorithms are not tractable when dealing with BigGraphs. Moreover, research for graphs analytics accompanied by machine learning has increasingly received more attention as the graphs have great convincing power and high interpretability. Recently, many Graph-based Deep Learning methods are proposed by researchers. Deep Learning has advanced many folds in a wide range of domains, such as image classification [2], Natural Language Processing (NLP) [3], acoustics [4], and is seen as a key method for various future applications. Graphs are an effective representation of the non-Euclidean data structure. Still, tasks like Link Prediction, Node Classification, Node Clustering, and Graph Classification are challenging. Thus, the analysis of the graph is the focus of many types of researches.

The main contributions of this paper are:

- We provide a detailed review of existing Graph-based Deep Learning models in BigGraph in four tasks: Link Prediction, Graph Classification, Node Classification, and Interpretation of Node. We also introduce three general frameworks for Graph-based Deep Learning: StellarGraph, PyTorch-BigGraph (PBG), and Spektral.
- We examined the different Deep Learning models for three tasks/challenges (Link Prediction, Graph Classification, and Node Classification) using different datasets and compared their performance for the training and validation process.

The rest of this paper is organized as follows. Section 2 briefly presents a background study of the Graph-based Deep Learning approaches. In Section 3, we define BigGraph and its characteristics. And we are introducing three general frameworks for BigGraphs. We define three main challenging problems in BigGraph. In Section 4, We elaborate on some methods of Graphs-based Deep Learning for BigGraph. In Section 5, we describe few popular Graph-based datasets. The experimental setup is presented in Section 6, and the results are shown in Section 7. Furthermore, we conclude the paper in Section 8.

2. Background

Recently, many researchers are working on Deep Learning-based approaches for BigGraphs. In 2017, Kipf and Welling proposed a semi-supervised learning approach for Graph-based data, Graph Convolution Networks (GCNs) [5]. In 2018, Zhang et al. [6] introduced Deep Graph Convolutional Neural Network (DGCNN), a novel neural network architecture for the Graph Classification inspired from [5]. Velivckovic et al. [7] proposed Graph attention networks (GATs), a novel neural network architecture for graph-structured data that uses masked self-attentional layers to overcome the limitations of prior approaches based on graph convolutions and their approximations. In 2019, Zhang et al. [8] proposed unified neural network architecture, Attri2vec, for node classification challenges in the attributed network embedding. In 2017, Hamilton et al. [9] proposed a well-established framework for solving the inductive node embedding problem, called GraphSAGE, and extended the work for heterogeneous graphs, also called it as HinSage. In 2016, Grovers and Leskovec founded Node2vec [10] algorithm, which was created to enable the learning of continuous network node representations. Dong et al. [11] in 2017 introduced Metapath2vec, a technique for dealing with network heterogeneity in heterogeneous networks with various nodes and ties. In 2018, Nguyen et al. [12] proposed a continuous dynamically built-in network embedding (CTDNE) framework which allows the incorporation of temporal information through network embedding methods. In 2016, Trouillon et al. [13] presented an approach for link prediction problem based on complex-valued embeddings, which is scalable for the large datasets also, called Complex Embeddings for Simple Link Prediction (ComplEx). In 2015, Yang et al. [14] introduces a DisMult model that uses the neural embedding methodology to describe entities and knowledge-bases (KBs). In 2017, Schlichtkrull et al. [15] presented an R-GCN model for tasks such as the prediction of links or the categorization of nodes. In 2019, Yu et al.

proposed simplifying graph convolutional networks (SGC) to reduce the excess complexity in the GCN [5] through successively removing nonlinearities and collapsing weight matrices between consecutive layers. In 2019, Klicpera et al. [16] presented improved Neural message passing algorithms using the relationship between graph GCN and PageRank for semi-supervised classification on graphs. And many more approaches and models. In Section 4, we elaborate on each Deep Learning-based architecture for graphs data in detail.

3. BigGraph
3.1 Characteristic of BigGraph

There are famous five V's that explain the characteristics of Big Data—Volume, Veracity, Variety, Value, and Velocity [17]. Similarly, BigGraph may also be defined by using the following four V's.

- Volume: The graph's volume is the sum of nodes and edges divided by the total number of nodes and edges. The research becomes more complicated as the graph volume expands continuously. If the graph volume is very high, the data-to-analysis time is too long.
- Velocity: The rate at which the graphs grow in size and complexity is called velocity. For example, in large graphs, the rate of streaming edges. Graph analytics becomes incredibly complicated due to these streaming edges. The steady stream of edges is too much for the memory to manage. As a result, computing a metric such as the shortest distance between two nodes or counting closely connected groups is difficult.
- Variety: The graphs are made up by combining the various sources and categories of data that are merged. The graphs data are often created through integration, such as document, XML/ JSON, graph-structured data, and relational. When different kinds of BigGraphs are combined, they can have different interpretations, which may make it more complex. Some examples are social networks, citation networks, ontologies, proteinprotein interaction networks, and linked data/semantic web.
- Valence: The degree of interdependence or connectedness is referred to as valence. The data elements are closely related if the valence is high. This relatedness is heavily exploited in graph analytics. For certain situations, the average distance between arbitrary node pairs decreases as the time valence grows, and the sections of the massive graph become denser.

3.2 BigGraph frameworks

3.2.1 StellarGraph

The StellarGraph [18] is an extensible, user-friendly, and modular Python library for machine learning which provides advanced algorithms for graph machine learning. It makes the graph structure data easier to discover and to answer questions. StellarGraph can analyze a wide range of graphs such as.

a)   Graph with just one type of nodes and links (homogeneous graph).

b)   Graph with several types of nodes and/or links (heterogeneous graph).

c)   Extreme heterogeneous graphs have many various forms of edges (knowledge graphs).

d)   Graphs with or without nodes associated data.

e)   Weighted graphs (graphs with edge weights).

In Graph-structured data, nodes or vertices represents entities, and the edges or links represent relationships between them. On Facebook, for example, persons represent nodes or vertices and their associations as connections, or edges, which are linked to data like a person's age and the date of the relationship. StellarGraph library can solve many challenges in BigGraph like link prediction, Graph Classification, Node classification, Node description interpretation, representation for nodes and edges, and numerous downstream jobs for displaying machines. StellarGraph is built by using TensorFlow 2 and its Keras high-level API, Pandas, and Numpy. The StellarGraph can train various models from the information graph embedding literature, including TransE, DistMult, RESCAL, and ComplEx.

3.2.2 PyTorch-BigGraph

The PyTorch - BigGraph (PBG) [19] is a large-scale graph embedding framework for graphs with billions of nodes and trillions of edges. PBG library is generous and extensible. PBG trains from an input graph's list of edges, each of which is identified by its source and target entities, as well as, potentially, a relation form. It generates a feature vector for each entity, attempting to place adjacent entities in the vector space while moving unrelated entities apart. Thus, the entities with identical distributions of neighbors would end up being close together. PBG can solve a wide range of graph problems using a combination of techniques, including.

a)   Graph partitioning, which prevents the model from being fully loaded into memory,

b)   Multithreaded computation on each machine,

c) Distributed execution across multiple machines, all operating on disjoint parts of the graph at the same time, and

d) Batched negative sampling, which allows for processing greater than 1 million edges/sec/machine with 100 vertices. PBG can train various models from the knowledge graph embedding literature, including TransE, Dist-Mult, RESCAL, and ComplEx.

3.2.3 Spektral

Spektral [20] is an open-source Python library for developing graph neural networks. Spektral is a simple and flexible framework for creating Graph Neural Networks (GNNs). Spektral is built on TensorFlow and the Keras application programming interface. Spektral provides a range of Deep Learning on graphs techniques, such as pooling operators and message-passing, and utilities for processing graphs and loading standard benchmark datasets. Spektral can classify social network users, create new graphs with GAN frameworks, predict molecular properties, cluster nodes, predict relations between the nodes, and perform any other tasks where graphs express data. Some of the most common layers for graph deep learning, Spektral can implement include—(i) Convolutional Graph Networks (GCN), (ii) GraphSAGE, (iii) Chebyshev convolutions, (iv) Approximated Personalized Propagation of Neural Predictions (APPNP), (v) ARMA convolutions, (vi) Diffusional Convolutions, (vii) Graph attention networks (GAT), (viii) Graph Isomorphism Networks (GIN), and so on.

3.3 Challenges and issues

Applying Deep Learning to the BigGraph is a challenging task as the ubiquitous graph data is nontrivial [21] Graphs are irregular in structure, so it is difficult to extend certain fundamental mathematical procedures to graphs. Graphs can be weighted or unweighted, signed or unsigned, and homogenous or heterogeneous by nature. As Heterogeneous graphs contain diverse types and properties, they are very complicated to compute. These diverse tasks, types, and properties need different model architectures to tackle specific problems. As the graphs are enormous, it is challenging to design models with a linear time complexity with respect to the graph size. Graphs are interdisciplinary by nature; they are often connected to different disciplines like social sciences, chemistry, and biology. So, integrating domain knowledge can complicate model designs Computer. Three main challenging applications in BigGraph.

3.3.1 Graph classification

Graph Classification is a type of problem in recognizing the graph class labels in a dataset. The problem includes real-world applications in various fields, including social analysis, cyber-security, chemoinformatics, and urban computing. For example, in domains like chemoinformatics, molecules can be represented as graphs where nodes correspond to atoms and edges indicate chemical bonds between pairs of atoms. The task then predicts each graph's class label, such as a molecule's anti-cancer activity, toxicity, or solubility. There are several approaches to resolving this issue. However, graphs in the real world can be both large and noisy. And the noisy graph, the significant subgraph patterns can be sparse and confined only to small neighborhoods within the graph. Some existing works on Graph Classification tasks are on Graph Classification—Niko-lentzos et al. [22] presented present two novel algorithms applicable for both labeled and unlabeled graphs for comparing pairs of graphs based on their global properties. Structure2vec model by Dai et al. [23] which is for structured data representation based on the concept of embedding latent variable models into feature spaces and studying those feature spaces using discriminative knowledge. Similarly, Niepert et al. [24]. Moreover, Johansson et al. [25] are also approaches for the neural network-based Graph Classification task.

3.3.2 Node classification

The problem of classifying each node into one of a set of predefined classes is known as node classification. For any given network structure with labels on some nodes, the problem of labeling for every node is the node classification problem [26]. In a social network analysis application, node classification is a critical task. For example, based on their connections, attributes, and activities, users of a social network can predict their political affiliation. Node classification may be in both transductive or inductive settings. In the transductive setting, the labels of few nodes are given, and the goal is to predict the labels of the remaining nodes in the graph. In the inductive setting, the label for new nodes not seen during training must predict. The dynamic case makes the node classification problem even more difficult because the label distribution may change over time. Node classification is performed by applying a classifier on the set of labeled node embedding for training. Wilson et al. [27], Han and Shen [28], Yao et al. [29], Wang et al. [30], and Pimentel et al. [31] are the some existing works.

3.3.3 Link prediction

The ability to predict linkage between data objects is known as link prediction that predicts edges in the network where edges appear over time [32,33]. Many data mining applications, such as social network analysis and product recommendation, rely heavily on link prediction. Link prediction algorithms are used in many data mining tasks, either implicitly or explicitly. Genetic prediction, Web hyperlink creation, record linkage problems, and protein–protein interactions are examples of explicit link prediction problems. The implicit link prediction problems include recommender systems which are the services that predict links between users and items in a user-item bipartite graph representing preferences or purchases. Information retrieval, which deals with the prediction of links between words and documents in a word document bipartite graph representing word occurrences, is also an essential example of implicit link prediction problems. Many researchers proposed various approaches for the link prediction tasks for both homogenous [10,31,34] and heterogenous [35,36] graphs.

4. Graphs-based deep learning models

4.1 Graph convolutional network

Graph Convolution Networks (GCN) [5] is a semi-supervised learning approach for Graph-based data, e.g., document type in a citation network. The approach is based on the efficient usage of CNNs, which work directly on graphs. The Convolutional architecture is based on spectral graph convolutions and choosing their local first-order approximations. The authors have considered the challenge of node(edge) classification in a graph where the label is only present for a small subset of nodes. The problem is framed as semi-supervised. The labels are spread over the whole graph using some forms of Graph-based regularization (authors have used Laplacian regularization in the loss function). The authors have shown the approach works well for efficiency and classification accuracy in semi-supervised learning against state-of-the-art methods and a baseline model [37]. They have also presented the limitations of the approach.

4.2 Deep graph convolutional neural network

Deep Graph Convolutional Neural Network (DGCNN) [6] architecture is for the graph classification tasks using the graph convolutional layers inspired from [5]which is an efficient variation of convolutions of neural networks which functions directly on graphs, and also a scalable solution to semisupervised learning on the graphic structure. DGCNN includes the graph convolutional layers and a new SortPooling layer that uses the representations learned for each node from a stack of graph convolutional layers to create a representation (also known as an embedding) for each given graph. The SortPooling layer's output is then fed into one-dimensional convolutional, max pooling, and dense layers, which acquire graph-level features that can be used to predict graph labels. Without the requirement to transform graphs to tensors, DGCNN receives graph data as input directly, enabling gradient-based end-to-end training. DGCNN also permits the study of global graph topology by sorting vertex features with the SortPooling layer. DGCNN outperforms the state-of-the-art methods on many benchmark datasets such as bioinformatics datasets (Proteins, NCI1, and D&D) and the social network datasets (IMDB-M, COLLAB, and IMDB-B).

4.3 Graph attention networks

Graph attention networks (GATs) [7] is a novel neural network architecture for graph-structured data that uses masked self-attentional layers to overcome the limitations of prior approaches based on graph convolutions and their approximations. GATs enable the specification of different weights to different nodes in a neighborhood implicitly by stacking layers in which nodes can attend over their neighborhoods' features without understanding the graph structure beforehand and without expensive matrix operations like inversion. GATs outperform the existing methods on all four well-established node classification benchmarks, both inductive graph benchmarks (like the PubMed Diabetes, Cite-Seer, and Cora citation network datasets) datasets and transductive benchmark datasets(like a protein–protein interaction dataset). Therefore, GATs can solve many challenges in spectral-based graph neural networks.

4.4 Attributed network embedding via subspace discovery

Attributed network embedding via subspace discovery (Attri2vec) [8] is a unified neural network architecture for node classification problems in the attributed network embedding. Through a network structure-guided transformation on the initial attribute space, Attri2vec discovers a latent node attribute subspace and learns network node representations.

By conducting a succession of linear and nonlinear mappings in node attributes, Attri2vec integrates network nodes into a structure-providing attribute subspace. The proposed system combined all network functionality and node material seamlessly to obtain insightful node representations. Based upon the node's accessible properties, Attrivec also presents a possible solution to the out-of-sample problem as it learns the new node's representations. In some benchmark networks, such as DBLP, CiteSeer, PubMed Diabetes, Flickr, and Facebook, Attri2vec outperforms the state-of-the-art architecture for many tasks such as node clustering, node classification, and out-of-sample relation prediction tasks.

4.5 GraphSAGE

GraphSAGE [9] is a well-established framework for solving the inductive node embedding problem, which uses node feature knowledge to create node embeddings for previously unseen data. Here, a function that produces embeddings by sampling and aggregating features from a node's local neighborhood rather than training individual embeddings for each node is learned. GraphSAGE outperforms the state-of-the-art architectures on following three inductive node classification benchmarks tasks such as.

a) Classifying Reddit articles as Deep Learning + BigGraph belonging to different communities.

b) Classifying research papers into different topics using the Web of Science citation dataset.

c) Classifying protein roles through multiple biological protein-protein interactions (PPI) graphs.

For additional functions like classification of node, clustering, and link prediction, GraphSAGE is quite beneficial.

4.6 Heterogeneous GraphSAGE

Previous approaches we have are all focused on homogenous graphs. Heterogeneous GraphSAGE (HinSage) [38] works on Heterogeneous graphs. The approach presents low dimensional embeddings for the graphs having a large number of nodes. The novelty is the approach is that it works on a small subset of nodes from the graph. Other approaches require having all the nodes present during the training of the embeddings. This approach leverages the node features' information (e.g., text attributes) for the efficient generation of embeddings for unseen data. They learn a function that samples and aggregates features of nodes in a local neighborhood to generate embeddings. Authors claim the approach performs better than strong baselines on node classification benchmarks (categorization of unseen nodes in citation and Reddit posts evolving data and multigraph protein interaction dataset).

4.7 Node2vec

Tasks to predict nodes and edges necessitate the cautious design of characteristics employed by algorithms for learning. Recent breakthroughs in the larger field of representation learning have resulted in fundamental advances in automating prediction by learning its features. Nonetheless, current learning methods are insufficiently descriptive to capture the many communication patterns seen in the networks—Node2vec [10], an algorithm for the continuous feature representations of network nodes. The algorithm learns node mappings to a low-dimensional space of features, which maximizes the probability of preservation for node network neighborhoods. Authors offer a flexible idea of a neighborhood of a node network and provide a random predictive approach to exploring different districts more effectively. Previous work focused on static notions of network neighborhoods, but the authors claim that the expanded versatility in discovering neighborhoods is the secret to deeper representations. Authors prove that Node2vec outperforms the latest multilabel classification, and Node2vec outperforms the relation estimation technologies in several real-world networks from diverse areas. The results reflect an innovative approach to studying current task-independent representations in dynamic networks.

4.8 Metapath2vec

Metapath2vec [11] is a technique for dealing with network heterogeneity in heterogeneous networks with various nodes and ties. Here, the meta-path-guided random walk approach, capable of catching both the structural and semantic associations of various types of nodes and connections in a heterogeneous network, is created. The method formalizes the node's heterogeneous neighborhood function to maximize the likelihood of skip-gram-based networks in the sense of several node types. In conclusion, a heterogeneous negative sampling methodology is devised for the accuracy and efficiency of optimization. The Metapath2vec not only outperforms state-of-the-art embedding models for a range of heterogeneous network mining tasks, such as link prediction, node classification, clustering, and search for similarities but can also differentiate between structural and semantic similarities in the two heterogeneous networking networks—the AMiner Computer Science and the database and Information Systems(DBIS) datasets.

4.9 Continuous-time dynamic network embeddings

Continuous-Time Dynamic Network Embeddings (CTDNE) [12] is a general framework for integrating temporal data into network embedding techniques. The framework provides a foundation for generalizing emerging random walk-based embedding methods for studying dynamic (time-dependent) network embeddings from continuous-time dynamic networks. The result is a more suitable time-dependent network representation that captures the continuous-time dynamic network's essential temporal properties. Initial temporal edge selection and temporal random walk are two key interchangeable components of the proposed continuous-time dynamic network embedding framework that enable the user to temporally bias the learning of time-dependent network representations. The proposed framework is a very generalized and practical dynamic network embedding approach for various real-world networks in temporal relation prediction.

4.10 Complex embeddings for simple link prediction

The complex Embeddings for Simple Link Prediction (ComplEx) [13] is a method for solving the problem of link prediction in Complex Embeddings. Complex valued embeddings are used in the method model. The composition of complex embeddings can handle a wide range of binary interactions, including symmetric and antisymmetric relationships. The approach based on complex embeddings is potentially simpler than state-of-the-art models such as Neural Tensor Network and Holographic Embeddings. It only uses the Hermitian dot product complex equivalent of the regular dot product between real vectors. In both simulated and actual datasets, the proposed model outperforms the current state-of-the-art. The synthetic dataset is built on symmetric or antisymmetric relations, while the actual datasets contain various types of relations in various, regular KBs.

4.11 DistMult

Using the neural-embedding technique, Yang et al. [14] analyze learning representations and relations in knowledge bases of entities. Authors say that most current models, including NTN [39] and TransE [40] can be used in a unified learning system in which entities are low-dimensional vectors learned by a neural network and the connections are linear and bilinear mapping functions. On the link prediction challenge, the authors use this model to compare various embedding models. They prove that a basic bilinear formulation results in new outcomes for the challenges. The authors present a novel method for logical mining rules using the learned relation embeddings. They claim that embeddings learned from the bilinear objective are particularly strong at catching relational semantics and that relation composition is captured by matrix multiplication. More intriguingly, they prove that their method of rule extraction surpasses an existing confidence-based rule mining approach in mining Horn rules involving compositional reasoning.

4.12 Relational graph convolutional networks

Despite the considerable effort put into their development and upkeep, even the largest are still incomplete (e.g., Wikidata, Yago). Knowledge graphs are used for many use-cases in information retrieval. Schlichtkrull et al., [15] presented Relational Graph Convolutional Networks (R-GCNs) and used it for two common completion tasks of knowledge bases: link estimates and the categorization of entities (recovery of missing entity attributes). R-GCN's linked to a new type of neural networks running on graphs and particularly designed to manage the extremely multirelated data in the actual world of knowledge. R-GCNs can be utilized for the entity classification as a stand-alone model. They also show that link estimation factorization models such as DistMult may be considerably improved by employing an R-GCN encoder model to gather evidence via numerous inference phases in the graph and achieve an improvement of 29.8% in FB15k-237 over a baseline decoder-only.

4.13 Simplifying graph convolutional networks

Simplifying graph convolutional networks (SGC) [41] is the simplest possible formulation of a graph convolutional model to grasp further and describe the dynamics of GCNs. The proposed method's node classification accuracy is evaluated on the Cora, CiteSeer, and PubMed Diabetes citation network datasets. On citation networks, SGC will equal the efficiency of GCN and other state-of-the-art graph networks. SGC outperforms GCN by around 1% on CiteSeer, as SGC having fewer parameters and therefore suffering less from overfitting. SGC outperforms the previous sampling-based GCN variants on Reddit by more than 1%. The proposed model is again evaluated to evaluate its applicability on five downstream applications: semi-supervised user geolocation, text classification, link extraction, Graph Classification, and zeroshot image classification. And, the proposed model outperforms the current state-of-the-art.

4.14 Personalized propagation of neural predictions and Approximate personalized propagation of *neural* predictions

Personalized propagation of neural predictions (PPNP) and its fast approximation, APPNP [16] is an improved prediction and propagation algorithm for semi-supervised classification on graphs. The proposed model is derived by using the relationship between GCN and PageRank and extending it to personalized PageRank. The model can also solve the small range problem implicit in various message-passing models without any additional parameters. The model uses the information from broad and customizable neighborhoods through the teleport probability for classifying each node. Moreover, one particular thing about the model is that—can be easily combined with the neural network. The model outperforms the state-of-the-art GCN-like models for semi-supervised classification on all the datasets like CiteSeer, Cora, PubMed Diabetes, and MICROSOFT ACADEMIC.

5. Graph-based datasets

5.1 Mutag

Mutag [42] is a widely used dataset that includes a collection of static graphs that describe chemical compounds, each of which has a binary label attached to it. The dataset is for evaluating various Graph Classification algorithms. There are 188 graphs with an average of 18 nodes and 20 edges in this dataset. Each graph defines a chemical compound, and the labels of the graph define its mutagenic effect on a particular bacteria of gram-negative. Graph nodes have seven labels, and each graph belongs to one of two classes.

5.1.1 Proteins

Proteins dataset [43] is a commonly used dataset made up of graphs, with a chemical molecule and a label as either or not an enzyme. In short, it is a dataset of Proteins classified as enzymes or non-enzymes. It comprises 1113 graphs, each with an average of 39 nodes and 73 edges. Each graph belongs to one of two groups, and each node has four attributes, each with a one-hot label encoding.

5.2 DBLP network data

DBLP network data: DBLP citation network is a subgraph form by extracting papers from four subjects(Data Mining, Database, Computer Vision, and Artificial Intelligence) from the DBLP-citation-network V3 based on their venue information. Here, the papers with no citation are removed. The DBLP subgraph is considered an undirected network by ignoring the citation direction. The dataset consists of 11,448 papers and 45,661 citation relations. In DBLP, 2476-dimensional binary node feature vectors are constructed from the paper titles. Each element represents the presence or absence of the corresponding word.

5.3 MovieLens

The MovieLens [44] are the commonly used datasets in many fields like research, education, and industry. These datasets are the products of participant participation in the MovieLens movie recommendation scheme, which has hosted several experiments since its establishment in 1997. The MovieLens 100 K dataset includes 100,000 reviews on 1682 films from 943 users.

5.4 Cora

The Cora dataset [45] contains 2708 scientific publications classifying into seven categories. There are 5429 connections in the citation network. Each paper is manually labeled into one of seven classes: Generic Algorithms, Case-Based, Neural Networks, Theory, Reinforcement Learning, Probabilistic Methods, and Rule Learning. A 0/1-valued word vector describes each publication in the dataset, showing the inclusion or absence of the corresponding word from the dictionary. There are 1433 unique words in the dictionary.

5.5 Blog Catalog 3 dataset

The Blog Catalog 3 dataset [46] is a heterogeneous network made up of two types of nodes: user and community. A friend edge type can connect two user nodes (i.e., two users are friends). The belongs edge type can connect a user to a party (i.e., the user belongs to the group). This dataset is a graph dataset representing a network of bloggers' social relationships described on the BlogCatalog website. There are 88,800 nodes and 2.1 million edges in the network.

5.6 IAEronEmployees

IAEnronEmployees [47] is a widely used dataset in many fields such as research, education, and many others. Here, Emails sent from one employee to another are represented by a set of edges. A total of 50,572 edges, each include time stamp data. Edges belong to a total of 151 different node IDs.

5.7 WN18

The WN18 dataset [48] is made up of triplets (synset, relation_type, triplet) from WordNet 3.0. There are 40,943 synsets in all, with 18 relation forms. There are 141,442 triplets for the training set, 5000 for the validation set, and 5000 for the testing set. Since the dataset includes several inverse relationships, only the published findings may be compared with this dataset.

5.8 FB15k

This FB15k dataset [48] is made up of triplets (synset, relation type, and triplet) derived from the Freebase website. The dataset consists of 14,951 mids and 1345 kinds of relationships among them. There are 483,142 triplets for the training set, 50,000 for the validation set, and 59,071 for the testing set. Since this dataset includes several inverse relationships, it can only be used to compare against published results.

5.9 AIFB dataset

The AIFB DataSet [49] is a Semantic Web (RDF) dataset used as a data mining benchmark containing information about the AIFB research institute's personnel, research groups, and publications. The collection includes 8 entities, 29 edges, and 45 various connections and edge kinds. There are 178 members of a study community of five separate research groups in the dataset.

5.10 CiteSeer dataset

The CiteSeer dataset [50] contains 3312 research articles classified into six categories: databases, agents, artificial intelligence, human–computer interaction, information retrieval, and machine learning. The citation network has 4732 connections, but only 4715 are included in the graph since 17 of them do not include a source or target publishing in the dataset. A 0/1-valued word vector identifies each publication in the dataset, showing the inclusion or absence of the corresponding word from the dictionary. There are 3703 different words in the dictionary of the dataset.

5.11 Pubmed diabetes dataset

The PubMed Diabetes dataset [51] contains 19,717 research articles on diabetes from the PubMed Diabetes website and classified into three categories: (i) Diabetes Mellitus, Experimental, (ii) Diabetes Mellitus Type 1, and (iii) Diabetes Mellitus Type 2. There are 44,338 connections in the citation network. A TF/IDF weighted word vector from a dictionary of 500 unique words is used to classify each publication in the dataset.

6. Experimental environment setup

The experiments were performed using the StellarGraph framework [18]. It provides an implementation of various Graph-based deep neural networks. Multiple algorithms are available as layers that can be incorporated into a custom model. This framework is based on Tensorflow [52] implementation, so it leverages all the technology of the Tensorflow. We have utilized the GPU acceleration of Tensorflow models. The experiments were conducted in the Google Colab environment. All the employed dataset for an experiment is split into three-way sets of 80, 10, and 10%. They represent the training, validation, and testing sets, respectively. For each experiment, a Graph-based Deep Learning model is trained for 100 epochs with "Adam" optimizer and "Accuracy" as the experiments' metric. Some algorithms require a particular setup of the graph data; those cases are handled as they were encountered. The optimizer's learning rate is also adjusted as needed for a specific experiment. Details of all these particular parameters can found here [53].

7. Experimental results and analysis

We began with the task of Graph Classification for our preliminary testing of Graph-based deep neural network architectures.

Fig. 1 shows the performance of GCN and DGCNN on the Mutag dataset. The dataset represents graphs of chemical compounds with labels representing the mutagenic effects of those compounds. As we see GCN algorithm performs reasonably well on both the training and validation split of the dataset. It is DGCNN that executes exceptionally well on both partitions.

A similar scenario is observed in Fig. 1 for the much complex dataset Proteins. Here too, DGCNN performs way better than GCN. This scenario shows the deep layers stacked one over each other in the DGCNN

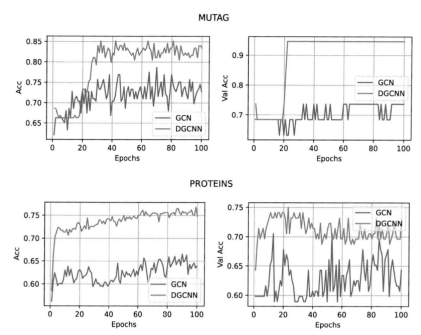

Fig. 1 Training and validation performance of GCN and DGCNN on Mutag and Proteins dataset for Graph Classification.

architecture learn enough patterns to generalize to new data. DGCNN gets an accuracy of 94.73% and 73.21%, while GCN gets 73.68% and 64.28% accuracies on a test split of the Mutag and Proteins datasets, respectively, as shown in Table 1.

The next set of experiments were for the task of Node Classification. Fig. 2 shows the performance of GAT, GCN, PPNP, RGCN, and SGC models on the four datasets: PubMed Diabetes, AIFB, Cora, and CiteSeer. As shown in the plots, GAT performance is the most reliable on all the datasets, with the highest accuracy of 86.96% for the test split of the PubMed Diabetes dataset. Like GAT's performance, GCN performs well on all the datasets and achieves the highest accuracy for the AIFB dataset with 99.44%. PPNP is another overall good performer. RGCN and SGC have unreliable performances; they do good for some datasets but plummets for a few.

We also experimented with some preliminary testing of Graph-based neural network architectures for the link prediction task on three commonly used datasets: Cora, PubMed Diabetes, and CiteSeer, using three well-performed models: Attri2vec, GraphSAGE, and HinSage. Fig. 3 shows each of the three models' performance on all of the datasets for training and validation splits. As we can see in the plots, the GraphSAGE model generally

Table 1 Performance of some models for Graph Classification, Link Prediction, and Node Classification applications on different datasets.

Graph-based task	Dataset	Algorithm	Accuracy (%)
Graph Classification	Mutag	GCN	73.68
		DCNN	**94.73**
	Proteins	GCN	64.28
		DCNN	**73.21**
Node Classification	AIFB	GAT	66.66
		GCN	**94.44**
		PPNP	66.66
		RGCN	**94.44**
		SGC	44.44
	CiteSeer	GAT	74.39
		GCN	75.60
		PPNP	**78.91**
		RGCN	23.19
		SGC	75.0
	Cora	GAT	80.81
		GCN	76.75
		PPNP	82.65
		RGCN	15.12
		SGC	**83.02**
	PubMed Diabetes	GAT	**86.96**
		GCN	72.16
		PPNP	60.09
		RGCN	39.35
		SGC	79.51
	CiteSeer	Attri2vec	58.49
		GraphSAGE	**82.59**
		HinSage	65.28

Table 1 Performance of some models for Graph Classification, Link Prediction, and Node Classification applications on different datasets.—cont'd

Graph-based task	Dataset	Algorithm	Accuracy (%)
	Cora	Attri2vec	56.91
		GraphSAGE	**74.26**
		HinSage	55.71
	PubMed Diabetes	Attri2vec	65.02
		GraphSAGE	**86.80**
		HinSage	76.05

Bold markings in the Accuracy column represent the best performer among models for a given graph dataset.

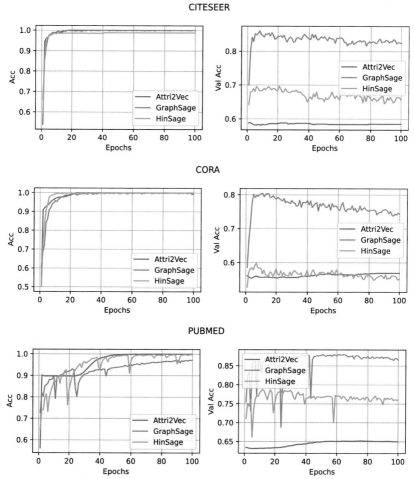

Fig. 2 Training and validation performance of Attri2vec, GraphSAGE, and HinSage on CiteSeer, Cora and PubMed Diabetes dataset for Link Prediction.

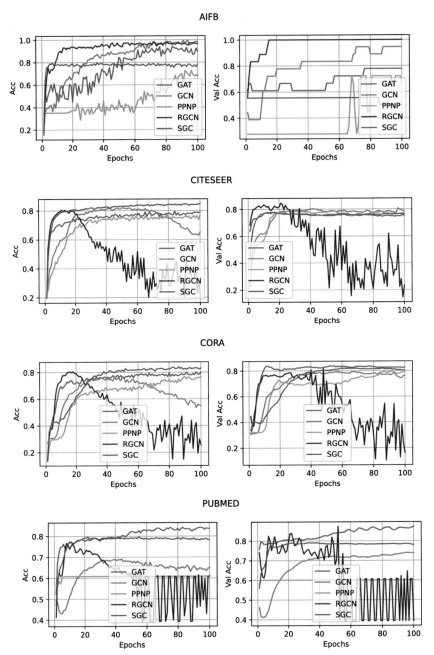

Fig. 3 Training and validation performance of GAT, GCN, PPNP, RGCN and SGC on AIFB, CiteSeer, Cora and PubMed Diabetes dataset for Node Classification.

performed well for all the datasets and best on for the PubMed Diabetes dataset with 86.80% accuracy on the test split of the dataset. For the CiteSeer and Cora datasets, GraphSAGE performance is the best among other models. For CiteSeer and Cora, it achieves 82.59% and 74.26% accuracies, respectively. As we see GraphSAGE algorithm performs reasonably well on both training and validation split for all the datasets over Attri2vec and HinSage algorithms.

8. Conclusion

There have been recent Deep Learning-based approaches to the challenges in Graph-based data. These approaches have achieved state-of-the-art performance on many graph data-based challenges. In this paper, we empirically investigate these algorithms on various tasks related to graph data. We compared various models of Graph-based deep neural network architectures such as GCN, DGCNN, GAT, PPNP, RGCN, SGC, Attri2vec, GraphSAGE, and HinSage on various datasets like Mutag, Proteins, PubMed Diabetes, AIFB, Cora, CiteSeer for three main applications of graph structure data: Graph Classification, Node Classification, and Link Prediction. The experiment results we got show that: For the Graph Classification task shows DGCNN perform exceptionally better perform on both datasets (Mutag and Proteins). For the Node Classification task, the performance of the models has differed with different datasets. For the PubMed Diabetes dataset, the GAT model outperforms the other four models (For the Node Classification task, we used GAT, GCN, PPNP, RGCN, and SGC). Similarly, the performance of the models is evaluated for other datasets. GraphSAGE outperforms the other two models (Attri2vec and HinSage) in the Link Prediction task. These experiments have also revealed the general trend of Deep Learning, where deep architecture network picks up more latent characteristics of the data, continues here too. Furthermore, the banes (overfitting) of Deep Learning are encounter along the way for some models. We conclude with this observation from these empirical results that advances in Deep Learning have benefitted how we design solutions for Graph-based problems. It has also brought new challenges along the way that were not explored before. All these experiments have demonstrated the power of Deep Learning in the context of graphs and new methods for tackling the challenges associated with this structured form of data.

References

[1] Backlinko.com, How Many People Use Facebook in 2021? 2021 (Online; accessed 14. Apr. 2021). https://backlinko.com/facebook-users.

[2] S. Wang, B. Kang, J. Ma, X. Zeng, M. Xiao, J. Guo, M. Cai, J. Yang, Y. Li, X. Meng, et al., A deep learning algorithm using ct images to screen for corona virus disease (covid-19), Eur. Radiol. (2021) 1–9.

[3] L. Deng, Y. Liu, Deep Learning in Natural Language Processing, Springer, 2018.

[4] M.J. Bianco, P. Gerstoft, J. Traer, E. Ozanich, M.A. Roch, S. Gannot, C.-A. Deledalle, Machine learning in acoustics: theory and applications, J. Acoust. Soc. Am. 146 (5) (2019) 3590–3628.

[5] T.N. Kipf, M. Welling, Semi-supervised classification with graph convolutional networks, arXiv (2016) (preprint arXiv:1609.02907).

[6] M. Zhang, Z. Cui, M. Neumann, Y. Chen, An end-to-end deep learning architecture for graph classification, in: Proceedings of the AAAI Conference on Artificial Intelligence, Vol. 32, 2018.

[7] P. Veličković, G. Cucurull, A. Casanova, A. Romero, P. Liò, Y. Bengio, Graph attention networks, arXiv (2017). arXiv:1710.10903. https://arxiv.org/abs/1710.10903v3.

[8] D. Zhang, J. Yin, X. Zhu, C. Zhang, Attributed network embedding via subspace discovery, Data Mining and Knowledge Discovery 33 (6) (2019) 1953–1980.

[9] W.L. Hamilton, R. Ying, J. Leskovec, Inductive representation learning on large graphs, arXiv (2017) (preprint arXiv:1706.02216).

[10] A. Grover, J. Leskovec, node2vec: scalable feature learning for networks, arXiv (2016) (arXiv:1607.00653).

[11] Y. Dong, N.V. Chawla, A. Swami, metapath2vec: scalable representation learning for heterogeneous networks, in: Proceedings of the 23rd ACM SIGKDD international conference on knowledge discovery and data mining, 2017, pp. 135–144.

[12] G.H. Nguyen, J.B. Lee, R.A. Rossi, N.K. Ahmed, E. Koh, S. Kim, Continuous-time dynamic network embeddings, in: Companion Proceedings of the The Web Conference, 2018, pp. 969–976.

[13] T. Trouillon, J. Welbl, S. Riedel, É. Gaussier, G. Bouchard, Complex embeddings for simple link prediction, in: International Conference on Machine Learning, PMLR, 2016, pp. 2071–2080.

[14] B. Yang, W.-t. Yih, X. He, J. Gao, L. Deng, Embedding entities and relations for learning and inference in knowledge bases, arXiv (2014) (preprint arXiv:1412.6575).

[15] M. Schlichtkrull, T.N. Kipf, P. Bloem, R. Van Den Berg, I. Titov, M. Welling, Modeling relational data with graph convolutional networks, in: European Semantic Web Conference, Springer, 2018, pp. 593–607.

[16] J. Klicpera, A. Bojchevski, S. Günnemann, Predict then propagate: graph neural networks meet personalized pagerank, arXiv (2018) (preprint arXiv:1810.05997).

[17] D.K. Singh, P.K.D. Pramanik, P. Choudhury, Big graph analytics: techniques, tools, challenges, and applications, Data Analytics: Concepts, Techniques, and Applications (2018) 171.

[18] Welcome to StellarGraph's Documentation! StellarGraph 1.2.1 Documentation, 2021 (Online; accessed 14. Apr. 2021). https://stellargraph.readthedocs.io/en/stable.

[19] A. Lerer, L. Wu, J. Shen, T. Lacroix, L. Wehrstedt, A. Bose, A. Peysakhovich, Pytorch-biggraph: a large-scale graph embedding system, arXiv (2019) (preprint arXiv:1903.12287).

[20] D. Grattarola, C. Alippi, Graph neural networks in tensorflow and keras with spektral, arXiv (2020) (preprint arXiv:2006.12138).

[21] Z. Zhang, P. Cui, W. Zhu, Deep learning on graphs: a survey, IEEE Transactions on Knowledge and Data Engineering (2018).

[22] G. Nikolentzos, P. Meladianos, M. Vazirgiannis, Matching node embeddings for graph similarity, in: Proceedings of the AAAI Conference on Artificial Intelligence, Vol. 31, 2017.

[23] H. Dai, B. Dai, L. Song, Discriminative embeddings of latent variable models for structured data, in: International conference on machine learning, PMLR, 2016, pp. 2702–2711.

[24] M. Niepert, M. Ahmed, K. Kutzkov, Learning convolutional neural networks for graphs, in: International conference on machine learning, PMLR, 2016, pp. 2014–2023.

[25] F.D. Johansson, D. Dubhashi, Learning with similarity functions on graphs using matchings of geometric embeddings, in: Proceedings of the 21th ACM SIGKDD International Conference on Knowledge Discovery and Data Mining, 2015, pp. 467–476.

[26] S. Bhagat, G. Cormode, S. Muthukrishnan, Node classification in social networks, in: Social Network Data Analytics, Springer, 2011, pp. 115–148.

[27] R.C. Wilson, E.R. Hancock, E. Pekalska, R.P. Duin, Spherical and hyperbolic embeddings of data, IEEE Trans. Pattern Anal. Mach. Intell. 36 (11) (2014) 2255–2269.

[28] Y. Han, Y. Shen, Partially supervised graph embedding for positive unlabelled feature selection, in: IJCAI, 2016, pp. 1548–1554.

[29] L. Yao, Y. Zhang, B. Wei, Z. Jin, R. Zhang, Y. Zhang, Q. Chen, Incorporating knowledge graph embeddings into topic modeling, in: Proceedings of the AAAI Conference on Artificial Intelligence, Vol. 31, 2017.

[30] X. Wang, P. Cui, J. Wang, J. Pei, W. Zhu, S. Yang, Community preserving network embedding, in: Proceedings of the AAAI Conference on Artificial Intelligence, Vol. 31, 2017.

[31] T. Pimentel, A. Veloso, N. Ziviani, Unsupervised and Scalable Algorithm for Learning Node Representations, 2017.

[32] Z. Huang, D.K. Lin, The time-series link prediction problem with applications in communication surveillance, INFORMS Journal on Computing 21 (2) (2009) 286–303.

[33] J. Kunegis, E.W. De Luca, S. Albayrak, The link prediction problem in bipartite networks, in: International Conference on Information Processing and Management of Uncertainty in Knowledge-based Systems, Springer, 2010, pp. 380–389.

[34] C. Zhou, Y. Liu, X. Liu, Z. Liu, J. Gao, Scalable graph embedding for asymmetric proximity, in: Proceedings of the AAAI Conference on Artificial Intelligence, Vol. 31, 2017.

[35] Z. Liu, V.W. Zheng, Z. Zhao, F. Zhu, K.C.-C. Chang, M. Wu, J. Ying, Semantic proximity search on heterogeneous graph by proximity embedding, in: Proceedings of the AAAI Conference on Artificial Intelligence, Vol. 31, 2017.

[36] Z. Liu, V. Zheng, Z. Zhao, F. Zhu, K. Chang, M. Wu, J. Ying, Distance-aware dag embedding for proximity search on heterogeneous graphs, in: Proceedings of the AAAI Conference on Artificial Intelligence, Vol. 32, 2018.

[37] Z. Yang, W. Cohen, R. Salakhudinov, Revisiting semi-supervised learning with graph embeddings, in: International conference on machine learning, PMLR, 2016, pp. 40–48.

[38] W.L. Hamilton, R. Ying, J. Leskovec, Inductive representation learning on large graphs, arXiv (2018) (1706.02216).

[39] R. Socher, D. Chen, C.D. Manning, A.Y. Ng, Reasoning with neural tensor networks for knowledge base completion, in: Proceedings of the 26th International Conference on Neural Information Processing Systems - Volume 1, NIPS'13, Curran Associates Inc., Red Hook, NY, USA, 2013, pp. 926–934.

[40] A. Bordes, N. Usunier, A. Garcia-Duran, J. Weston, O. Yakhnenko, Translating embeddings for modeling multi-relational data, in: Proceedings of the 26th International Conference on Neural Information Processing Systems - Volume 2, NIPS'13, Curran Associates Inc., Red Hook, NY, USA, 2013, pp. 2787–2795.

[41] F. Wu, A. Souza, T. Zhang, C. Fifty, T. Yu, K. Weinberger, Simplifying graph convolutional networks, in: International conference on machine learning, PMLR, 2019, pp. 6861–6871.
[42] A.K. Debnath, R.L. Lopez de Compadre, G. Debnath, A.J. Shusterman, C. Hansch, Structure-activity relationship of mutagenic aromatic and heteroaromatic nitro compounds. Correlation with molecular orbital energies and hydrophobicity, J. Med. Chem. 34 (2) (1991) 786–797. https://doi.org/10.1021/jm00106a046.
[43] K.M. Borgwardt, C.S. Ong, S. Schonauer, S.V.N. Vishwanathan, A.J. Smola, H.-P. Kriegel, Protein function prediction via graph kernels, Bioinformatics 21 (suppl_1) (2005) i47–i56. https://doi.org/10.1093/bioinformatics/bti1007.
[44] MovieLens, 100K Dataset, 2015. https://grouplens.org/datasets/movielens/100k/.
[45] Dataset, 2021. Online; accessed 14. Apr. 2021. https://relational.fit.cvut.cz/dataset/CORA.
[46] BlogCatalog, dataset, 2020, Online; accessed 14. Apr. 2021. https://figshare.com/articles/dataset/BlogCatalog_dataset/11923611.
[47] N.K.A. Ryan, A. Rossi, Ia-Enron-Employees | Dynamic Networks | Network Data Repository, 2021, Online; accessed 14. Apr. 2021. http://networkrepository.com/ia-enron-employees.php.
[48] en:transe, Everest, 2021, Online; accessed 14. Apr. 2021. https://everest.hds.utc.fr/doku.php?id=en:transe.
[49] AIFB, DataSet, 2013, Online; accessed 14. Apr. 2021. https://figshare.com/articles/dataset/AIFB_DataSet/745364.
[50] N.K.A. Ryan, A. Rossi, Citeseer | Labeled Networks | Network Data Repository, 2021, Online; accessed 14. Apr. 2021. http://networkrepository.com/citeseer.php.
[51] djoharap, Datasets | LINQS, 2017, Online; accessed 14. Apr. 2021. https://linqs.soe.ucsc.edu/data.
[52] TensorFlow, 2021. Online; accessed 14. Apr. 2021. https://www.tensorflow.org.
[53] Graph-Based Deep Learning, 2021. Online; accessed 14. Apr. 2021. https://github.com/lilapati/deepgraph.

About the authors

Ms. Lilapati Waikhom is pursuing her Ph.D. degree at the Department of Computer Science & Engineering, National Institute of Technology Silchar, Assam, India. She obtained her M.Tech. from the Department of Computer Science & Engineering, National Institute of Technology Arunachal Pradesh, India. Her research interest spans from Machine Learning, Big Graph, and Network Sciences.

Dr. Ripon Patgiri has received his Bachelor Degree from Institution of Electronics and Telecommunication Engineers, New Delhi in 2009. He has received his M.Tech. degree from Indian Institute of Technology Guwahati in 2012. He has received his Doctor of Philosophy from National Institute of Technology Silchar in 2019. After M. Tech. degree, he has joined as Assistant Professor at the Department of Computer Science & Engineering, National Institute of Technology Silchar in 2013. He has published numerous papers in reputed journals, conferences, and books. His research interests include distributed systems, file systems, Hadoop and MapReduce, big data, bloom filter, storage systems, and data-intensive computing. He is a senior member of IEEE. He is a member of ACM and EAI. He is a lifetime member of ACCS, India. Also, he is an associate member of IETE. He was General Chair of 6th International Conference on Advanced Computing, Networking, and Informatics (ICACNI 2018, http://www.icacni.com) and International Conference on Big Data, Machine Learning and Applications (BigDML 2019, http://bigdml.nits.ac.in). He is Organizing Chair of 25th International Symposium on Frontiers of Research in Speech and Music (*FRSM 2020*, http://frsm2020.nits.ac.in) and International Conference on Modelling, Simulations and Applications (CoMSO 2020, http://comso.nits.ac.in). He is convenor, Organizing Chair and Program Chair of 26th annual International Conference on Advanced Computing and Communications (ADCOM 2020). He is guest editor in special issue "Big Data: Exascale computation and beyond" of EAI Endorsed Transactions on Scalable Information Systems. He is also an editor in a multi-authored book, title *"Health Informatics: A Computational Perspective in Healthcare"*, in the book series of *Studies in Computational Intelligence*, Springer. Also, he is writing a monograph book, titled *"Bloom Filter: A Data Structure for Computer Networking, Big Data, Cloud Computing, Internet of Things, Bioinformatics and Beyond"*, Elsevier

CHAPTER FOUR

Analyzing correlation between quality and accuracy of graph clustering

Soumita Das and Anupam Biswas
Department of Computer Science and Engineering, National Institute of Technology Silchar, Silchar, Assam, India

Contents

Abstract

In this chapter, a model is proposed to establish correlation between the quality and accuracy metrics that are used during graph clustering evaluation. Earlier works have discussed trade-off between accuracy and quality, but it has never been studied extensively to derive correlation between the two and to ensure accuracy alternatively. The experimental analysis also shows such trade-off between accuracy and quality metrics. The proposed model has addressed a solution to the trade-off between quality and accuracy by establishing correlation between two via number of clusters. We have performed empirical analysis to validate the existence of correlation between quality and accuracy via number of clusters. The analysis indicated that the number of clusters plays significant role in assurance of quality as well as accuracy of the clustering.

Advances in Computers, Volume 128
ISSN 0065-2458
https://doi.org/10.1016/bs.adcom.2021.09.008

1. Introduction

Modern data systems are growing very fast and becoming very big and complex. Handling and maintenance of such big data are very challenging. Discovery of knowledge from those big data systems is even more challenging. Graphical knowledge representation of a complex system makes the system easily understandable, and it has the capability to analyze data intelligently. The graphical representations of big data systems also yield huge graphs. Hence, study and analysis of big data in entirety is a very tedious task. Often big data is partitioned into certain groups or clusters. From the view point of expert systems, the process of clustering is treated as unsupersvised learning of clusters based similarity measure. The main objective of such clustering is to maximize the similarity within the clusters, while minimize similarity among different clusters. With the similar notion, huge graphs are also partitioned into subgraphs or clusters where, nodes within the cluster are connected densely but nodes of different clusters are connected sparsely. Partitioning of the graph into subgraphs is referred as graph clustering.

The graph clustering algorithms are vastly employed in the applications of diverse domains [1–4]. The effectiveness of these algorithms is generally evaluated in terms of quality and accuracy of the clustering predicted by the algorithm. To account the importance and significance of quality and accuracy of a clustering, several metrics [5–7] have been proposed to evaluate performance of clustering algorithm. Quality of graph clustering deals only with connectivity whereas measuring accuracy requires ground truth. Measuring quality is a unsupervised approach that completely depends on the edges. On the contrary, measuring accuracy is a supervised approach that measures similarity between the real clustering and the predicted clustering. Thus, accuracy depends only on nodes. This fundamental difference between quality and accuracy has led to a potential conflict. There has no direct correlation between the two aspects of clustering evaluation. The trade-off between the two is often pointed out in several literature [8–10]. Though such trade-off is obvious, it has never been studied or analyzed extensively to draw correlation between accuracy metric and quality metric.

The two major issues emerged with the clustering evaluation are the trade-off between quality and accuracy, and the inability to ensure accuracy in most of the networks. Importance of effective clustering evaluation grows a lot in these days with problems such as big data analysis and study of complex systems [11, 12]. Quality and accuracy of clustering are very important and challenging factors in such cases. Important decisions [12, 13] regarding various applications in complex and bigger systems are made based on the information gathered from the clustering. Therefore, accuracy of the clustering has to be ensured along with the quality. As mentioned above, accuracy cannot be ensured directly for unknown networks so it has to be ensured indirectly without knowing ground truth. Thus, one of the simplistic ways to ensure accuracy alternatively would be through quality metric, since it does not require ground truth to evaluate clustering. This approach requires well established correlation between quality and accuracy. The derivation of a correlation between quality and accuracy metrics will not only helpful to avoid trade-off but also to ensure accuracy through quality. In this chapter, we put our primary focus on this direction.

A model is proposed to establish correlation between accuracy metric and quality metric via number of clustering. The clustering generation process is directed with a predetermined specific framework suitable for defining functional relationships such as one-to-one and inverse functions. Various properties are derived by extrapolating the functional relationships to establish the correlation between quality and accuracy via number of clusters. A methodology is designed with the incorporation of multiple-objective optimization and scalarization techniques to validate the proposed model. The empirical analysis shows the significance role of number of clusters in indicating both accuracy and quality.

Rest of the chapter is organized as follows. Section 2 briefs preliminary definitions. Section 3 discusses various metrics and different methodology used to evaluate graph clustering. Section 4 explains the proposed model to establish correlation between accuracy and quality. Section 5 illustrates the methodology for verifying the proposed model. Section 6 briefs widely used quality and accuracy metrics for evaluating graph clustering. Section 7 analyze the empirical results incorporating the proposed model. Section 8 concludes this works.

2. Preliminary definitions

An undirected weighted graph G is a pair (V, E) of a finite set V of nodes and a finite set $E = V \times V$ of edges where a function $\delta_E :$ $\{V \times V \to \Re, \Re \geq 0\}$ defines the strengths or weights of edges such that $\delta_E(u, v) = \delta_E(v, u)$ for all $(u, v) \in E$. If edge set is not specified then the strength of edge (u, v) is represented simply with $\delta(u, v)$. For unweighted graphs, function δ_E is defined as $\delta_E : V \times V \rightarrow \{0, 1\}$, where 0 represents absence and 1 represents presence of edge. Self loops are allowed in both weighted and unweighted graphs. Note that only difference between weighted and unweighted graph is presence of edges are marked only with 1 in unweighted graph, whereas it can be any $\Re > 0$ in weighted graph.

Definition 1 Clustering

A nonoverlapping clustering C of graph $G = (V, E)$ is partitioning of G into a nonempty set of subgraphs, known as clusters, such that

- any $C_i \in C$ is a graph $C_i = (V_i, E_i)$ with $V_i \subseteq V$ nodes and $E_i \subseteq E$ edges,
- for all $u, v \in V_i$, if there exists a edge $(u, v) \in E$ then $(u, v) \in E_i$,
- $\delta_{E_i}(u, v) = \delta_E(u, v)$ for all $(u, v) \in E_i$,
- $\bigcup_{i=1}^{k} V_i = V$, and
- $V_i \cap V_j \neq \phi$ if and only if $V_i = V_j$, for all $C_i, C_j \in C$.

If any node $u \in V_i$ is a member of cluster $C_i \in C$, then we write it simply $u \in C_i$, otherwise $u \notin C_i$. When two nodes u and v are in same cluster $C_i \in C$, then we write $u \in_{C_i} v$. Otherwise we write $u \notin_{C_i} v$.

Definition 2 Ground truth

Ground truth G^T of graph $G(V, E)$ is a triplet $(\Gamma, \Sigma, f_\Gamma)$ of a finite set Γ of properties, a finite set Σ of different measures and function $f_\Gamma :$ $\{\gamma \rightarrow w | \forall \gamma \in \Gamma \& w \subseteq \Sigma\}$ where $w_i : \{w_i \rightarrow \Re | \forall w_i \in w\}$ is a function defined for the measures in the subset w. For instance, the property friendship in social network is associated with measures such as closeness, number of common friends etc. These measures will have distinct values that are observed during on site study or survey.

Definition 3 Real clustering

Clustering C^r of graph $G(V, E)$ is referred as *real clustering* if with respect to ground truth $G^T(\Gamma, \Sigma, f_\Gamma)$, any property $\gamma \in \Gamma$ holds on for all $u \in C_i$ in G and this is true for all $C_i \in C^r$. The real clustering is nothing but the clustering that is defined based on ground truth.

Definition 4 Modeled clustering

Clustering C^m of graph $G(V,E)$ is referred as *modeled clustering* if it is predicted with any graph clustering algorithm or if it is generated with graph generating algorithms. The graph clustering algorithms are applicable to both real world networks and already generated synthetic networks. However, graph generating algorithms are applicable only to synthetic networks for defining modeled clustering during graph generation. Modeled clustering have no direct correlation with ground truth $G^T(\Gamma, \Sigma, f_\Gamma)$. If in G holds any property

$\gamma \in \Gamma$ on any node $u \in V$ with respect to G^T then that γ may not hold in $u \in C_i$ for any cluster $C_i \in C^m$.

Definition 5 Accuracy metrics

An accuracy metric A is a function $A : C^r \times C^m \to \Re$, where C^r is real clustering of any graph G and C^m is a modeled clustering of G.

Definition 6 Quality metrics

Graph clustering quality metric Q is a function $Q : G \times C^m \to \Re$, where G is any graph and C^m is any modeled clustering of G.

3. Related work

Assurance of accuracy and quality of clustering is an important task for study and big data analysis of modern day complex data systems. Numerous methods have developed to evaluate clustering obtained with any clustering algorithm. In Ref. [14], proposed data mining process model based approach by considering multiple clustering algorithms with different parameter settings and prespecified objectives. Clemencon [15] proposed a U–process based statistical framework for studying the performance of clustering methods. Kou et al. [16] proposed a MCDM based method to evaluate clustering. Kriegel et al. [17] designed mathematical model for comparing different clusters. Kremer et al. [18] proposed cluster mapping measure (CMM) for dynamic or stream data clustering. Delling et al. [19] used some unit-test strategies for clustering. In Refs.[9, 20, 21] defined framework for evaluating clusters by referencing to previous results obtained with different algorithms. Hascoët et al. [22] proposed multilayer interaction model to compare clustering visually. Brandes et al. [6] also proposed graphical approach to analyze clusters visually.

With the above mentioned methods, clusters are measured in terms of either quality metrics or accuracy metric or both at the same time. Several metrics have been designed both in perspective of accuracy as well as quality of the clustering. Starting from the very popular quality metric modularity [5] which was proposed specially to measure quality of clustering. Another popular quality measure is coverage defined in Ref. [4]. With the same quest as of quality there has a deep concern about accuracy of clustering. Several accuracy metrics were proposed over the decades accounting from information theoretic approach to mathematical indexing. Some popular accuracy measure such as adjusted rand index (ARI) [23], normalized mutual information (NMI) [24], F-measure, and purity [25] are still in the front line of measuring accuracy of clustering. Jo and Lee [26] also proposed quality metric called clustering index (CI). Lin et al. [27] defined several distance based quality metrics. Meilă [28] has proposed information theory based accuracy metric variation of information (VI). In Ref. [29] defines quality metric based on the ratio of compactness to separability of clusters. Reichart and Rappoport [30] further improved VI by normalizing and referred it as normalized variation of information (NVI). Foggia et al. [31] also discussed several accuracy metrics. Recently, Zaidi et al. [32] proposed path length based metric cluster path length (CPL). Leskovec et al. [33] proposed quality metric network community profile (NCP), which considers cluster sizes in background. Aldecoa and Marin [7] proposed probabilistic metric surprise to measure quality of clustering.

Apart from defining new metric or improvement of existing metric several researcher have studied properties, shortcomings of present metrics and illustrated correlation with clustering processes. Orman et al. [9] discussed several quality metrics and they have noted community distribution has very important role in overall community structure. Almeida et al. [8] also discussed quality metrics and discovers several shortcomings of them. They have showed quality metrics have biases toward good quality clustering but they fail to assure accuracy. They also showed that quality metric does not behave as usual when cluster structure is different from the traditional one. Delling et al. [19] discussed centrality and duality of quality metrics. With these two properties they have demonstrated that random graphs can be generated corresponding to a quality metric or vice-versa. They have also showed that clustering algorithms, quality metrics and random graphs are nothing but the different aspects of one and the same problem. We have extended further such study in the direction of exploring correlation through a generic model considering both kinds of metrics, modeled clustering and parameters of modeled clustering generation process.

4. Proposed model

The objective of the proposed model is to derive correlation between accuracy and quality based on the modeled clustering obtained with different parameters of algorithms associated with the clustering generation process (i.e., the parameters of clustering algorithm and graph generating algorithm). Consider P is the set of parameters used in any algorithm to obtain modeled clustering C^m. The ordered variation O_i of any parameter $P_i \in P$ is the set of values obtained by varying P_i from r_s^i to r_e^i with interval d that are arranged in increasing order. Each of the ordered variations O_i obtained for each $P_i \in P$ is stored in a matrix $O^{[M \times N]}$. The number of columns in the matrix O is defined as follows:

$$N = |O_c| \text{ such that } c = \underset{i}{\operatorname{argmax}} |O_i| \qquad (1)$$

where, c is the index of parameter that have maximum number of values in ordered variation, $|O_i|$ is the number of values in ordered variation O_i of parameter P_i and O_c is the ordered variation that have maximum number of values. The number of rows in the matrix O is defined as follows:

$$M = |P| \times \prod_{i=1 \& i \neq c}^{|P|} |O_i| \qquad (2)$$

where, $|P|$ is the number of parameters and $|O_i|$ is the number of values in ordered variation O_i of parameter P_i. The matrix O is filled with the values in ordered variations of all parameters as follows. Every $(j \times |P| + 1)^{th}$ row is filled with O_c, where $j = \{0, 1, \ldots, (\prod_{i=1 \& i \neq c}^{n} |O_i| - 1)\}$. Every $(j \times |P| + i \times k)^{th}$ row is filled with R_i^k, where $i = \{1, 2, \ldots, (|P| - 1)\}, j = \{0, 1, \ldots, (|O_i - 1|)\}$, $k = \{1, 2, \ldots, (|O_i|)\}$ and R_i^k is the row matrix for k^{th} value $O_i(k)$ in ordered variation O_i of parameter $P_i \in P$, $i \neq c$ where each element of R_i^k have value $O_i(k)$. A set P_{val} on the matrix O is defined. Each element P_{val}^i in P_{val} represents a combination of all parameter values as in ordered variations. Thus, P_{val}^i on matrix O is expressed as follows:

$$P_{val}^i = \{O_{jk} | \ j = \{(m \times |P| + 1), (m \times |P| + 2), \ldots, \\ (m \times |P| + |P|)\}, \ m = \left\lfloor \frac{i-1}{N} \right\rfloor \& k = \text{ mod } (i, N)\} \qquad (3)$$

where, $i = \{1, 2, \ldots, \frac{M}{|P|}\}$ and $|P_{val}| = |P| \times \prod_i^{|P|} |O_i|$. For all such P_{val}^i of parameter set P used in any algorithm will result in a modeled clustering C^m. Let, $\Omega = \{C^m | m = \{1, 2, 3, \ldots, \frac{M}{|P|}\}\}$ is the set of modeled clustering for all the elements in P_{val}. Thus, we have function $f_r : P_{val} \to \Omega$. Again, for each clustering $C^m \in \Omega$ will result in exactly one accuracy metric value and quality metric value. Let, $A = \{A_i | i = \{1, 2, \ldots, \frac{M}{|P|}\}\}$ and $Q = \{Q_i | i = \{1, 2, \ldots, \frac{M}{|P|}\}\}$ are the sets of all accuracy metric values and quality metric values respectively for all modeled clustering $C^m \in \Omega$. This implies two functions $f_a : \Omega \to A$ and $f_q : \Omega \to Q$.

Lemma 1. *Given $f_{ab} : A \rightarrow B$ and $f_{bc} : B \rightarrow C$ are two functions from A to B and from B to C respectively. Then, we have function $f_{ac} : A \rightarrow C$ from A to C.*
Proof. By definition of function, every element in A will have exactly one image in B, i.e., for all $a_i \in A$ there exist $b_i \in B$ such that $f_{ab}^{-1}(b_i) = a_i$. similarly, for all $b_i \in B$ there exist $c_i \in C$ such that $f_{bc}^{-1}(c_i) = b_i$. This implies, for all $a_i \in A$ alternatively associated with some images $c_i \in C$ i.e., for all $a_i \in A$ there exist $c_i \in C$ such that $f_{ab}^{-1}(f_{bc}^{-1}(c_i)) = a_i$. Now, we have to show $f_{ab}^{-1}(f_{bc}^{-1}(c_i)) \neq f_{ab}^{-1}(f_{bc}^{-1}(c_j))$. By definition of function $f_{bc}^{-1}(c_i) \neq f_{bc}^{-1}(c_j)$, since f_{bc} is function from B to C. Suppose, $f_{bc}^{-1}(c_i) = b_i$ and $f_{bc}^{-1}(c_j) = b_j$. Thus, we have $f_{ab}^{-1}(f_{bc}^{-1}(c_i)) = f_{ab}^{-1}(b_i)$ and $f_{ab}^{-1}(f_{bc}^{-1}(c_j)) = f_{ab}^{-1}(b_j)$. Since, f_{ab} is function from A to B, it has to satisfy $f_{ab}^{-1}(b_i) \neq f_{ab}^{-1}(b_j)$. Hence, we have $f_{ab}^{-1}(f_{bc}^{-1}(c_i)) \neq f_{ab}^{-1}(f_{bc}^{-1}(c_j))$. Since, from A to C we have exactly one image in C for all elements in A, we have $f_{ac} : A \rightarrow C$ from A to C. □

Theorem 1. *Given P_{val}, is the set of all combinations of parameters values obtained with ordered variation, Ω is the set of all clustering obtained with respect to combinations in P_{val} and, A and Q are the set of accuracy metric and quality metric resulted with respect to all clustering in Ω. Then, we have functions $\delta_A : P_{val} \rightarrow A$, from P_{val} to A and $\delta_Q : P_{val} \rightarrow Q$, from P_{val} to Q.*
Proof. We have functions $f_r : P_{val} \rightarrow \Omega$, $f_a : \Omega \rightarrow A$ and $f_q : \Omega \rightarrow Q$. With Lemma 1 we have function from P_{val} to A as $\delta_A : P_{val} \rightarrow A$ and from P_{val} to Q as $\delta_Q : P_{val} \rightarrow Q$. □

Now, we define indirect correlation between A and Q. We have functions from P_{val} to A and from P_{val} to Q, i.e., $\delta_A : P_{val} \rightarrow A$ and $\delta_Q : P_{val} \rightarrow Q$ respectively. Since, there has no direct correlation between A and Q by their definitions, the correlation between A and Q has to be established through such an entity X, which by definition is common to both and determined by same entity Ω, and the X determines both A and Q. Thus, we have following theorem for defining indirect correlation between A and Q via X.

Theorem 2. *If there exist a function $\delta_X : P_{val} \rightarrow X$ then indirect correlation between A and Q via X is defined by the function $\delta_{cor} : (\delta_A, \delta_Q) \rightarrow \delta_X$ such that $\delta_X(\delta_A^{-1}(A)) = \delta_X(\delta_Q^{-1}(Q))$.*
Proof. Since, X is determined by Ω, we have function $f_x : \Omega \rightarrow X$, from Ω to X. Again, we have $f_r : P_{val} \rightarrow \Omega$ and hence, by Lemma 1 we have $\delta_X : P_{val} \rightarrow X$. Condition of necessity $\delta_X(\delta_A^{-1}(A)) = \delta_X(\delta_Q^{-1}(Q))$ is proved by contradiction. Assume $\delta_X(\delta_A^{-1}(A)) \neq \delta_X(\delta_Q^{-1}(Q))$ for the purpose of contradiction. Since A and Q are the sets of accuracy metric values and quality metric values obtained for clustering set Ω, some clustering can have same

accuracy values or quality values or X values. Thus, either all $p_i \in P$ have distinct images in A, Q, and X or some of the $p_i \in P$ have same image in A or Q or X. We have to show the assumption contradicts in all these cases.

Let, all $p_i \in P$ have distinct images in A, Q and X. Now, for any a_i, $a_j \in A$ we have $\delta^{-1}(a_i) = p_i$, $\delta^{-1}(a_j) = p_j$ and $p_i \neq p_j$, $p_i, p_j \in P_{val}$. Suppose, for $p_i, p_j \in P_{val}$ we have $q_k, q_l \in Q$, then we have $\delta^{-1}(q_k) = p_i$ and $\delta^{-1}(q_l) = p_j$. Since $\delta_X : P_{val} \to X$, P_{val} must have exactly one image, i.e., we cannot have $\delta_X(p_i) = x_m$ and $\delta_X(p_i) = x_n$, $x_m, x_n \in X$. Thus, if for p_i, $p_j \in P_{val}$ we have $x_m, x_n \in X$ then $\delta_X(\delta_A^{-1}(a_i)) = x_m = \delta_X(\delta_Q^{-1}(q_k))$ and $\delta_X(\delta_A^{-1}(a_j)) = x_n = \delta_X(\delta_Q^{-1}(q_l))$, contradicting the assumption.

Let, some of the $p_i \in P$ have same image in A or Q or X. Suppose, $a_i \in A$ is the image of both $p_i, p_j \in P_{val}$, then we have $\delta^{-1}(a_i) = p_i$ and $\delta^{-1}(a_i) = p_j$. Suppose, $p_i, p_j \in P$ have distinct images in both Q and X, then we have $\delta^{-1}(q_k) = p_i$ and $\delta^{-1}(q_l) = p_j$. Since $\delta_X : P_{val} \to X$, if for $p_i, p_j \in P_{val}$ we have x_m, $x_n \in X$ then $\delta_X(\delta_A^{-1}(a_i)) = x_m = \delta_X(\delta_Q^{-1}(q_k))$ and $\delta_X(\delta_A^{-1}(a_i)) = x_n = \delta_X(\delta_Q^{-1}(q_l))$, again contradicting the assumption. Similarly, we can show for all other cases the assumption contradicts. Thus, $\delta_X(\delta_A^{-1}(A)) = \delta_X(\delta_Q^{-1}(Q))$, which implies that for all pairs of (a_i, q_i), $a_i \in A$ and $q_i \in Q$ that are images of same $p_i \in P_{val}$ will have same $x_i \in X$. Thus, we have function $\delta_{cor} : (\delta_A, \delta_Q) \to \delta_X$. □

5. Validation of proposed model

For any modeled clustering generated with any clustering algorithm, both accuracy and quality have to be optimal. Thus, in the correlation function $\delta_{cor} : (\delta_A, \delta_Q) \to \delta_X$, both A and Q have to be optimized and get the corresponding X values. Let us assume multiple accuracy metrics and multiple quality metrics are on board. In this case, we have objectives in two levels. First one is individual level, where all the members of each group have the objective to optimize their own values. Second one is group level, where each group has the objective to optimize the values of the group as a whole. This is a multiobjective optimization problem. There will be Pareto optimal solutions, where the solution cannot be improved without degrading at least one objective. Note that the individual level objective for each group member is also multiobjective problem. For accuracy metrics, the multiobjective optimization function is defined as follows:

$$\text{maximize } (A_1(C^r, C^m), A_2(C^r, C^m),$$
$$..., A_k(C^r, C^m)) \text{ s. t. } C^m \in \Omega \tag{4}$$

where, C^r is the real clustering. Linear scalarizing the multiobjective function in Eq. (4) a single-objective is formulated as follows:

$$\text{maximize } F_A \left(\sum_{i=1}^{k} w_i A_i(C^r, C^m) \right) \text{ s. t. } C^m \in \Omega \tag{5}$$

Similarly, the multiobjective optimization function for quality metrics is defined as follows:

$$\text{maximize } (Q_1(C^m), Q_2(C^m),$$
$$..., Q_k(C^m)) \text{ s. t. } C^m \in \Omega \tag{6}$$

Linear scalarizing the multiobjective function in Eq. (6) a single-objective is formulated as follows:

$$\text{maximize } F_Q \left(\sum_{i=1}^{k} w_i Q_i(C^m) \right) \text{ s. t. } C^m \in \Omega \tag{7}$$

Now, we have two single-objective functions, i.e., Eqs. (5) and (7) for the accuracy and quality metrics groups respectively. The group level multiobjective function is defined with these two single-objective functions as follows:

$$\text{maximize } (F_A, F_Q) \text{ s. t. } \delta_X(\delta_A^{-1}(A)) = \delta_X(\delta_Q^{-1}(Q)) \tag{8}$$

Here, functions F_A and F_Q both are linear scalarization of multiobjective functions. Since, the Eq. (8) is a multiobjective function, there will be Pareto optimal solutions. These Pareto optimal solutions will result in multiple X values corresponding to each solution. Let, X_{opt} is the Pareto optimal solutions' set and, A_{opt} and Q_{opt} are the optimal values of A and Q in Pareto optimal solutions' set respectively. A range $[X_{min}, X_{max}]$ is defined on the X_{opt}, where X_{min} and X_{max} are respectively the upper and lower bound of X values. Let, X_g is the ground truth value of X. The range $[X_{min}, X_{max}]$ is defined as $R_1 = [X_g - d, X_g + d]$, where d is the minimum deviation from the actual X_g for the values in X_{opt}. The value of d is obtained as follows:

$$d = min(|X_g - x_i|), \forall x_i \in X_{opt} \tag{9}$$

For the range R_1, A_{exp} and Q_{exp} are the maximum expected values of A and Q respectively such that $A_{exp} = \alpha\%$ of A_{opt} and $Q_{exp} = \beta\%$ of Q_{opt}.

In the above case, ground truth is considered to define the range of X values. The networks for which ground truth is unavailable the range has to be defined alternatively. Also note that, accuracy cannot be measured directly due unavailability of ground truth. Therefore, we have to rely only on quality metrics' values. In that case, we only have the single-objective function in Eq. (7) for quality metrics. The range $Q_r = [Q_{mean} - SD, Q_{mean} + SD]$ is defined on Q values obtained with Eq. (7). The range of X values $R_2 = [X_{min}, X_{max}]$ is defined using Q_r as follows. Obtain the all X values corresponding to the Q values that are within the range Q_r, and minimum and maximum values of these X values are considered as X_{min} and X_{max}, respectively.

We follow two-way validation in the networks where ground truth is available. In the forward validation, ground truth X_g is considered to examine the A and Q values with respect to the range R_1. In the backward validation, we assume ground truth X_g is not available and define the range R_2 based on the Q values. Compare the A and Q values for the R_2 with the values obtained for R_1. There by, we show how the level of accuracy corresponding to the quality via X can be ensured.

In order to validate the proposed model empirically, there has to be a practical X that is associated with the clustering. The Theorem 2 states that there has to be a function $\delta_X : P_{val} \rightarrow X$ for defining correlation between accuracy and quality via X. The following theorem proves that the number of clusters can be considered as one of the X, through which correlation between accuracy and quality metric can be established.

Theorem 3. *Correlation between accuracy and quality metrics through number of clusters (NoC) is valid. For given sets P_{val}, Ω, A, Q and, if NoC is the set of all numbers of clusters obtained for all clustering $C^m \in \Omega$, then we have $\delta_{cor} : (\delta_A, \delta_Q) \rightarrow \delta_{NoC}$.*

Proof. Since for a modeled clustering $C^m \in \Omega$ we can have exactly one NoC value, we have function $\delta_{NoC} : \Omega \rightarrow NoC$. Again, we have $f_r : P_{val} \rightarrow \Omega$ and hence, by Lemma 1 we have $\delta_{NoC} : P_{val} \rightarrow NoC$. By Theorem 2, we have $\delta_{NoC}(\delta_A^{-1}(A)) = \delta_{NoC}(\delta_Q^{-1}(Q))$, and hence we have $\delta_{cor} : (\delta_A, \delta_Q) \rightarrow \delta_{NoC}.\square$

By Theorem 3, we have the correlation between accuracy metrics and quality metrics via NoC. The methodology discussed above for validating the proposed model is used in the empirical analysis (Section 7).

6. Clustering measures

Several metrics have been designed to evaluate clustering as discussed in Section 3. In this section, we have briefed very popular and widely accepted measures for evaluation modeled clustering. We have selected four accuracy metrics and three quality metrics for our discussion. Significance in terms of modeled clustering and mathematical definition of these metrics are explained below.

6.1 Accuracy metrics

Consider any modeled clustering $C^m = \{C_1^m, C_2^m, ..., C_k^m\}$ obtained with a clustering algorithm on any network of n nodes. Let $C^r = \{C_1^r, C_2^r, ..., C_l^r\}$ be the real clustering. Here, C_k^m and C_l^r are interpreted as set of nodes in the respective clusters. Overlapping of nodes in C^m and C^r is summarized with a contingency table as presented in Table 1. Contingency table is also presented with 2×2 matrix as shown in Table 2. Entries in the tables are decision pairs of two different nodes. With respect to any pair of nodes, if they are predicted as in the same cluster C_i^m and these two nodes are actually lies in same cluster C_j^r then entry for this pair of nodes will be true positive (TP). In this case cluster labels of C_i^m and C_j^r must be equivalent (i.e., $i \equiv j$) or represent the same cluster (i.e., $C_i^m \equiv C_j^r$). If nodes pair is predicted to lie in same cluster, but they actually lied in different cluster the entry will be false positive (FP). If nodes pair is predicted to lie in different cluster and they actually lied in different cluster the entry will be true negative (TN). If nodes pair is predicted to lie in different cluster, but they actually lied

Table 1 Contingency table type 1.

$C^r \searrow C^m$	C_1^m	C_2^m	...	C_k^m	Sums
C_1^r	n_{11}	n_{12}	...	n_{1k}	a_1
C_2^r	n_{21}	n_{22}	...	n_{2k}	a_2
⋮	⋮	⋮	⋱	⋮	⋮
C_l^r	n_{l1}	n_{l2}	...	n_{lk}	a_l
Sums	b_1	b_2	...	b_k	$\sum_{ij} n_{ij} = n$

Here, each n_{ij} denotes the number of nodes in common between cluster C_i^m and C_j^r : $n_{ij} = |C_i^m \cap C_j^r|$.

Table 2 Contingency table type 2.

Real↘Predicted	Same	Different	Total
Same	TP	FN	P
Different	FP	TN	N
Total	$\hat{P}$	$\hat{N}$	P+N

in the same cluster the entry will be false negative (FN). With these assumptions accuracy metrics are defined as follows:

6.1.1 ARI

ARI is corrected version of rand index [23] which measures degree of overlapping between two partitions. Rand index [34] suffers scaling problem so [35] proposed a corrected version as in the form of Eq. (10).

$$AdjustedIndex = \frac{Index - ExpectedIndex}{MaxIndex - ExpectedIndex} \tag{10}$$

More specifically, with overlapping entries of different clusters of modeled clustering C^m and real clustering C^r in contingency table as shown in Table 1. Considering entries of this table, ARI can be computed as follows:

$$ARI(C^r, C^m) = \frac{\Sigma_{ij}\binom{n_{ij}}{2} - \frac{\left[\Sigma_i\binom{a_i}{2}\Sigma_j\binom{b_j}{2}\right]}{\binom{n}{2}}}{\frac{1}{2}\left[\Sigma_i\binom{a_i}{2} + \Sigma_j\binom{b_j}{2}\right] - \frac{\left[\Sigma_i\binom{a_i}{2}\Sigma_j\binom{b_j}{2}\right]}{\binom{n}{2}}} \tag{11}$$

where, n_{ij} is the number of nodes that are present in both cluster C_i^m and C_j^r, a_i is the summation of all n_{ij} corresponding to any C_j^r of C^r and all C_i^m of C^m, and b_j is the summation of all n_{ij} corresponding to any C_i^m of C^m and all C_j^r of C^r. Rand index can take a value between the range $[0, 1]$, while the ARI may take negative values if the index is less than the expected index [36]. Even though ARI takes negative values, generally considered range for ARI is $[0, 1]$. ARI value 0 indicates real and modeled clustering do not agree on pairing, ARI value 1 indicates real and modeled clustering both represent the same clusters.

6.1.2 Normalized mutual information (NMI)

NMI [24] is an information theoretic approach to measure shared information between two data distribution. NMI between real clustering C^r and modeled clustering C^m can be defined as:

$$NMI(C^r, C^m) = \frac{2 \times I(C^r, C^m)}{H(C^r) + H(C^m)} \tag{12}$$

where, $I(C^r, C^m)$ is mutual information share between C^r and C^m, $H(C^r)$ and $H(C^m)$ entropy of real clustering C^r and modeled clustering C^m respectively. NMI takes values in the range [0, 1]. Value 1 indicates maximum mutual information share between C^r and C^m or in other words both represents same clustering. Value 0 indicates no information sharing between C^r and C^m.

6.1.3 Purity

Purity is very simple and clear measure. It only considers correctly assigned nodes to any cluster. Each cluster is assigned a cluster label which is most frequent, and then correctly assigned nodes are counted with respect to real clustering. This number is then divided by the total number of nodes in the network. Purity can be calculated as follows [25]:

$$Purity(C^r, C^m) = \frac{1}{n} \sum_i max_j(C_i^r \cap C_j^m) \tag{13}$$

Actually, purity is the average of total correctly assigned nodes to different clusters. Purity takes values in the range [0, 1]. Higher purity value indicates high accuracy and lower value indicates bad clusters.

6.1.4 F-measure

Harmonic mean of precision and recall is referred as F-measure, also known as F-score or F_1-score.

$$F - measure = 2 \times \frac{Precision \times Recall}{Precision + Recall} \tag{14}$$

F-measure takes its best values from the range [0,1]. Precision is a measure of exactness of the modeled clustering with respect to real clustering. It evaluates the proportion of true positives against all the positive results. Precision can be expressed in terms of elements of contingency table (Table 2) as follows:

$$Precision = \frac{TP}{TP+FP} = \frac{TP}{\hat{P}} \qquad (15)$$

Recall is a measure of completeness of the modeled clustering with respect to real clustering. Recall also known as the true positive rate, or the recall rate, or sensitivity. It evaluates the proportion of true positives against actual true positives results. Recall can be expressed in terms of elements of contingency table (Table 2) as follows:

$$Recall = \frac{TP}{TP+FN} = \frac{TP}{P} \qquad (16)$$

6.2 Quality metrics

6.2.1 Modularity

Modularity (Q) [5] is the most widely used metric designed specially for the purpose of measuring quality of modeled clustering. This metric has got wide acceptance over the years for evaluating clustering when ground truth is not known. Modularity can be computed as follows. Let us consider a modeled clustering C^m with k clusters for a network. Define a $k \times k$ symmetric matrix e whose element e_{ij} is the fraction of all edges in the network that connect nodes in the cluster i to nodes in cluster j. The trace of this matrix $tr(e) = \sum_i e_{ii}$ gives the fraction of edges in the network that connect nodes in the same cluster. Whereas, the row (or column) sums $a_i = \sum_j e_{ij}$ represents the fraction of edges that connect to nodes in cluster i. With these $tr(e)$ and a_i modularity can be defined as:

$$Q(C^m) = \sum_i \left(e_{ii} - a_i^2\right) = tr(e) - \left\|e^2\right\| \qquad (17)$$

where, $\|x\|$ represents the sum of the elements of matrix x. Thus, Q effectively measures the fraction of edges in the network that connect nodes in the same cluster by subtracting the expected value of this quantity from it if the edges were placed at random. If the number of edges between nodes of same cluster is less than random, we will get $Q = 0$ and value approaches $Q = 1$ if edges between nodes of same cluster is higher.

6.2.2 Coverage

Coverage [4] of modeled clustering C^m is the fraction of edges that connect two nodes of same cluster within the total no of edges present in the network. Coverage can be computed as follows:

$$Coverage(C^m) = \frac{\sum_{i=1}^{k} \sum_{j=1, i=j}^{k} e_{ij}}{m} \tag{18}$$

where, k is the number of cluster present in C^m and m is total number of edges present in the network. Intuitively, higher value of coverage indicates better quality of clustering. Coverage will become 0 if all clusters predicted have only one node. Coverage takes value 1 if number of cluster is one, since all edges of the network will lie within the cluster.

6.2.3 External density
External density of a clustering is defined as the ratio of edges that connect two different clusters to the maximum number of edges possible that connect two different clusters. It is the ratio of intercluster edges to the maximum number of intercluster edges possible. External density can be computed as follows:

$$ExtDensity(C^m) = \frac{\left\{ (u,v) | u \in C_i^m, v \in C_j^m, i \neq j \right\}}{n(n-1) - \sum_{i=1}^{k} \left(|C_i^m| (|C_i^m| - 1) \right)} \tag{19}$$

where, u and v are any pair of nodes, C_i^m and C_j^m are clusters, n is the number of nodes in the network, C^m is the modeled clustering which comprises k clusters. Lesser value of external density indicates better quality clustering.

7. Empirical analysis

The details about experimental setup such as synthetic network generation, accuracy and quality metrics considered, and clustering algorithms are briefed in this section. The methodology discussed for validating the proposed model empirically in Section 5. The empirical analysis validates the existence of correlation between quality and accuracy via *NoC*.

7.1 Experimental setup
We have considered four accuracy metrics ARI, NMI, Purity and F-measure, and three quality metrics Modularity, coverage and external density for our experiments. Both quality and accuracy metrics have to be effective irrespective of domain so we have considered algorithms of different domains. We have considered three clustering algorithms HC-PIN

[1], LICOD [2], and SCAN [3]. Modeled clustering from synthetic network is generated in two stages. In first stage, Brandes's method [6] is used to generate synthetic network. The Brandes's method has parameter set $P = \{Pin, Pout\}$. We have considered variation of Pout within the range [0.05, 0.95] with interval of 0.05 keeping Pin constant with value 0.8. We have considered 1000 nodes with prepartitioned into eight clusters for generating synthetic networks corresponding to all pairs of Pin and Pout. In second stage, clustering algorithm is used to generate modeled clustering from the synthetic network. The clustering algorithms are applied to the synthetic networks generated with ordered variation of *Pout* for obtaining the modeled clustering. We have considered nonoverlapping clustering for our experiments. Therefore, we have initialized the parameters related to overlapping- *os* for HC-PIN and ϵ for LICOD to the value of 0.0. Best performing values of other parameters suggested in Refs. [1–3] for all the algorithms are considered.

7.2 Result analysis

The various fronts obtained for the multiobjective function in Eq. (8) are presented in Fig. 1. The Pareto optimal solutions and corresponding NoC_{opt} are presented in Fig. 2. The trade-off between quality and accuracy can be noticed in each of the fronts in the clustering obtained with all the algorithms. The metric values are normalized to the range [0,1], where 0 and 1 are the worst and best values for the metrics. The deviations from the ground truth $NoC_g = 8$, A_{exp}, Q_{exp} and other values for the forward verification is presented in the Table 3. Clearly the maximum quality value we can expect is moderate for the NoC range R_1 where accuracy is higher. Even the optimal values of quality are also not the best values. Although the maximum expected quality is less than 100% of the optimal quality values, the maximum expected accuracy is 100% of optimal accuracy. That means we are certain that we will be able to get 100% of the optimal accuracy within the NoC range R_1 with moderate quality values. The deviation from the NoC_g is very low, particularly for SCAN and HC-PIN it is 0. This is because SCAN and HC-PIN algorithms produce some clustering that have $NoC = NoC_g = 8$. Outside the NoC range R_1, specifically for the higher quality values $NoC \ll NoC_g$ and accuracy is also very low. That is higher quality does not imply higher accuracy, especially if NoC is deviated far from the NoC_g. This shows the significant role of NoC in indicating both quality and accuracy.

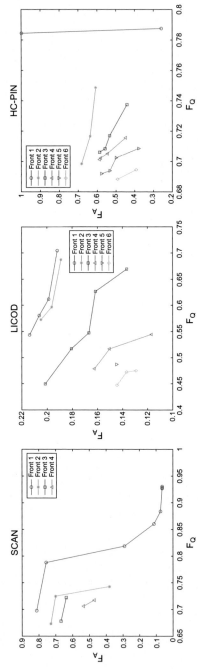

Fig. 1 Different fronts obtained corresponding to SCAN, LICOD and HC-PIN algorithms.

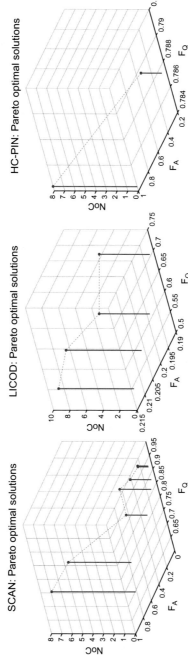

Fig. 2 Pareto optimal solutions and corresponding *NoC*.

Table 3 The ranges of NoC, deviations of NoC and expected accuracy and quality levels.

Algorithm	d	R_1	A_{exp}	Q_{exp}	A_{opt}	Q_{opt}	α	β
SCAN	0	[8, 8]	0.8152	0.6974	0.8152	0.9302	100	74.96
LICOD	1	[7, 9]	0.2138	0.5805	0.2138	0.7044	100	82.41
HC-PIN	0	[8, 8]	1	0.7843	1	0.7875	100	99.59

Table 4 The ranges of NoC, and expected accuracy and quality levels obtained based on mean and standard deviation of quality metrics values.

Algorithm	Q_{mean}	SD	R_2	A_{exp}	Q_{exp}	A_{opt}	Q_{opt}	α	β
SCAN	0.8163	0.1036	[2,10]	0.8152	0.8835	0.8152	0.9302	100	94.98
LICOD	0.5542	0.0794	[6,11]	0.2138	0.7044	0.2138	0.7044	100	100
HC-PIN	0.7163	0.0287	[3, 5]	0.6828	0.7487	1	0.7875	68.28	95.06

The range of NoC, A_{exp}, Q_{exp} and other values for the backward verification is presented in the Table 4. The mean quality and standard deviation values indicates that most of the cases SCAN and HC-PIN produce high quality clustering. However, the NoC range R_2 obtained are quite different. The range R_2 for SCAN covers wide range of NoC values including the NoC_g. On the contrary, the range R_2 for HC-PIN is narrow, it includes only three NoC values. The results indicates that if the predicted NoC range R_2 includes the NoC_g, we are certain to have expected accuracy 100% of optimal accuracy and expected quality more than 90% of optimal quality. However, if the NoC range R_2 does not include NoC_g, the maximum expected accuracy drastically degrades. For the HC-PIN algorithm, the NoC range is $R_2 = [3, 5]$, which excludes NoC_g and accordingly one can notice expected accuracy is 68% of optimal accuracy. However, the optimal accuracy of HC-PIN is high in comparison to LICOD, which show expected accuracy 100% of optimal accuracy. Therefore, the level of accuracy for HC-PIN will be higher than LICOD but level of expected accuracy would be lower.

There is no way to know the actual accuracy of clustering without measuring it. However, it can be explained whether the algorithm will produce accurate clusters or the level of accuracy with the proposed model as follows. An algorithm can show 100% of optimal accuracy (be it best or worst) and more than 90% of optimal quality if and only if R_2 includes NoC_g. However, it does not mean that accuracy will have best value 1. For instance, in the case of LICOD algorithm, maximum expected accuracy is 100% of optimal accuracy, but the optimal accuracy is very low. Thus, the significance of α depends on optimal accuracy value. If the optimal accuracy value is high it indicates the algorithm mostly produces accurate clusters, and optimal accuracy will be low only if the algorithm mostly produce inaccurate clusters. Therefore, with the proposed model, if an algorithm show higher optimal accuracy and it is capable of showing maximum expected accuracy 100% of optimal accuracy in both forward and backward verification then we are certain to have highly accurate clustering with that algorithm.

7.3 Analysis with real world network

To perform analysis on real world network, we have considered very popular and widely used four data sets: Dolphin [37], Karate [38], Strike [39], and Football [40]. Ground truth clustering of all these data sets are known and are available online.[a] Dolphin network has 62 nodes, 159 edges, and 2 clusters. Karate network has 34 nodes, 78 edges and 2 clusters. Strike network has 24 nodes, 34 edges, and 3 clusters. Football network has 115 nodes, 613 edges, and 12 clusters. Real world networks are supposed to retain their actual structure, i.e., number of nodes and edges remains unchanged. Therefore, modeled clustering with real world network is obtained without changing *edges* and *the total number of nodes*. Accounting both these constraints, different modeled clustering are generated with ordered variation of parameters of same clustering algorithm.

The values of quality and accuracy metrics for the modeled clustering generated with HC-PIN algorithm on all four networks are presented in Fig. 3. For Dolphin network, due to smaller λ values, initially number of cluster produced is very high compared to the real clustering. As a result, all the accuracy metrics showed lower values. Quality metrics coverage and modularity also showed comparatively lower values but external density showed higher values, which indicate presence of large number of inter-cluster edges. With increment of λ value, number of clusters reduced and all accuracy metrics increased. In this case, number of cluster reduction actually implied convergence toward the real clustering so accuracy metrics showed higher values. The reduction may go up to a single cluster or less than real clustering. In that case, accuracy metrics' values will degrade as modeled clustering diverse from the real clustering. Clearly, λ values 2.5–5.5 produced same number of clusters as in the real clustering of

---

[a] http://vldao.fmf.uni-lj.sj/pub/networks/pajek.

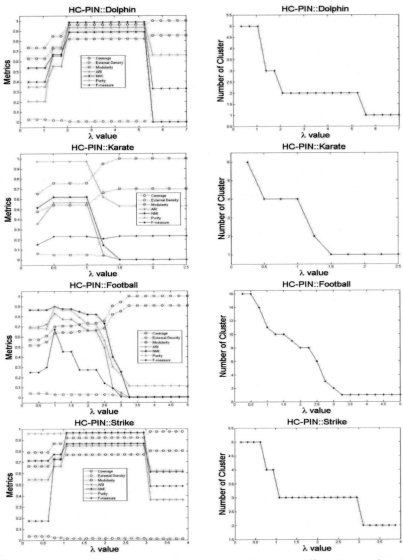

Fig. 3 Various quality and accuracy metrics' values obtained on different networks with HC-PIN algorithm with variation of λ parameter.

Dolphin network and accordingly accuracy metrics showed highest values. When number of clusters become smaller than the real clustering, all accuracy metrics' values degraded. As described earlier, when number of clusters decreases, intercluster edges also decrease so quality metric

External Density reduces and other two increases (i.e., all quality metrics improve). Interestingly, all the quality metrics improved even when number of clusters become less than the real clustering. Similar characteristics can also be noted for remaining three networks: Karate, Football, and Strike. Disturbance created during clustering with ordered variation of parameter λ also showed similar characteristics of metrics as noted in synthetic network. These implied clear impact of the number of clusters on different metrics.

Influence of number of clusters on quality and accuracy metrics can also be noticed in modeled clustering generated with LICOD algorithm as shown in Fig. 4. Comparatively very large number of clusters produced by LICOD than the real clustering has caused degradation of both quality and accuracy metrics values. Clearly, for all variation of σ from 0.05 to 1, LICOD produced very large number of clusters on Dolphin, Karate and Football networks. Specially for Football network, number of clusters are much large compared to number of clusters present in real clustering, which was 12. However, for Dolphin and Karate network at the end (i.e., $\sigma = 1$) produces similar number of clusters as in original network, as a result we can see spikes in both quality and accuracy. For Strike network, both quality and accuracy metrics showed similar characteristics as in the case of HC-PIN.

The values of quality and accuracy metrics for the modeled clustering generated with SCAN algorithm is presented in Fig. 5. With variation of ϵ, instead of continuously decreasing number of clusters as was in HC-PIN and LICOD, here number of clusters initially increased and then decreased. More interestingly for all four networks, number of clusters remains constant with single cluster for lower and higher values of ϵ, which resulted in least accuracy metrics' values. When number of clusters increased accuracy metrics also increased and quality metrics degraded. Again, when number of clusters decreased accuracy metrics also decreased and quality metric improved. Note that SCAN produces hubs and outliers along with the clusters. We have considered hubs and outliers as individual clusters, so number of clusters showed more than actual clusters produced. Despite this fact, one can notice that the accuracy metrics and quality metrics have clear correlation with the number of clusters.

The common characteristics noticed in all four networks is that when number of clusters converges toward the real clustering, both quality and accuracy improves. Larger number of clusters than real clustering cause both

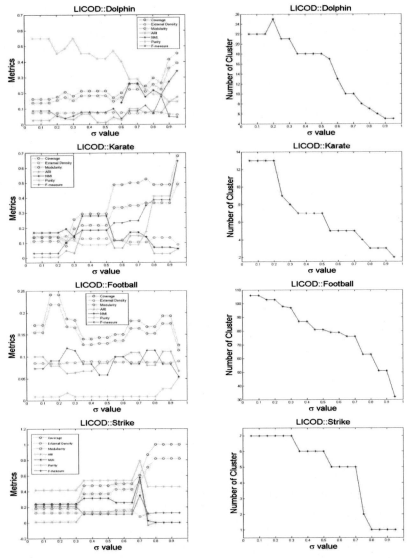

Fig. 4 Various quality and accuracy metrics' values obtained on different networks with LICOD algorithm with variation of σ parameter.

quality and accuracy metrics to acquire worst values. Smaller number of clusters than real clustering results in improvement of quality metrics, whereas accuracy metrics deteriorates. When number of clusters are similar to the real clustering, accuracy metrics gains best values but quality metrics do not attain best values instead it attains moderate best values.

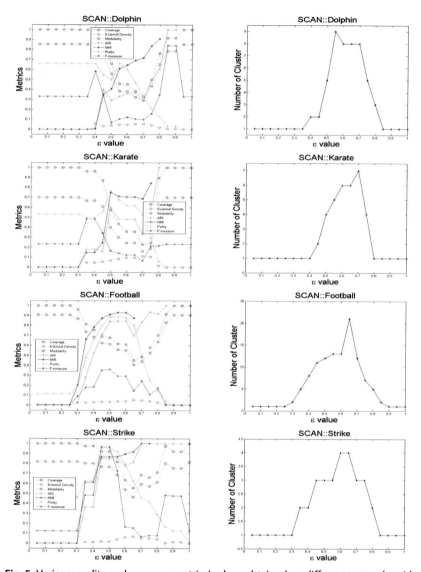

Fig. 5 Various quality and accuracy metrics' values obtained on different networks with SCAN algorithm with variation of ϵ parameter.

8. Conclusion

In this chapter, we have proposed a model to define correlation between quality and accuracy metrics via number of clusters. The model is validated empirically. The modeled clustering generated with ordered variation of different parameters. The analysis shows highest quality metrics values showed very inaccurate clustering. Quality and accuracy metrics contradict one another with the exception of when the number of clusters of modeled clustering is almost similar to the real clustering. We have also noticed that the quality and accuracy metrics cannot attain optimal values at the same time. Neither the best nor the worst values of quality metrics justify accuracy. However, moderate optimal values of quality metrics can assure reasonable level of accuracy if the number of clusters deviate far from the ground truth.

References

[1] J. Wang, M. Li, J. Chen, Y. Pan, A fast hierarchical clustering algorithm for functional modules discovery in protein interaction networks, IEEE/ACM Trans. Comput. Biol. Bioinform. 8 (3) (2011) 607–620, https://doi.org/10.1109/TCBB.2010.75.

[2] R. Kanawati, LICOD: leaders identification for community detection in complex networks, in: 2011 IEEE Third International Conference on Privacy, Security, Risk and Trust (passat) and 2011 IEEE Third International Conference on Social Computing (socialcom), October, IEEE, 2011, pp. 577–582, https://doi.org/10.1109/PASSAT/SocialCom.2011.206.

[3] X. Xu, N. Yuruk, Z. Feng, T.A.J. Schweiger, SCAN: a structural clustering algorithm for networks, in: Proceedings of the 13th ACM SIGKDD International Conference on Knowledge Discovery and Data Mining, ACM, New York, NY, USA, 2007, pp. 824–833, https://doi.org/10.1145/1281192.1281280.

[4] U. Brandes, M. Gaertler, D. Wagner, Experiments on graph clustering algorithms, in: G. Di Battista, U. Zwick (Eds.), Algorithms–ESA 2003, Lecture Notes in Computer Science, vol. 2832, Springer Berlin Heidelberg, 2003, pp. 568–579, ISBN: 978-3-540-20064-2, https://doi.org/10.1007/978-3-540-39658-1_52.

[5] M.E.J. Newman, M. Girvan, Finding and evaluating community structure in networks, Phys. Rev. E 69 (2) (2004) 026113.

[6] U. Brandes, M. Gaertler, D. Wagner, Engineering graph clustering, J. Exp. Algorithmics 12 (1) (2008) 26, https://doi.org/10.1145/1227161.1227162.

[7] R. Aldecoa, I. Marin, Surprise maximization reveals the community structure of complex networks, Sci. Rep. 3 (1060) (2013) 1–9.

[8] H. Almeida, D. Guedes, W. Meira Jr, M.J. Zaki, Is there a best quality metric for graph clusters? in: Machine Learning and Knowledge Discovery in Databases, Springer, 2011, pp. 44–59.

[9] G.K. Orman, V. Labatut, H. Cherifi, Comparative evaluation of community detection algorithms: a topological approach, J. Stat. Mech. Theory Exp. 2012 (08) (2012) P08001, https://doi.org/10.1088/1742-5468/2012/08/P08001.

[10] A. Biswas, B. Biswas, Investigating community structure in perspective of ego network, Expert Syst. Appl. 42 (20) (2015) 6913–6934, https://doi.org/10.1016/j.eswa.2015.05.009.

[11] M. Dayarathna, T. Suzumura, Towards emulation of large scale complex network workloads on graph databases with XGDBench, in: 2014 IEEE International Congress on Big Data (BigData Congress), June, 2014, pp. 748–755, https://doi.org/10.1109/BigData.Congress.2014.140.

[12] H.-H. Shuai, D.-N. Yang, P. Yu, C.-Y. Shen, M.-S. Chen, On pattern preserving graph generation, in: 2013 IEEE 13th International Conference on Data Mining (ICDM), December, 2013, pp. 677–686, https://doi.org/10.1109/ICDM.2013.14.

[13] M. Hua, M.K. Lau, J. Pei, K. Wu, Continuous K-means monitoring with low reporting cost in sensor networks, IEEE Trans. Knowl. Data Eng. 21 (12) (2009) 1679–1691, https://doi.org/10.1109/TKDE.2009.41.

[14] K.M. Osei-Bryson, Towards supporting expert evaluation of clustering results using a data mining process model, Inf. Sci. 180 (3) (2010) 414–431, https://doi.org/10.1016/j.ins.2009.09.019.

[15] S. Clemencon, A statistical view of clustering performance through the theory of processes, J. Multivar. Anal. 124 (2014) 42–56, https://doi.org/10.1016/j.jmva.2013.10.001.

[16] G. Kou, Y. Peng, G. Wang, Evaluation of clustering algorithms for financial risk analysis using MCDM methods, Inf. Sci. 275 (2014) 1–12, https://doi.org/10.1016/j.ins.2014.02.137.

[17] H.-P. Kriegel, E. Schubert, A. Zimek, Evaluation of multiple clustering solutions, in: MultiClust@ ECML/PKDD, Citeseer, 2011, pp. 55–66.

[18] H. Kremer, P. Kranen, T. Jansen, T. Seidl, A. Bifet, G. Holmes, B. Pfahringer, An effective evaluation measure for clustering on evolving data streams, in: Proceedings of the 17th ACM SIGKDD International Conference on Knowledge Discovery and Data Mining, ACM, New York, NY, USA, 2011, pp. 868–876, https://doi.org/10.1145/2020408.2020555.

[19] D. Delling, M. Gaertler, R. Görke, Z. Nikoloski, D. Wagner, How to Evaluate Clustering Techniques, Univ., Fak. für Informatik, Bibliothek, 2006.

[20] S. Datta, S. Datta, Methods for evaluating clustering algorithms for gene expression data using a reference set of functional classes, BMC Bioinformatics 7 (1) (2006) 397.

[21] M. Emmanuel, A. Stephan, T. Seidl, A framework for evaluation and exploration of clustering algorithms in subspaces of high dimensional databases, in: BTW, 2011, pp. 347–366.

[22] M. Hascoët, M. Cedex, P. Dragicevic, Interactive graph matching and visual comparison of graphs and clustered graphs, in: Proceedings of the International Working Conference on Advanced Visual Interfaces, ACM, 2012, pp. 522–529.

[23] W.M. Rand, Objective criteria for the evaluation of clustering methods, J. Am. Stat. Assoc. 66 (336) (1971) 846–850.

[24] A. Strehl, J. Ghosh, Cluster ensembles–a knowledge reuse framework for combining multiple partitions, J. Mach. Learn. Res. 3 (2002) 583–617.

[25] C.D. Manning, P. Raghavan, H. Schütze, Introduction to Information Retrieval, first ed., Cambridge University Press, 2008 (July).

[26] T. Jo, M. Lee, The evaluation measure of text clustering for the variable number of clusters, in: Advances in Neural Networks-ISNN 2007, Springer, 2007, pp. 871–879.

[27] C. Lin, Y.-r. Cho, W.-c. Hwang, P. Pei, A. Zhang, Clustering methods in protein-protein interaction network, in: Knowledge Discovery in Bioinformatics: Techniques, Methods and Application, John Wiley & Sons, Inc, 2007, pp. 1–35.

[28] M. Meilă, Comparing clusterings–an information based distance, J. Multivar. Anal. 98 (5) (2007) 873–895.

[29] A. Celikyilmaz, I. Burhan Türkşen, Validation criteria for enhanced fuzzy clustering, Pattern Recogn. Lett. 29 (2) (2008) 97–108, https://doi.org/10.1016/j.patrec.2007.08.017.

[30] R. Reichart, A. Rappoport, The NVI clustering evaluation measure, in: Proceedings of the Thirteenth Conference on Computational Natural Language Learning, Association for Computational Linguistics, 2009, pp. 165–173.

[31] P. Foggia, G. Percannella, C. Sansone, M. Vento, Benchmarking graph-based clustering algorithms, Image Vision Comput. 27 (7) (2009) 979–988.

[32] F. Zaidi, D. Archambault, G. Melançon, Evaluating the quality of clustering algorithms using cluster path lengths, in: Advances in Data Mining. Applications and Theoretical Aspects, Springer, 2010, pp. 42–56.

[33] J. Leskovec, K.J. Lang, M. Mahoney, Empirical comparison of algorithms for network community detection, in: Proceedings of the 19th International Conference on World Wide Web, ACM, 2010, pp. 631–640.

[34] K. Steinhaeuser, N.V. Chawla, Identifying and evaluating community structure in complex networks, Pattern Recogn. Lett. 31 (5) (2010) 413–421, https://doi.org/10.1016/j.patrec.2009.11.001.

[35] L. Hubert, P. Arabie, Comparing partitions, J. Classif. 2 (1) (1985) 193–218, https://doi.org/10.1007/BF01908075.

[36] S. Wagner, D. Wagner, Comparing Clusterings: An Overview, Universität Karlsruhe, Fakultät für Informatik Karlsruhe, 2007.

[37] D. Lusseau, K. Schneider, O.J. Boisseau, P. Haase, E. Slooten, S.M. Dawson, The bottlenose dolphin community of doubtful sound features a large proportion of long-lasting associations, Behav. Ecol. Sociobiol. 54 (4) (2003) 396–405.

[38] W.W. Zachary, An information flow model for conflict and fission in small groups, J. Anthropol. Res. 33 (1977) 452–473.

[39] J.H. Michael, Labor dispute reconciliation in a forest products manufacturing facility, For. Prod. J. 47 (11–12) (1997) 41–45.

[40] M. Girvan, M.E.J. Newman, Community structure in social and biological networks, Proc. Natl. Acad. Sci. 99 (12) (2002) 7821–7826.

About the authors

Soumita Das received the B.Tech. degree in information technology from North Eastern Hill University (N.E.H.U), Shillong, India, in 2015 and the M.Tech. degree in Computer Sience and Engineering from Tezpur University in 2018. She is currently pursuing the Ph.D. degree with the Department of Computer Science and Engineering, National Institute of Technology Silchar (NITs), Silchar, India. Her research interests include social network analysis.

Dr. Anupam Biswas is currently working as an Assistant Professor with the Department of Computer Science and Engineering, National Institute of Technology Silchar, Silchar, India. He has received the B.E. degree in Computer Science and Engineering from the Jorhat Engineering College, Jorhat, India, in 2011, M.Tech. degree in Computer Science and Engineering from the Nehru National Institute of Technology Allahabad, Prayagraj, India, in 2013, and the Ph.D. degree in Computer Science and Engineering from IIT (BHU) Varanasi, Varanasi, India, in 2017. He has authored or coauthored several research articles in reputed international journals, conference, and book chapters. His research interests include machine learning, deep learning, computational music, information retrieval, social networks, and evolutionary computation. He has served as a Program Chair for the International Conference on Big Data, Machine Learning and Applications (BigDML 2019). He has served as a General Chair for the 25th International Symposium on Frontiers of Research in Speech and Music (FRSM 2020) and co-edited proceedings of FRSM 2020 published as book volume in Springer AISC Series. He has edited three books titled Health Informatics: A Computational Perspective in Healthcare, Principles of Social Networking: The New Horizon and Emerging Challenges in different Springer book series.

CHAPTER FIVE

genebF: Filtering protein-coded gene graph data using Bloom filter

Sabuzima Nayak and Ripon Patgiri
Department of Computer Science and Engineering, National Institute of Technology Silchar, Silchar, Cachar, Assam, India

Contents

Abstract

Today's digital era is facing an enormous obstacle called Big data. Every digital device is producing a huge volume of data. The exponential increment of the number of digital devices is leading to an exponential increment of generated data. One of the domains of Big data is Big graph, for instance, DNA structure. Currently, Bioinformatics is focusing on a probabilistic membership checking data structure called Bloom Filter. Bloom Filter has a simple architecture with approximately constant time complexity operations. It proved to be efficient for handling DNA data, but Bloom Filter takes a single input. Thus, Bloom Filter cannot be used for the processing of paired data such as

Advances in Computers, Volume 128
ISSN 0065-2458
https://doi.org/10.1016/bs.adcom.2021.09.009

protein-coding genes. The protein-coding genes provide the association of a gene with a protein. This information helps in the reconstruction of DNA sequences. In this chapter, a novel Bloom Filter called geneBF is proposed for the processing of paired data. It has a simple architecture and uses arithmetic operations for insertion and query. Our experimental results prove that geneBF is suitable for Big data and paired data such as protein-coding genes. The insertion of a 300 MB file takes less than 6 s. Similarly, query operation on a 300 MB file takes approximately 4 s with 0.0024 false positive probability. geneBF has zero false negative probability. Finally, geneBF is experimented with a 3.3 MB protein-coding gene dataset which gives 0.0000585 false positive probability.

1. Introduction

Big data is a pressing issue in every domain. One such domain where Big data is a major issue is networking, for example, Big graph. Some examples of Big graphs are communication networks, social graphs, endorsement graphs, protein-coding genes, etc. [1]. With each passing day, these graphs are interlinking and increasing the complexity of their structure. For example, a social media user is adding new friends or new DNA sequences are determined along with its complex relationship with other DNA and proteins. Even though there is a advent of many data mining tools, but they are unable to manage the Big data efficiently. Moreover, accessing the database containing this complex Big data is very tedious. Thus, many applications are focusing on a simple data structure called Bloom Filter.

Bloom Filter is a simple data structure for membership checking. Its simple architecture and faster operation speed, along with the incapability of storing the original data, which in turn provide high security for data, has proved to be an excellent choice for handling Big data. Bloom Filter is used in many domains such as Networking [2–4], Network Security [5–7], and Bioinformatics [8, 9]. The time complexity of Bloom Filter operations is $O(1)$. Every Bloom Filter operation first performs hashing of the input element by k number of hash functions. The hashed values give the bit locations of the Bloom Filter array. In insertion operation, all these bit locations are set to 1. In query operation, all bit locations are checked whether all have bit value 1; if true, then the element is present otherwise absent. However, Bloom Filter has two issues, namely, false positive and false negative. A response is called false positive when Bloom Filter returns true for an element that is not inserted into the Bloom Filter. Whereas a response is called false negative when the Bloom Filter returns false for an element that is inserted into the Bloom Filter. Although these issues are present in Bloom

Filter, but this data structure is preferred because the occurrence of these cases is very low and can be ignored. On the contrary, Bloom Filter is not appropriate for Big graphs.

The data of the Big graphs are mostly paired data, i.e., they have two values, in other words, two vertices. But, Bloom Filter stores a single data. This motivated us to propose geneBF for the storage of paired data. geneBF is a novel Bloom Filter that is capable of processing Big data within few seconds. geneBF uses an r-dimensional Bloom Filter [10] which uses only a single hash function with faster execution of operations and high storage capacity. geneBF performs two additional hashing operations to store the paired data.

geneBF is a Bloom Filter which is using rDBF for faster processing of Big Data. Hence, Section 2 precisely elaborates about both Bloom Filter and rDBF. Then, Section 3 helps to understand the architecture and algorithm of the proposed system, i.e., geneBF. We have mentioned earlier, geneBF is faster and proficient in handling Big Data. To prove this statement we have used two different datasets whose description is provided in Section 4. In addition, Section 5 presents the experimentation and the analysis of the obtained result. In Section 6, we have further clarified some important points and future works about geneBF. Finally, Section 7 concludes our chapter.

2. Related works

There are two main concepts, in other words, two data structures that are closely connected to geneBF, i.e., Bloom Filter and rDBF. In this section, these data structures are explained precisely.

2.1 Bloom filter

Bloom Filter [11] is a probabilistic data structure used mainly for membership checking. It is a bit array where each slot is of one-bit length. Initially, all bits or slots of the Bloom Filter are set to 0. In insertion operation, the input element is hashed by k number of hash functions. Each hash function gives a hash value which is a slot location of the Bloom Filter array. The bit of those k locations are set to 1. Again in query operation, the queried element is hashed by the same k hash functions. The slot locations given by the hash values are checked. In case all the bit values are 1, then the Bloom Filter returns true. However, if any one location has a value 0, then it is considered that the element is absent, and Bloom Filter returns false as a response.

Many elements have some overlapping locations after hashing. Hence, for some elements, even if it is not inserted, their bit locations are already set to 1. Such an element, when queried, returns true. Such a response is called false positive, which is an issue in the Bloom Filter. Similarly, Bloom Filter has another issue called false negative. In the delete operation of Bloom Filter, the bit locations are set to 0. But, when the delete operation is executed, some other inserted elements with one or more common bit locations with the deleted element are set to 0. Thus, when those inserted element is queried, then Bloom Filter gives false as a response, i.e., an inserted element is absent in Bloom Filter without deletion. Such a response is called false negative. Therefore, Bloom Filter responses are classified into four types, (i) True positive, (ii) False positive, (iii) True negative, and (iv) False negative.

Let BF is a Bloom Filter and S be a set of elements, where $S \in BF$
Let q is an queried element.
If $q \in S$ and $q \in BF$, then BF response is "True positive"
If $q \notin S$ and $q \in BF$, then BF response is "False positive"
If $q \notin S$ and $q \notin BF$, then BF response is "True negative"
If $q \in S$ and $q \notin BF$, then BF response is "False negative"
Between false positive and false negative, a false negative is more damaging. Let's consider the situation of password databases. An unregistered user gave its username and password. Due to the false positive response of Bloom Filter, the user is granted access, but the account is empty because its account is not created. In another scenario, a registered user is not given access due to a false negative response. Thus, false negative has a more negative impact on the application compared to false positive. Moreover, identifying a response as a false negative is difficult. Therefore, Bloom Filter does not permit delete operation to prevent false negative probability. The time complexity of insertion and query operations of Bloom Filter is $O(k)$ where k is number of hash functions. Thus, Bloom Filter is dependent on the number of hash functions.

2.2 rDBF

r–Dimensional Bloom Filter (rDBF) [10] is a novel Bloom Filter proposed for massive scale membership query. rDBF uses a single hash function for its operations which makes it independent of number of hash functions. Furthermore, each element is hashed to a single bit which increases the element insertion capacity drastically while remaining below the tolerable false positive probability. rDBF considers prime numbers for dimensions to reduce the false positive probability. Moreover, rDBF executes bitwise operations to enhance performance. Initially, each cell of rDBF is set to 0. In the insertion operation, the element is hashed by one hash function. The bit location, which indicates the presence of the element, is found by performing the modulus operation between hash value and the dimensions of rDBF. Then using OR operation, the bit location is modified to insert the new element into rDBF. Similar to the insertion operation in query operation, the bit location is calculated by performing modulus operation. Then, an AND operation is performed to check the presence of the element in the rDBF. rDBF has many variants, namely, 2DBF, 3DBF, 4DBF, and 5DBF. Increasing the dimension of the rDBF increases the operation execution time, whereas decreases the false positive probability. In our experimentation, we have used 2DBF for faster execution of the operations.

3. geneBF: The proposed system

geneBF is a novel Bloom Filter for the storage of paired data. It performs fewer arithmetic operations to store the paired data. geneBF uses 2DBF for faster operations. The 2DBF uses a single hash function and achieves a highly efficient Bloom Filter. Whereas by adding two more hash functions, we were able to achieve another highly efficient Bloom Filter that is capable of storing paired data. Let, $\mathbb{B}_{x,y}$ be the 2DBF where x and y are the dimensions of the Bloom Filter. x and y are prime numbers where $x \neq y$. When a 2DBF takes composite numbers as its dimensions, then the false positive probability increases. geneBF has a simple algorithm and gives excellent execution speed. Let, *Key–Value* be a paired data where the first value of the data is called *Key* and the second value of the data is called *Value*. *Key–Value* is given as input to geneBF. First, both *Key* and *Value* are hashed separately. Then, both the hashed values are concatenated, where the *Key* hash value is followed by the *Value* hash value. This concatenated string is given as input to the 2DBF for insertion and query operation.

3.1 Insertion operation

Let, $\mathbb{B}_{x,y}$ be a 2DBF and *Key–Value* be a paired data. Let, $\mathcal{H}()$ be a hash function. We use the murmur hash function [12]. In the murmur hash function, different seed values create a different hash value for the same key. We have given separate seed values to hash *Key* and *Value*. Algorithm 1 demonstrates the insertion operation of geneBF. In 2DBF, the array is of unsigned long int, which makes each cell of 63 bits. Hence, in the algorithm, the value of the bits is 63.

ALGORITHM 1 Insertion of a *Key–Value*($\mathcal{K}, \mathcal{V}$) into geneBF.

1: **procedure** IgeneBF($\mathbb{B}_{x,y}$, $\mathcal{K}$, $\mathcal{V}$)
2:     $h_k = \mathcal{H}(\mathcal{K}, Seed_1)$
3:     $h_v = \mathcal{H}(\mathcal{V}, Seed_2)$
4:     $C_{kv} = STRCAT(h_k, h_v)$ ▷ Concatenating both hash values
5:     $h_{kv} = \mathcal{H}(C_{kv}, Seed_3)$
6:     $i = h_{kv} \% x$
7:     $j = h_{kv} \% y$
8:     $l = h_{kv} \% bits$
9:     $d = \mathbb{B}_{i,j}$
10:    $p = 1 << pos$
11:    $\mathbb{B}_{i,j} = d \ OR \ p$
12: **end procedure**

3.2 Query operation

The initial steps of hashing the *Key* and *Value* are similar in both the operations. The *Key* and *Value* are hashed, and the hash values are concatenated. The concatenated value is again hashed and given as a query to the 2DBF. Algorithm 2 demonstrates the query operation of geneBF.

ALGORITHM 2 Query a *Key–Value*($\mathcal{K}, \mathcal{V}$) into geneBF.

1: **procedure** QgeneBF($\mathbb{B}_{x,y}$, $\mathcal{K}$, $\mathcal{V}$)
2:     $h_k = \mathcal{H}(\mathcal{K}, Seed_1)$
3:     $h_v = \mathcal{H}(\mathcal{V}, Seed_2)$
4:     $C_{kv} = STRCAT(h_k, h_v)$ ▷ Concatenating both hash values
5:     $h_{kv} = \mathcal{H}(C_{kv}, Seed_3)$
6:     $i = h_{kv} \% x$
7:     $j = h_{kv} \% y$
8:     $l = h_{kv} \% bits$
9:     $d = \mathbb{B}_{i,j}$
10:    $p = 1 << pos$
11:    $r = d \oplus p$
12:    $t = r \ AND \ p$
13:    **if** $(t = 0)$ *AND* $(d \neq 0))$ **then**
14:            **return** true
15:    **else**
16:            **return** false
17:    **end if**
18: **end procedure**

4. Data description

For the experimentation on geneBF we have considered two datasets, namely, *Key–Value* and Protein-coding Gene. In this section, some details regarding the dataset and the procedure for preprocessing the data are provided. For our experimentation, we have generated three different files, which are explained using an example:-

Let, $M = \{(1, i), (2, ii), (3, iii), (4, iv), (5, v), ..., (10, x)\}$

$Same = \{(1, i), (2, ii), (3, iii), (4, iv), (5, v), ..., (10, x)\}$ and $Same = M$

$Disjoint = \{(11, xi), (12, xii), (13, xiii), (14, xiv), (15, xv), ..., (20, xx)\}$

$Mixed = \{(5, v), (6, vi), (7, vii), (8, viii), (9, ix), ..., (15, xv)\}$

$(a, b) \in \{M, Same, Mixed, Disjoint\}$ where a is *Key* and b is *Value*.
Same set is equal to M and it is not generated separately. The mixed set is a set where some inputs are present in geneBF whereas the Disjoint set has inputs absent in the geneBF. M is the file that is inserted into the geneBF. The same, mixed, and disjoint sets are given as input for the query operation. Same set is given as input to determine the false negative probability of geneBF. Similarly, the mixed and disjoint set determines the false positive probability of geneBF. The mixed set and disjoint set are created to test geneBF by two different cases.

4.1 Key–value dataset

The *Key–Value* dataset consists of two values, namely, *Key* and *Value*. *Key* is a unique number assigned in incremental order. All *Keys* are of a fixed length of eight digits. The *Value* is a nodeID in the ego network [13]. The nodeID belongs to the network of Facebook or Twitter. Also, all *Values* are of a fixed length of 21 characters. In the ego network [13], both *Key* and *Value* are both nodeID. However, we were unable to determine the unique lines. Thus, to simplify, the *Keys* are assigned in incremental order, and *Values* are nodeID from ego network dataset. We generated two files following this procedure, namely, M and *Disjoint* set. Some *Key–Value* pairs from M and *Disjoint* set are combined to generate *Mixed* set.

We wanted to test the performance of geneBF against Big Data, i.e., big sized files. Therefore, we generated the M, Same, Mixed, and Disjoint sets of three different sizes, namely, 100 MB, 200 MB, and 300 MB. However, the exact 100 MB, 200 MB, and 300 MB files are not created; these files are marginally bigger. Table 1 presents the exact file size. In the following sections, we have used 100 MB, 200 MB, and 300 MB to refer to the file size

Table 1 Notation and file size of *Key–V alue* dataset.

Notation	Denotation	File size (MB)
M_{100}	M set of size 100 MB approximately	108.4
$Same_{100}$	M_{100}	108.4
$Mixed_{100}$	Mixed set of size 100 MB approximately	116.0
$Disjoint_{100}$	Disjoint set of size 100 MB approximately	110.8
M_{200}	M set of size 200 MB approximately	214.3
$Same_{200}$	M_{200}	214.3
$Mixed_{200}$	Mixed set of size 200 MB approximately	212.6
$Disjoint_{200}$	Disjoint set of size 200 MB approximately	212.0
M_{300}	M set of size 300 MB approximately	325.4
$Same_{300}$	M_{300}	325.4
$Mixed_{300}$	Mixed set of size 300 MB approximately	324.4
$Disjoint_{300}$	Disjoint set of size 300 MB approximately	323.5

for simplicity and more clear understanding. Table 1 lists the notation used in the following sections. An important point to note is $Same_n$ where $n \in \{100, 200, 300\}$ is another name of M_n where $n \in \{100, 200, 300\}$ and $Same_n$ is not constructed separately.

4.2 Protein-coding gene dataset

We have selected the protein-coding gene for experimentation on a real dataset. Protein-coding genes are DNA sequences that code for proteins. Every gene has the information to inform the biological cell to combine the building blocks to construct a certain protein (excluding some exceptions such as isoforms). The protein-coding genes help in translating the information present in the genomic sequence into valuable biological knowledge.

Similar to the *Key–V alue* dataset, we have generated two more datasets to determine the false positive probability, namely, Mixed and Disjoint set. Table 2 presents the notation to refer to the files used for the experimentation. M_{PG} is the protein-coding genes dataset [14] of size 3.3 MB. The file has two columns, the first column lists the genes and the second column lists the protein. A single row represents the connected gene and protein pair. $Same_{PG}$ is M_{PG}, which is given as a query to geneBF to evaluate the false

Table 2 Notation and file size of protein-coding gene dataset.

Notation	Denotation	File size (MB)
M_{PG}	Protein-coding gene dataset	3.3
$Same_{PG}$	M_{PG}	3.3
$Mixed_{PG}$	Mixed set	3.3
$Disjoint_{PG}$	Disjoint set	3.3

negative probability. Similarly, $Mixed_{PG}$ and $Disjoint_{PG}$ are the Mixed and Disjoint set of protein-coding genes, respectively. To generate the Disjoint set, we have taken the genes as numbers in incremental order and the protein column contains the proteins copied from M_{PG}. The length of the number taken as gene is same as the length of the gene in M_{PG} file. $Mixed_{PG}$ file is constructed by merging half of the content from M_{PG} and half from the $Disjoint_{PG}$. All the files have the same size. An important point to note is $Same_{PG}$ is another name of M_{PG} and $Same_{PG}$ is not constructed separately.

5. Experimental results and analysis

We have conducted extensive experimentation on geneBF. This experimentation proves that geneBF is faster and efficient for membership checking of paired data. We have conducted our experimentation using Linux operating system with 8GB RAM. Table 3 demonstrates the configuration of the experimental environment. We use the C programming language to conduct our experiments on geneBF.

5.1 Buffer size

In our experiment on geneBF, we have to read a file with a maximum size of 300 MB; hence, we first determine the buffer size for reading continuous blocks of data to speed up our insertion and query operation. We experimented to decide the buffer size using static and dynamic memory allocation. The file sizes considered for this experimentation are 100 MB and 200 MB. Figs. 1 and 2 illustrates the time taken by buffer of different memory sizes allocated statically and dynamically to read 100 MB and 200 MB files, respectively. It is already known that static memory allocation is faster than dynamic memory allocation. The system allocates static memory in the stack and dynamic memory in the heap. The stack (static memory

Table 3 Experimental environment.

Name	Features
CPU	Intel(R) Core(TM) i5-8500 CPU @ 3.00GHz
Number of CPU	6
L1d cache	32K
L1i cache	32K
L2 cache	256K
L3 cache	9216K
RAM	8 GB
Operating system	Ubuntu 18.04.5 LTS
HDD	500 GB
Programming language	C
Version of the language	gcc (Ubuntu 7.5.0-3ubuntu1 18.04) 7.5.0

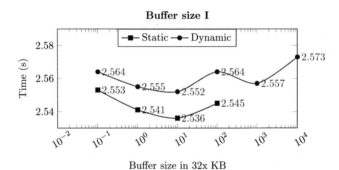

Buffer size in 32x KB

Fig. 1 Time taken to read 100 MB file by buffer of different memory size allocated statically and dynamically. Lower value is better. Buffer size is multiple of 32 KB.

allocation) is faster than the heap (dynamic memory allocation). The allocation and deallocation of memory in the stack is a minor task due to its access pattern. Whereas heap maintains complex bookkeeping for allocation and deallocation of memory which results in slower operations compared to static memory allocation. Knowing this fact, we still considered experimenting both static and dynamic memory allocated buffer because a bigger sized dynamic allocated buffer may take less time for reading the whole files. The aim of the experiment was to find the time taken by geneBF for executing the insertion operation while varying the buffer size.

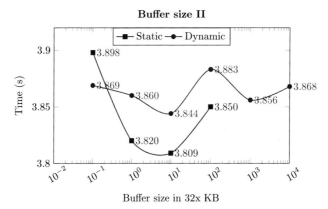

Fig. 2 Time taken to read 200 MB file by buffer of different memory size allocated statically and dynamically. Lower value is better. Buffer size is multiple of 32 KB.

Instead of taking a fixed buffer size, we have considered the number of lines read from the files to reduce the complexity of the experiment. Thus, the buffer size is a multiple of 32. The number of characters in a line, including the length of *Key* and *Value*, space, and a newline is 32. An extra fact we came to learn while conducting the experiment is in Windows OS, the number of characters for a newline is 1, but for Linux OS, it is 2. When the static allocation buffer size is 32 MB, then the core is dumped by the system. Whereas the dynamic allocation is unable to allocate memory when the buffer size is 3.2 GB. A 320 KB buffer in both static and dynamic allocation takes the least time for insertion operation compared to other buffer sizes. An increase in buffer size decreases the insertion time; however, when the buffer size is higher than 320 KB, the insertion time increases. The reason is the cache is unable to accommodate the whole buffer in its memory. Figs. 1 and 2 clearly illustrate that the buffer size 320 KB has the least time in both the 100 MB and 200 MB files. Moreover, the static allocated buffer was faster than the dynamically allocated buffer. Therefore, in further experiments of geneBF, we have considered 320 KB buffer size.

5.2 Insertion operation

The experiment aims to find the time taken by geneBF for inserting *Key–Value* inputs and observing whether geneBF is capable of handling big sized files. Fig. 3 presents the result of the experiment. The input files are M_{100}, M_{200}, and M_{300} which are inserted separately into geneBF. It is clear from Fig. 3 that geneBF is faster because inserting a 300 MB file takes less than 6 s.

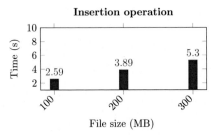

Fig. 3 Comparison of time taken by geneBF to insert *Key–V alue* from files having different sizes. Lower value is better.

Furthermore, the increment of file size by 100 MB, increases the time taken by geneBF by 1.455 s on average. Therefore, geneBF is faster and efficient for Big Data applications.

5.3 Query operation

This experiment has many aims such as: to determine (i) time taken by geneBF for executing the query operation on big sized files, (ii) false negative probability of geneBF, and (iii) false positive probability of geneBF. The experiment about false positive and false negative are explained in the Section 5.4. The input file sizes that are queried to geneBF are 100 MB, 200 MB, and 300 MB (approximately). Fig. 3 represents the time taken by the geneBF for performing query operations. After the insertion of the inputs from M_{100}, the input of *Same*$_{100}$, *Mixed*$_{100}$, and *Disjoint*$_{100}$ are given as queries to geneBF separately. The information regarding these files is explained in Section 4. The time taken by geneBF for the execution of the query operation is recorded. Similarly, M_{200} and M_{300} are inserted into the geneBF separately, and their respective Same, Mixed, and Disjoint set is given as queries. The geneBF has a fast query speed; 100 MB files take less than 2 s, 200 MB files take less than 3 s, and 300 MB files take approximately 4 s. All input files have some difference in size; hence, the time taken varies by some fraction of the second. Another important point to notice is the increment of file size by 100 MB, increases the time taken by geneBF for query operation (considered Mixed set) by 1.26 s on average as shown in Fig. 4.

5.4 False negative and false positive

After inserting M_n, the *Same*$_n$, *Mixed*$_n$, and *Disjoint*$_n$ where $n \in \{100, 200, 300\}$ sets are given as queries to determine the false negative and false positive probability.

Query operation

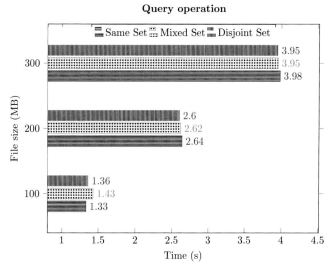

Fig. 4 Comparison of time taken by geneBF to query *Key–V alue* from files having different sizes. Lower value is better.

5.4.1 False negative

The Same set is given as queries to determine the false negative probability of geneBF. Fig. 5 illustrates the false negative probability of geneBF. In all three cases, i.e., the query of $Same_{100}$, $Same_{200}$ and $Same_{300}$ files gives zero false negative response. The figure is used to emphasize that geneBF has zero false negative probability because false negative is more harmful than false positive. False positives can be rectified but rectifying false negatives is impossible. Thus, the Bloom Filter with zero false negative probability is more efficient.

5.4.2 False positive

Mixed and Disjoint sets are given as queries to determine the false positive probability of geneBF. Fig. 5 illustrates the false positive probability of geneBF. The Mixed sets have more false positive probability compared to their respective Disjoint set because original size of the Mixed set is more compared to the original size of their respective Disjoint set. For instance, the original size of the Mixed set of 100 MB is 116 MB, and the Disjoint set of 100 MB is 110.8 MB. The size of other files is mentioned in Table 1. Furthermore, an increase in the file size increases the false positive probability. However, an important point to notice is giving a 300 MB sized file as queries have less than 0.0025 false positive probability.

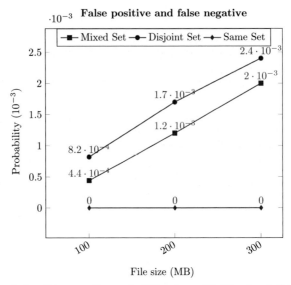

Fig. 5 Same set gives false negative probability of geneBF. Mixed and disjoint set gives false positive probability of geneBF. Lower value is better.

5.5 Operation speed

Fig. 6 illustrates the operation speed of insertion and query operation of geneBF. Operation speed provides the efficiency of geneBF. It is calculated as Megabyte processed per second. The query operation speed is calculated using Mixed set. From Fig. 6, we can notice the query operation has higher efficiency than insertion operation. Moreover, the query operation speed is approximately constant for all three files, whereas an increase in file size increases the insertion operation speed.

5.6 Protein-coding gene

In this chapter, the real dataset considered is protein-coding gene. The details of the dataset construction and other details are provided in Section 4. The length of a single row in all files (i.e., M_{PG}, $Same_{PG}$, $Mixed_{PG}$, and $Disjoint_{PG}$) is 32 characters, including space and newline. Thus, the buffer size taken in the experimentation based on this dataset is similar to the previous Key–$Value$ dataset, i.e., 320 KB.

The experiment aims to determine whether geneBF is working accurately and efficiently on the real dataset. Fig. 7 illustrates the time taken by the insertion and query operations of geneBF giving protein-coding genes as a dataset. The protein-coding gene dataset, i.e., M_{PG} of size

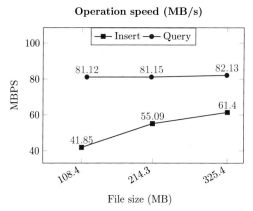

Fig. 6 The operation speed, i.e., MB/s. The mixed set is used for query. Higher value is better.

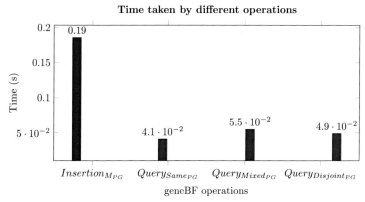

Fig. 7 Time taken by the insertion and query operations giving protein-coding gene as input. Lowest is better. File size: 3.3 MB

3.3 MB, is inserted into geneBF, and the time taken by the operation is recorded. After inserting M_{PG}; $Same_{PG}$, $Mixed_{PG}$ and $Disjoint_{PG}$ are given as queries to geneBF separately. Fig. 7 illustrates the time taken by geneBF for the execution of various operations when protein-coding genes are given as input. $Same_{PG}$ has such big time difference compared to $Mixed_{PG}$ and $Disjoint_{PG}$ is due to locality of reference in cache. Another important point to note is geneBF is not biased toward any specific elements or data. In the experimentation, insertion of protein–coding gene took nearly 0.2 s, whereas query operation took less than 0.06 s. Therefore, geneBF is qualified to handle both Big data and DNA datasets.

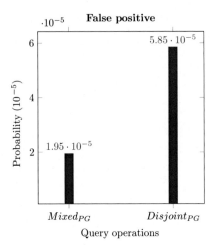

Fig. 8 False Positive Probability of geneBF. Lowest is better. File size: 3.3 MB.

This experiment aims to calculate the false negative and false positive probability of geneBF after giving protein-coding genes as input. After inserting M_{PG} into geneBF, the $Same_{PG}$ is given as queries. geneBF gave zero false negative responses. Thus, the false negative probability of geneBF is again determined as zero. Similarly, $Mixed_{PG}$ and $Disjoint_{PG}$ are given as queries to geneBF to find the false positive probability. Fig. 8 illustrates the false positive probability of geneBF when the Protein-coding Gene dataset is given as queries. The false positive probability is coming in the range of 10^{-5}. Thus, the false positive probability of geneBF is very low compared to a megabyte file.

6. Discussion

The Big graph is more complicated compared to Big data due to its complex structure. The increase in focus toward Bloom Filter to filter and membership check the Big data to reduce the data processing complexity of an application leads to introducing Bloom Filter in Big graph. However, the paired nature of the data is a hindrance for implementing Bloom Filter in Big graph because a single Bloom Filter only takes single input. Thus, geneBF is the first step toward handling Big graph data. geneBF executes arithmetic operations to increase the operational speed. But, its only demerit is it has fixed dimensions, and the dimensions need to be prime numbers to have a low false positive probability. One significant advantage of geneBF is its zero false negative probability. False positive responses are later fixed by the application; however, it is difficult to determine the false negative response, so rectifying is even more impossible. Thus, zero false negative probability and high operational speed is making geneBF ideal for Big graph applications.

7. Conclusion

We proposed a simple and efficient Bloom Filter, called geneBF, for processing paired data. It takes a constant time for insertion and query operation. We experimented with big sized files which are in the range of megabytes. Our experimental result shows that geneBF is a faster Bloom Filter that takes 5.3 s for insertion and approximately 4 s for query operation for a 300 MB file. Through experimentation, we also determined that geneBF has zero false negative probability. Furthermore, the false positive probability for processing a 300 MB file is 0.0024. We further experimented on geneBF with a real dataset which is the protein-coding genes. geneBF again gave zero false negative probability and 0.0000585 false positive probability. This novel Bloom Filter is ideal for paired data. In the future, we will perform further experimentation on other Big graphs, for instance, social graphs.

References

[1] D.K. Singh, R. Patgiri, Big graph: tools, techniques, issues, challenges and future directions, in: Sixth International Conference on Advances in Computing and Information Technology (ACITY 2016), 2016, pp. 119–128.

[2] R. Patgiri, S. Nayak, S.K. Borgohain, Hunting the pertinency of bloom filter in computer networking and beyond: a survey, J. Comput. Netw. Commun. 2019 (2019) 1–10, https://doi.org/10.1155/2019/2712417.

[3] A. Singh, S. Garg, S. Batra, N. Kumar, J.J.P.C. Rodrigues, Bloom filter based optimization scheme for massive data handling in IoT environment, Future Gener. Comput. Syst. 82 (2018) 440–449.

[4] S. Nayak, R. Patgiri, A. Borah, A review on impact of Bloom filter on named data networking: the future internet architecture, Computer Networks, 196, Elsevier B. V., 2020, pp. 1–24. https://doi.org/10.1016/j.comnet.2021.108232.

[5] R. Patgiri, S. Nayak, S. Borgohain, Preventing DDoS using Bloom filter: a survey, ICST Trans. Scalable Inform. Syst. 5 (19) (2018) 155865, https://doi.org/10.4108/eai.19-6-2018.155865.

[6] S. Geravand, M. Ahmadi, Bloom filter applications in network security: a state-of-the-art survey, Comput. Netw. 57 (18) (2013) 4047–4064.

[7] R. Patgiri, S. Nayak, N.B. Muppalaneni, Is Bloom filter a bad choice for security and privacy? in: 2021 International Conference on Information Networking (ICOIN), 2021, pp. 648–653, https://doi.org/10.1109/ICOIN50884.2021.9333950.

[8] S. Nayak, R. Patgiri, A review on role of Bloom filter on DNA assembly, IEEE Access 7 (2019) 66939–66954, https://doi.org/10.1109/ACCESS.2019.2910180.

[9] G. Benoit, C. Lemaitre, D. Lavenier, E. Drezen, T. Dayris, R. Uricaru, G. Rizk, Reference-free compression of high throughput sequencing data with a probabilistic de Bruijn graph, BMC Bioinformatics 16 (1) (2015) 1–14.

[10] R. Patgiri, S. Nayak, S.K. Borgohain, rDBF: a r-dimensional Bloom filter for massive scale membership query, J. Netw. Comput. Appl. 136 (2019) 100–113, https://doi.org/10.1016/j.jnca.2019.03.004.

[11] Burton Howard Bloom, Space/time trade-offs in hash coding with allowable errors, Commun. ACM 13 (7) (1970) 422–426.

[12] A. Appleby, Murmur Hashing, 2010. Retrieved from: https://sites.google.com/site/murmurhash/.

[13] J.J. McAuley, J. Leskovec, Learning to discover social circles in ego networks, in: NIPS, vol. 2012, Citeseer, 2012, pp. 548–556.

[14] M. Zitnik, R. Sosič, S. Maheshwari, J. Leskovec, BioSNAP Datasets: Stanford Biomedical Network Dataset Collection, 2018. (August). http://snap.stanford.edu/biodata.

About the authors

Sabuzima Nayak is a research scholar in the Department of Computer Science & Engineering, National Institute of Technology Silchar. She has published several journal articles, conferences papers and book chapters. Her research interest is Bloom Filter, Computer Networking, and Bioinformatics.

Dr. Ripon Patgiri has received his Bachelor Degree from Institution of Electronics and Telecommunication Engineers, New Delhi in 2009. He has received his M.Tech. degree from Indian Institute of Technology Guwahati in 2012. He has received his Doctor of Philosophy from National Institute of Technology Silchar in 2019. After M.Tech. degree, he has joined as Assistant Professor at the Department of Computer Science & Engineering, National Institute of Technology Silchar in 2013. He has published numerous papers in reputed journals, conferences, and books. His research interests include distributed systems, file systems, Hadoop and MapReduce, big data, bloom filter, storage systems, and data-intensive computing. He is a senior member of IEEE. He is a member of ACM and EAI. He is a lifetime member of ACCS, India. Also, he is an associate member of IETE. He was General Chair of 6th International Conference on Advanced Computing, Networking, and Informatics (ICACNI 2018, http://www.icacni.com) and International Conference on Big Data, Machine Learning and Applications (BigDML 2019, http://bigdml.nits.ac.in). He is Organizing Chair of 25th International Symposium on Frontiers of Research in Speech and Music (*FRSM 2020, http://frsm2020.nits.ac.in*) and International Conference on Modeling, Simulations and Applications (CoMSO 2020, http://comso. nits.ac.in). He is convenor, Organizing Chair and Program Chair of 26th annual International Conference on Advanced Computing and Communications (ADCOM 2020). He is guest editor in special issue "Big Data: Exascale computation and beyond" of EAI Endorsed Transactions on Scalable Information Systems. He is also an editor in a multi-authored book, title "*Health Informatics: A Computational Perspective in Healthcare*", in the book series of *Studies in Computational Intelligence*, Springer. Also, he is writing a monograph book, titled "*Bloom Filter: A Data Structure for Computer Networking, Big Data, Cloud Computing, Internet of Things, Bioinformatics and Beyond*", Elsevier.

CHAPTER SIX

Processing large graphs with an alternative representation

Ravi Kishore Devarapalli and Anupam Biswas
Department of Computer Science and Engineering, National Institute of Technology Silchar, Silchar, Assam, India

Contents

Abstract

This chapter investigates the processing of large graphs with a novel graph representation scheme, the centered subgraph data matrix (CESDAM). The graph is represented in multiple CESDAMs, each containing subgraphs of the graph. Processing time and memory requirement are the two major issues while dealing with the processing of large graphs. The representation of graph is crucial for handling both these issues. The efficiency of CESDAM scheme in terms of time requirement is analyzed by incorporating the breadth-first searching, which is the backbone of most graph processing algorithms. The analysis shows, breadth-first searching on CESDAM requires time $O(n \times B \times m^2)$, where n is the number of nodes in the graph, B is the branching factor of the meta-graph prepared on top of CESDAM ids, and m is the size of each CESDAM. The widely used adjacency matrix is becoming inefficient with enormously growing sizes of modern-day graphs, which requires very large amount of space for such graphs. Representation of graph in parts with CESDAM reduces the memory requirement significantly in contrast to adjacency matrix, which requires space $O(k \times m^2)$ only, where k is the number of CESDAMs.

Advances in Computers, Volume 128
ISSN 0065-2458
https://doi.org/10.1016/bs.adcom.2021.10.001

1. Introduction

Graphical modeling of relationships among different objects that are present within the data is common to several application domains [1–4], such as social networks in Sociology [5], protein-interaction networks in Biology [6, 7], food supply chain networks in Ecology [8, 9], chemical compounds in Chemistry [10, 11], attribute graphs in image processing [12] and so on. Numerous techniques have been developed for representing such graphs [13], which include edge list, adjacency matrix, adjacency list etc. The adjacency matrix representation is a very natural and widely used approach for representing graphs [14]. However, with enormously growing sizes of modern-day real world graphs, matrix-based approach such as adjacency matrix representation is becoming inefficient from the perspective of both memory requirement and processing time requirement. The adjacency matrix also requires large amount of memory even when the graph is sparse, where most of the matrix entries are zeros. Often, matrix with mostly zero entries is referred as sparse matrix. Various formats such as compressed sparse row (CSR), compressed sparse column (CSC), and block sparse row (BSR) are evolved to reduce memory requirement of sparse matrix [15–17]. However, these sparse matrix storage formats are beneficial only when the matrix is significantly sparse, otherwise they cost more for their different data access patterns [18, 19]. To overcome these issues, a matrix-based scheme, the centered subgraph data matrix (CESDAM) is proposed for representing large graphs.

The proposal of CESDAM scheme is inspired by the notion of data partitioning. Resource limitation is the key constraint in dealing with both memory requirement and processing time requirement while dealing with large graphs [3, 20]. Often, large data that are larger than the memory is processed with external memory algorithms [21, 22]. Large data are partitioned into smaller parts to fit them into the memory. External memory algorithms process those smaller parts of the data individually and output final result by accumulating all. Primary hurdle with graph in utilizing these algorithms is partitioning of graph requires processing of entire graph [23]. The preparation of sparse matrix formats also requires processing of entire adjacency matrix. The adjacency matrix of large graph cannot be loaded into the memory in its entirety. Thus, it is not possible to process a large graph with adjacency matrix representation within limited memory. Although, adjacency matrix can be processed in parts to prepare sparse matrix formats, such processing is not suitable for external memory algorithms specially when the graph is not too sparse. Simply division of adjacency matrix is unfruitful for external memory algorithms since it will result in processing of disconnected portions of graph, which have different meaning in the undivided graph. The parallel graph processing approaches such as GPU processing [24, 25], Pregel [20], Seraph [26], etc. are also confronting the same problem, since they also require partitioning of large graph. Therefore, effective mechanism is necessary to divide the graphs for processing those graphs with both external memory algorithms and parallel processing algorithms; and also to overcome memory limitation. Moreover, there has huge potential to improve the performance of parallel graph processing systems with topology-aware prepartitioning of graphs as raised in Ref. [20]. The prepartitioning of input graph based on topology will reduce the intermachine message passing when different parts of the graph will be processed in multiple processing units.

In CESDAM scheme, graph $G(V, E)$ is divided into k small overlapping subgraphs, each having $m_1, m_2, ..., m_k$ number of nodes. Each subgraph $g_u(u, V_u, E_u)$ is represented as a special kind of *Matrix* called CESDAM. The subgraph g_u is represented in the CESDAM has an identifier that is the central node u of the subgraph g_u. In general, entire graph is represented as a single entity, i.e., as an adjacency matrix or as an edge list, etc. For these representations, entire graph is the representing element. On the contrary, for the CESDAM representation, subgraphs of the graph are the representing elements. Thus, with the proposed approach a single large graph is represented with multiple CESDAMs. Moreover, the sizes, i.e., number of rows and columns of CESDAMs are dependent on connectivity pattern among the nodes in graph G, whereas the size of adjacency matrix is dependent on number of nodes.

This work is focused primarily on theoretical analysis of large graph processing with CESDAM representation from the perspective of breadth-first searching. The space requirement is also analyzed theoretically in contrast to adjacency matrix. Main contributions of the chapter are summarized as follows:

- The proposed CESDAM representation scheme is illustrated with suitable examples (Section 2).
- The breadth-first searching and breadth-first-tree (BFT) generation with adjacency matrix representation is discussed. The branching factor of graph with respect to BFT is briefed (Section 3).
- The processing of graph is analyzed in perspective of breadth-first searching. In this context, an algorithm is designed for breadth-first-searching

with CESDAM scheme (Section 4.2). To deal with the links among CESDAMs and for accumulating interpretations of multiple CESDAMs correctly during processing of graph meta-graph is introduced (Section 4.1).

- Theoretically proved that with meta-graph, breadth-first searching accumulates correctly the interpretation of CESDAMs during processing of graph (Section 4.3).
- Space complexity of CESDAM scheme is analyzed theoretically and showed that CESDAM requires less space than the widely used adjacency matrix (Section 5). In addition, derived several important properties with respect to the number of CESDAMs required for effective graph representation with CESDAM scheme.

2. The CESDAM representation

The essence of graph representation is to avail the information about graph as much as possible in an efficient way. Nodes and associated edges are the elementary information utilized during graph processing. This notion can be viewed as grouping of nodes and edges. As many nodes cover in the representation of graph through grouping, it becomes more suitable for processing. For instance, the adjacency matrix is more efficient for processing than edge list. Adjacency matrix covers all neighbors of a node in representation whereas edge list covers only one edges, i.e., one neighbor. The notion of grouping of nodes in graph representation can be further broadened from the perceptive of graph processing. Searching of nodes or edges is vital in sequential processing. Breadth-first-search (BFS) and depth-first-search (DFS) are the two widely used approaches in graph. Generally, BFS is preferred for sequential processing. Most of the BFS processing requires processing the neighbor nodes of a node. Any node along with associated connections with neighbors is simply a personal network of the node. Such personal network is referred as ego network in social network terminology [27]. Ego network is nothing but a graph property that is determined by the central node. We incorporate the notion of both ego network and grouping of nodes in the CESDAM representation scheme.

The CESDAM scheme represents large graph by dividing it into smaller subgraphs. Any large graph $G(V, E)$ is divided into smaller subgraphs $g(u, V_u, E_u)$, where $u \in V$ is the center of the subgraph. The division of graph is done based on the social network property called ego network [27, 28].

The subgraph identified around any specific node u constituting its neighbors is referred as ego network with u as a central node or ego. Depending on the distance from the central node u, defines different levels of ego networks such as level 1.0 ego network, level 1.5 ego network, level 2.0 ego networks and so on. The CESDAM scheme considers level 2.5 ego network. That is, each subgraph $g(u, V_u, E_u)$ of graph G represents a level 2.5 ego network. Various levels of ego network in G with respect to central node u is defined as follows:

Definition 1 Level 1.0 ego network

In graph $G(V, E)$, the level 1.0 ego network is defined as subgraph $g(u, V_u, E_u)$ with respect to ego node u such that if $\exists v \in V$ then $v \in V_u$ iff $\exists(u, v) \in E$ and $\forall(x, y) \in E_u$ satisfies following conditions:
1. $x = u$
2. $y \in V_u$

Definition 2 Level 1.5 ego network

In graph $G(V, E)$, the level 1.5 ego network is defined as subgraph $g(u, V_u, E_u)$ with respect to ego node u such that if $\exists v \in V$ then $v \in V_u$ iff $\exists(u, v) \in E$ and $\forall(x, y) \in E_u$ satisfies following conditions:
1. $(x = u)$ or $(x \in V_u)$
2. $y \in V_u$

Definition 3 Level 2.0 ego network

In graph $G(V, E)$, the level 2.0 ego network is defined as subgraph $g(u, V_u, E_u)$ with respect to ego node u such that if $\exists v \in V$ then $v \in V_u$ iff $\exists(u, v) \in E$ and satisfies following conditions:
1. $\forall(x, y) \in E_u \Rightarrow \{(x = u)$ or $(x \in V_u)\}$ and $y \in V_u$
2. $V'_u = \bigcup_{\forall y \in V_u}(V_u, V_y)$ and $E'_u = \bigcup_{\forall y \in V_u}(E_u, E_y)$, where $g(y, V_y, E_y)$ is the Level 1.0 ego network
3. $V_u = V'_u$ and $E_u = E'_u$

Definition 4 Level 2.5 ego network

In graph $G(V, E)$, the level 2.5 ego network is defined as subgraph $g(u, V_u, E_u)$ with respect to ego node u such that if $\exists v \in V$ then $v \in V_u$ iff $\exists(u, v) \in E$ and satisfies following three conditions:
1. $\forall(x, y) \in E_u \Rightarrow \{(x = u)$ or $(x \in V_u)\}$ and $y \in V_u$
2. $V'_u = \bigcup_{\forall y \in V_u}(V_u, V_y)$ and $E'_u = \bigcup_{\forall y \in V_u}(E_u, E_y)$, where $g(y, V_y, E_y)$ is the Level 1.5 ego network
3. $V_u = V'_u$ and $E_u = E'_u$

In ego network, we refer central node as level 0 node, since distance from the central node is 0. Similarly, the nodes having minimum distance from central node 1 is referred as level 1 nodes and so on. Now, we define external nodes with respect to level 2 ego network.

Definition 5 Level 1 external node

Given a level 2.0 ego network $g_u(V_u, E_u)$, then the neighbor nodes of level 1 nodes that have no edge with level 0 node are referred as level 1 external nodes.

Definition 6 Level 2 external node

Given a level 2.0 ego network $g_u(V_u, E_u)$, then the neighbor nodes of level 2 nodes that have no edge with level 1 nodes are referred as level 2 external nodes.

A generic model of CESDAM for representing any subgraph is shown in Fig. 1. The data structure used in CESDAM representation is a matrix $A = [\ ]_{(m+2)\times d}$, where m is the number of nodes included in subgraph and $d = max(|row\ a_i|)$. The value of d depends on the number of neighbors of nodes included in subgraph, the number of level 1 external nodes, and the number of level 2 external nodes. The CESDAM data model contains three parts as follows:

External level 1: All level 1 external nodes are included into this part. It covers first row of the matrix excluding first column.

External level 2: All level 2 external nodes are included into this part. It covers second row of the matrix excluding first column.

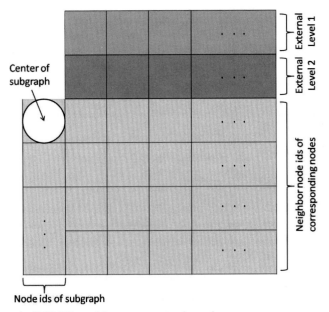

Fig. 1 Generic CESDAM model to represent subgraphs.

Main subgraph: All nodes associated with the subgraph are included into this part. It covers row 3 onwards rest of the matrix. First element in row 3, i.e., $a_{3,1}$ is the center of the subgraph. With reference to the center node, level 2.5 ego network is constructed. Therefore, the center node is considered as an identifier for the subgraph. Once center node is known, corresponding CESDAM for the subgraph can be constructed easily. For all nodes included in the subgraph (excluding level 2 external nodes) there is an entry in the first column (entry in one of $a_{i,1}$ where $i > 2$). If node u is entered in row i (i.e., in $a_{i,\ 1}$) then all neighbor nodes node u are entered in row i in successive columns (i.e., entered in one of $a_{i,j}$ where $j > 1$).

Since, the entire graph is viewed as multiple overlapped subgraphs representing each subgraph as a level 2.5 ego network. The entire graph is represented with multiple CESDAMs, where each CESDAM is considered as a subgraph. An example graph is shown in Fig. 2 indicating levels of nodes in probable subgraph having center at 1. The CESDAMs for the example graph is shown in Fig. 3.

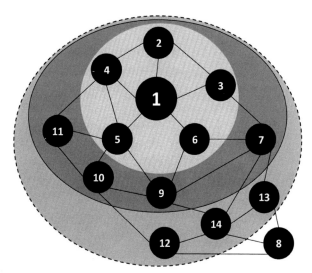

Fig. 2 Example graph showing various levels in the subgraph centered at node 1. Nodes inside inner white circle other than 1 are the level 1 nodes. Nodes outside inner *white circle* and inside *dark boundary* are level 1 external nodes or level 2 nodes. Nodes inside the *dark boundary* are strongly part of subgraph. Nodes outside *dark boundary* and inside *dotted boundary* are not part of subgraph that are level 2 external nodes.

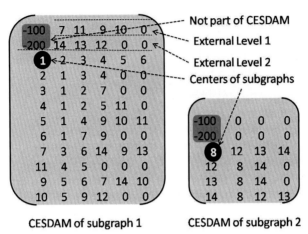

CESDAM of subgraph 1   CESDAM of subgraph 2

Fig. 3 CESDAMs prepared from the example graph in Fig. 2. Note that the CESDAM representation requires only $O(12 \times 6 + 6 \times 4)$ space, which is much smaller than that of $O(14 \times 14)$ and $O(14 \times 27)$ space required by the matrix-based approaches such as adjacency matrix and incidence matrix, respectively.

3. BFS in adjacency matrix

BFS is a very simple graph searching algorithm. The notion of BFS has been utilized widely in many graph related algorithms. Given a graph $G(V, E)$ and a prespecified start node s, BFS systematically explores the edges of G to discover every node that is reachable from s. Basically, the BFS does two tasks in the course of searching. Firstly, it computes the shortest distance from s to all reachable nodes. Second, it also produces a BFT with root s that covers all reachable nodes from s. The algorithm discovers all nodes at distance d from s before discovering any node at distance $d + 1$. To keep track of progress, BFS uses an indicator *visited* for each node. The indicator tells whether the node and all its neighbors are discovered or not. Initially, all nodes start with value 0 and become 1 after completion of discovery of all neighbor nodes. We say node is visited only if all its neighbors are discovered, otherwise not visited. Thus, at any time in-between starting and completion of visiting all nodes we have two frontiers: the nodes that are not visited and the nodes that are visited. Besides, these two frontiers there is another intermediate frontier, i.e., the nodes that are discovered visited partially, i.e., all neighbors of the node are not discovered.

3.1 Time complexity analysis

Algorithm 1 assumes that the input graph $G(V, E)$ is represented using adjacency matrix. It maintains several additional data structures with each node in the graph. The indicator for each node $u \in V$ is stored in variable $visited[u]$, the predecessor of u is stored in the variable $p[u]$ and the distance from start node s to u is stored in variable $d[u]$. If u has no predecessor (for instance, if $u = s$ or u has not been discovered), then $p[u] = NIL$. A queue Q is also used in the algorithm to maintain the intermediate frontier nodes. The time complexity of the BFS algorithm as analyzed as follows. After initialization of $d[s] = 0$, non of the nodes $u \in V - \{s\}$ is ever changed to 0, and thus the line 15 ensures that each node is enqueued at most once, and hence dequeued at most once. The operations of enqueuing and dequeuing take O(1) time, so the total time devoted to queue operations is O(n), where $n = |V|$ is the total

ALGORITHM 1 BFS(G, s).
```
1:  Input: G, s
2:  Output: d, p
3:  for all u ∈ V [G] do
4:      visited[u] ← 0 // here, 0 indicates not visited.
5:      d[u] ←∞ //distance from s to all nodes is initialized.
6:      p[u] ← NIL //parent of all nodes is initialized.
7:  end for
8:  d[s] ← 0
9:  Q ← create empty queue.
10: ENQUEUE(Q, s)
11: while Q is not empty do
12:     u ← DEQUEUE(Q)
13:     N_u ← getNeighbor(G, u)
14:     for all v ∈ N_u do
15:         if d[v] = ∞ then
16:             d[v] = d[u] + 1
17:             p[v] = u
18:             ENQUEUE(Q, v)
19:         end if
20:     end for
21:     visited[u] ← 1 // here, 1 indicates visited.
22: end while
23: return d, p
```

number of nodes. After dequeue operation in line 13, adjacency matrix is scanned for discovering neighbors. The neighbor searching in adjacency matrix requires scanning of a row so it requires $\Theta(n)$. The neighbor searching has to be done for all of the n nodes so total time spend on scanning adjacency matrix is $\Theta(n^2)$. The overhead for initialization is $O(n)$, and total running time is $O(n) + \Theta(n^2)$. Hence, the run time complexity experienced by the BFS algorithm with adjacency matrix is $\Theta(n^2)$. In the case of adjacency list, the time complexity of BFS would be $O(n + e)$, where $e = |E|$ is the total number of edges. Note that the number of edges of any graph varies from 0 to n^2, assuming no self-loops and no multiedge between two nodes. Thus, worst case time complexity of BFS algorithm with adjacency list would be $O(n + n^2)$, i.e., $O(n^2)$.

3.2 BFT

The BFS algorithm constructs a BFT in the course of node discovery. Initially, the BFT contains only its root, which is the start node s. Whenever a node v is discovered during the scanning of neighbors of an already discovered node u but not visited, the edge (u, v) connecting nodes u and v are added to the tree. In this case, u is referred as the predecessor or parent of v in the BFT. Since a node is discovered at most once so it has at most one parent. Ancestor and descendant relationships in the BFT are defined relative to the root s. If any node u is on a path in the tree from the root s to node v, then u is an ancestor of v and v is a descendant of u. Note that, even if the graph is undirected, the ancestry and descendent relationships from root node to leaf node impose pseudo-directed edges among the nodes in BFT.

3.3 Branching factor

The branching factor is property of tree that defines the number of children at each node. The tree data structure is categorized broadly in two categories: the tree having same maximal number of children, i.e., the tree is bound by the maximal number of children and the tree with no prespecified number of children. For the first category, number of children at each level starting from root node s towards the leaf nodes grows uniformly. For instance, number of child grows in the binary tree by factor of two. However, for the second category, the growth of number of children is not uniform since each node can have different number of nodes. Thus, branching factor is defined as average number of children per node in this case. The BFT generated in BFS algorithm is of second kind since number of children depends on start node s and the graph G. Branching factor of BFT obtained with respect to any start node s is computed as follows. We compute branching factors of nodes on each level, i.e., depth with respect to s and average of all gives the final branching factor. The branching factor of nodes of BFT that are at depth d is evaluated as below:

$$b_d(BFT, s) = \frac{T_c}{T_d} \qquad (1)$$

where, T_c is the total number of children of nodes at depth d in BFT and T_d is the total number of nonleaf nodes at depth d in BFT. If D is the depth of BFT having root s then the branching factor of BFT with respect to s is computed as follows:

$$b(BFT, s) = \frac{\sum_{d=0}^{D} b_d(BFT, s)}{D} \qquad (2)$$

For different start node s, BFS generates different BFT with different number of nonleaf nodes result in different branching factors on the same graph G. Therefore, we define branching factor of graph with respect to BFT obtained in BFS for any start node s. Similar process is followed as in case of BFT to define branching factor of graph. The depth of any node in the graph G is considered in reference to BFT obtained for start node s. The branching factors of nodes on each depth in reference to the BFT and average of all gives the final branching factor. The branching factor of nodes of graph G that are at depth d is evaluated as below:

$$b_d(G, s) = \frac{T_n}{T_d} \qquad (3)$$

where, T_n is the total number of neighbors of nodes at depth d in G and T_d is the total number of nonleaf nodes at depth d in BFT. If D is the depth of BFT having root s then the branching factor of graph G with respect to s is computed as follows:

$$b(G, s) = \frac{\sum_{d=0}^{D} b_d(G, s)}{D} \qquad (4)$$

To work with large graphs that are too large to store, time complexity of BFS is defined in terms of edge traversal required for the nodes at depth x as $O(b^{(x+1)})$, where b is the branching factor. Generally, branching factor of graph is defined as average number of neighbors. However, in practical, BFS will definitely not experience branching factor of G to reach nodes at depth $x < D$, instead it will experience branching factor covering nodes up to depth x. Thus, average number of neighbors cannot serve as branching factor. On the contrary, the branching factor defined in Eq. (4) is depends on the BFS. It gives privilege to compute branching factor of graph covering nodes up to any depth $x < D$, simply by replacing D with x in Eq. (4). Thus, edge traversal required for the nodes at depth x can give more precise value with Eq. (4).

4. Graph processing with CESDAM

In CESDAM representation scheme, graph is divided into multiple overlapping subgraphs. Each subgraph is represented in one CESDAM. Therefore, all the CESDAMs for subgraphs as a whole represent the entire graph. Thus, the processing of graph with CESDAM scheme should accumulate all the subgraphs represented in multiple CESDAMs. In addition, correctness of graph processing with CESDAM should also be ensured. Therefore, graph processing with CESDAM have to address two objectives: accumulation of subgraphs and correctness of the accumulation. Both these objectives are addressed in this section with detailed theoretical foundation. Moreover, correctness of accumulation of subgraphs represented in CESDAMs is also analyzed and proved that the processing of graph with the CESDAM results same as in the undivided graph.

4.1 Meta-graph preparation

The accumulation of subgraphs requires to draw links among CESDAMs. Since subgraphs represented in CESDAMs are overlapped, some of the nodes of one CESDAM will common with other CESDAMs. These shared nodes are the backbone for drawing links among CESDAMs. Through these links, dependency among CESDAMs is realized. If CESDAM A has shared nodes with CESDAM B, we say CESDAM A is dependent on CESDAM B for accumulation. Any CESDAM can have multiple shared nodes with different CESDAMs. Thus, a CESDAM can have multiple dependency on other CESDAMs. The CESDAMs is a single entity that represents a subgraph identifiable with its center, i.e., the id of CESDAM. Therefore, we interpret each CESDAM as node with their ids and dependency among CESDAMs as edges. Thus, we have a meta-graph on top of CESDAM ids expressing the dependency among CESDAMs.

Definition 7 Meta-graph

Given a graph $G(V, E)$ and V_M is the set of all CESDAM ids, i.e., centers u of all subgraphs $g_u(u, V_u, E_u)$, the meta-graph $G_M(V_M, E_M)$ is a graph such that for any pair of u, $v \in V_M$, $(u, v) \in E_M$ iff there exists $x \in V_u$ and $y \in V_v$ for subgraphs $g_u(u, V_u, E_u)$ and $g_v(v, V_v, E_v)$, respectively.

The meta-graph accumulates all the subgraphs represented in CESDAMs. The meta-graph can be represented as edge list or adjacency matrix or even meta-CESDAMs. We presume that the number of links among CESDAMs would be less so meta graph will be a sparse graph. Therefore, it would be better to consider edge list instead of adjacency matrix. As far as meta-CESDAMs are concerned, this work is not focused in that direction. Nevertheless, meta-CESDAMs are completely feasible and we will definitely work on it near future. We represent meta-graph with an edge list. Algorithm 2 presents the procedure for preparation of edge list representing meta-graph. Input to the algorithm is the list of all CESDAMs indexed with their ids in *SGlist* and outputs *EdgeList*.

4.1.1 Time complexity analysis

In Algorithm 2, for each of the CESDAM, its dependency has to be checked for all other CESDAMs. Thus, for k CESDAMs require $k \times k$ checking at meta-graph level. The dependency has to be matched at node level, i.e., nodes of subgraph. Therefore, nodes in one CESDAM has to be matched with nodes in other CESDAMs. Hence, in worst case (if node is in last position of CESDAM or no match) every node in one CESDAM has to be

---

ALGORITHM 2 getEdgeList(*SGlist*).

1: Input: *SGlist{#SGid}[CESDAM]* // List of subgraphs, here *#SGid* is numeric subgraph id.
2: Output: *EdgeList*
3: **for all** i^{th} CESDAM $\in$ *SGlist* with id *#SGid$_i$* **do**
4:    **for all** j^{th} CESDAM $\in$ *SGlist* with id *#SGid$_j$* such that $i \neq j$ **do**
5:       **if** there exists a common node between i^{th} and j^{th} CESDAMs **then**
6:          add an edge between *#SGid$_i$* and *#SGid$_j$* into *EdgeList* for i^{th} CESDAM and j^{th} CESDAM.
7:       **end if**
8:    **end for**
9: **end for**
10: **return** *EdgeList*

matched with all nodes in other CESDAMs. Suppose, there are k CESDAMs of sizes $m_1 \times m_1, m_2 \times m_2, \ldots, m_k \times m_k$, considering number of rows and columns are equal in all CESDAMs. In worst case requires $m_i^2 \times m_j^2$ comparisons for each pair of CESDAMs. If all CESDAMs are equal, i.e., $m_1 = m_2 = \ldots = m_k = m$ (say) then we have m^4 number of comparisons. This will be the case if we consider entire matrix, i.e., CESDAM. However, CESDAM scheme avails the privilege to search the nodes within the CESDAM. All nodes that are strongly part of the CESDAM, i.e., level 1 nodes and level 2 or external level 1 nodes, has an entry in column 1. All external level 2 nodes are stored in row 2. Therefore, checking of column 1 and row 2 of CESDAM would be enough. Thus, in worst case scenario requires only $4 \times m^2$ comparisons. For all pairs of CESDAMs requires $4 \times k \times k \times m^2$ comparisons. With Corollary 1, we have $k \times m^2 < n^2$ (detailed in Section 5). Hence, Algorithm 2 would experience worst case time complexity of $O(k \times n^2)$. In best case implies all shared node should be in first place so requires only $k \times k$ comparisons. Thus, Algorithm 2 would experience best case time complexity of $O(k^2)$.

4.2 Meta-breadth-first-search (MBFS)

Here, we address second objective, i.e., the correctness of accumulation of CESDAMs during processing of graph G with its representation as per CESDAM scheme. Sequential processing of large graphs is the essence of designing of CESDAM representation scheme. Most of the algorithms that follows sequential processing are inspired by BFS. Therefore, it is very much essential for the CESDAM scheme to ensure correctness of graph processing on BFS to smoothly operable in most the sequential algorithms. Unlike, conventional representation schemes such as adjacency matrix, adjacency list or edge list, the CESDAM scheme has multiple CESDAMs for different subgraphs of the graph G. Neighbors of a node are spread over multiple CESDAMs and, hence requires processing of multiple CESDAMs to accumulate all the neighbors of a node. Generic BFS cannot handle the links among CESDAMs when requires to scan neighbors. With meta-graph, the links among CESDAMs are manageable very effectively. We have designed MBFS by utilizing meta-graph to deal with the links among CESDAMs. The procedure for MBFS is presented in Algorithm 3. Algorithm 3 takes the input graph $G(V, E)$ is represented using CESDAM scheme. It maintains several additional data structures with each node in the graph. The indicator for each node $u \in V$ is stored in variable

ALGORITHM 3 MBFS(*EdgeList, SGlist, s*).

1: Input: *EdgeList* /*List of edges of meta-graph prepared on top of CESDAM ids.*/
   SGlist{#SGid}[CESDAM] /* List of subgraphs, *#SGid* is subgraph id.*/
2: Output: *d, p*
3: **for all** $u \in V$ **do**
4:   *visited[u]* ← 0 // here, 0 indicates not visited.
5:   *d[u]* ← −∞ //distance from *s* to all nodes is initialized.
6:   *p[u]* ← NIL //parent of all nodes is initialized.
7: **end for**
8: *d[s]* ← 0 //*s* is an arbitrary start node.
9: *Q* ← create empty queue. /*each element in *Q* contains node and CESDAM id
   where the node belongs*/
10: *ENQUEUE(Q, {s, #SGid_i})* /*#SGid_i is the ith CESDAM where *s* belongs.*/
11: **while** *Q* is not empty **do**
12:   {*cnode, #SGid_i*}← *DEQUEUE(Q)* /*cnode is the current node to be processed
      and *#SGid_i* is the ith CESDAM where *conde* belongs. */
13:   *NSGid* ← *getMetaNeighbors(EdgeList, #SGid_i)* /*get neighbors of CESDAM id
      #sg_i with respect to meta-graph.*/
14:   $NSGid^+$ ← with addition of *#SGid_i* into *NSGid*.
15:   **for all** j^{th} CESDAM id in $NSGid^+$ **do**
16:     SG_j ← *SGlist{#SGid_j}[]* /*j^{th} CESDAM having id *#SGid_j*.*/
17:     *flag* ← 0 //for indicating presence of *cnode* in SG_j.
18:     **if** *cnode* $\in SG_j(2, k)$, *k*=all columns **then**
19:       //*cnode* is external level 2 node of *sg_j*
20:       *flag* ← 1
21:       *rows* ← *getNodeRows(SG_j, cnode)*
22:       remove row 2 from *rows*
23:       *cnn* ← empty list. /*neighbor nodes of *cnode* with respect to SG_j.*/
24:       **for all** $r \in rows$ **do**
25:         add $SG_j(r, 1)$ to *cnn*
26:       **end for**
27:     **else if** *cnode* $\in SG_j(k, 1)$, *k*=all rows **then**
28:       *flag* ← 1
29:       *r* ← *getNodeRows(SG_j, cnode)* /* r is the row containing *cnode*.*/
30:       *columns* ← all elements of rth row of SG_j excluding *cnode*.
31:       **for all** $c \in columns$ **do**
32:         add $SG_j(r, c)$ to *cnn*
33:       **end for**
34:     **end if**
35:     **if** *flag* = 1 **then**
36:       **for all** *nnode* $\in cnn$ **do**
37:         **if** *d[nnode]* = ∞ **then**

Continued

ALGORITHM 3 MBFS(*EdgeList, SGlist, s*).—cont'd

38:        $d[nnode] \leftarrow d[cnode] + 1$
39:        $p[nnode] \leftarrow cnode$
40:        $ENQUEUE(Q, \{nnode, \#SGid_j\})$
41:        **end if**
42:     **end for**
43:   **end if**
44: **end for**
45: $visited[cnode] \leftarrow 1$
46: **end while**
47: **return** *d, p*

visited[*u*], the predecessor of *u* is stored in the variable *p*[*u*] and the distance from start node *s* to *u* is stored in variable *d*[*u*]. If *u* has no predecessor (for instance, if *u* = *s* or *u* has not been discovered), then *p*[*u*] = *NIL*. A queue *Q* is also used in the algorithm to maintain the intermediate frontier nodes.

4.2.1 Time complexity analysis

After initialization of *d*[*s*] = 0, non of the nodes $u \in V - \{s\}$ is ever changed to 0, and thus the line 37 ensures that each node is enqueued at most once, and hence dequeued at most once. The operations of enqueuing and dequeuing take $O(1)$ time, so the total time devoted to queue operations is $O(n)$, where $n = |V|$ is the total number of nodes. After dequeue operation in line 12, CESDAMs are scanned for discovering neighbors. The neighbor searching of any node *u* in CESDAM requires scanning of all the CESDAMs that are linked to the CESDAM where *u* belongs. As shown above, considering all CESDAM equal and have equal number of rows and columns, neighbor searching for any node on a CESDAM requires $O(2 \times m)$, where *m* is number of rows or columns. However, nodes that are external level 2 node in CESDAM require $O(2 \times m^2)$ as in line 21, since they requires to find the rows where they belongs. Algorithm 4 finds rows in the CESDAM where an external level 2 node belongs. The MBFS also generates a meta-breadth-first-tree (MBFT) with respect to start node *s*. Suppose, the branching factor of meta-graph is *B*, i.e., the *B* number of CESDAMs has to be scanned for identifying neighbors of each node, and hence requires $O(2 \times B \times m^2)$. The neighbor searching has to be done for all of the

ALGORITHM 4 getNodeRows(SG_j, cnode).
1: Input: SG_j // CESDAM with id #$SGid_j$. cnode // current node under processing.
2: Output: rows //list of row numbers where cnode exists.
3: rows ← empty list
4: **for all** row of SG_j **do**
5:      **for all** column of SG_j **do**
6:          **if** SG_j(row, column) = cnode **then**
7:              add rows into rows.
8:          **end if**
9:      **end for**
10: **end for**
11: **return** rows

$O(2 \times n \times B \times m^2)$ nodes so total time spend for scanning CESDAMs is $O(n \times B \times m^2)$. Hence, running time complexity of the MBFS algorithm is $O(n \times B \times m^2)$.

4.3 Correctness of graph processing

The MBFS algorithm finds the distances to all nodes reachable from the start node s along with searching of nodes. The distance between two nodes is the number of edges appears in the path between them. The minimum number of edges appeared in any path between two nodes is referred as shortest path. The the shortest-path distance $\delta(s, v)$ from s to v is the minimum number of edges in the shortest path from s to v. If there is no path between s and v then $\delta(s, v) = \infty$. Before showing that MBFS actually computes shortest-path distances, we investigate an important property of shortest-path distances.

Lemma 1. *Given a graph $G(V, E)$, and $s \in V$ be an arbitrary node. For any edge $(u, v) \in E$ and $u, v \in V$, if u is reachable from s then $\delta(s, v) = \delta(s, u) + 1$.*

Proof. The edge $(u, v) \in E$ implies that if u is reachable from s, then v is also reachable from s. In this case, the shortest path from s to v cannot be longer than the shortest path from s to u followed by the edge (u, v), and thus $\delta(s, v) = \delta(s, u) + 1$ holds. In case u is not reachable from s, then $\delta(s, u) = \infty$, and $\delta(s, v) = \infty$ since v would also not reachable from s so $\delta(s, v) = \delta(s, u) + 1$ holds.$\square$

Our objective is to show that use of CESDAM representation does not alters the essence of BFS in the MBFS, so we want to show that MBFS also properly computes $d[v] = \delta(s, v)$ for each node $v \in V$. We first show that $d[v]$ $\geq \delta(s, v)$ in the following lemma.

Lemma 2. *Given a graph G(V, E) represented with CESDAM, and suppose that MBFS is run on CESDAM representation of G from a given start node $s \in V$. Then upon termination, for each node $v \in V$, the value d[v] computed by MBFS satisfies d [v] $\geq \delta(s, v)$.*

Proof. We use induction on the ENQUEUE operations of MBFS. The inductive hypothesis is that $d[v] \geq \delta(s, v)$ for all $v \in V$. In MBFS, when s is enqueued the line 10 the hypothesis hold, because $d[s] = 0 = \delta(s, s)$ and $d[v] = \infty \geq \delta(s, v)$ for all $v \in V - \{s\}$. Now, consider a node v that is discovered during the searching neighbor from node u. With induction hypothesis we have $d[u] \geq \delta(s, u)$. After discovering v, i.e., *nnode* in Line 38, we have assignment $d[nnode] \leftarrow d[cnode] + 1$. From the assignment performed in the line 38 and from Lemma 1, we have

$$d[v] = d[u] + 1$$
$$\geq \delta(s, u) + 1$$
$$\geq \delta(s, v).$$

The node v is then enqueued, and it is never enqueued again because $d[nnode]$ is no longer ∞ and it never again becomes ∞. Thus, the value of $d[v]$ never changes again. Therefore, the inductive hypothesis is maintained as long as MBFS correctly discovers v while scanning for neighbors of u, i.e., *cnode*. This is ensured through lines 16–34, where each jth CESDAM that is linked with the CESDAM in which *cnode* belongs is searched. Links among all CESDAMs are defined in the meta-graph, which is represented in edge list. With CESDAM scheme, requires only to check row 2 and column 1 for detecting presence of any node, which is done with line 18 and line 27, respectively. Thus, neighbors are identified correctly with meta-graph and all linked CESDAMs and, hence the inductive hypothesis is maintained. □

To prove that $d[v] = \delta(s, v)$, we should first examine more precisely to show how the queue Q operates during execution of MBFS. The next lemma shows that how d values of nodes present in the Q at any time changes after ENQUEUE and DEQUEUE operations.

Lemma 3. *Suppose, at any time during the execution of MBFS on a graph G(V, E) represented with CESDAM, the queue Q contains the nodes $v_1, v_2, ..., v_r$, where v_1 is the font of Q and v_r is the rear. Then, $d[v_r] \leq d[v_1] + 1$ and $d[v_i] \leq d[v_{i+1}]$ for $i = 1, 2, ..., r - 1$.*

Proof. The proof is by induction on the number of queue operations. Initially, when the queue contains only s, the lemma certainly holds. For the inductive step, we should show that the lemma holds after both dequeuing and enqueuing a node. If the font v_1 of the queue is dequeued, v_2 becomes the new font. (If the queue becomes empty, then the lemma holds vacuously.) By the inductive hypothesis, $d[v_1] \leq d[v_2]$. In that case, we also have $d[v_r] \leq d[v_1] + 1 \leq d[v_2] + 1$, and the remaining inequalities will remain unaffected. Thus, the lemma follows with v_2 as the font. Thus, the lemma follows when nodes are dequeued.

Again, when we enqueue a node v, i.e., *nnode* in line 40 of the MBFS, it becomes new rear v_{r+1}. At that time, we have already removed node u, i.e., *cnode*, whose neighbors are currently being scanned from the queue Q. Therefore, the font v_1 has $d[v_1] \geq d[u]$. In that case, $d[v_{r+1}] = d[v] = d[u] + 1 \leq d[v_1] + 1$. Thus, the lemma follows when nodes are enqueued. □

During the course of MBFS, all nodes can have at most two distinct d values, i.e., ∞ (if node is not reachable from s) and the value obtained after discovery of the node (if node is reachable from s). The following lemma shows that the d values are monotonically increasing over time depending on the time when the nodes are enqueued.

Lemma 4. *Consider nodes v_i and v_j are enqueued during the execution of MBFS, and v_i is enqueued before v_j. Then, $d[v_i] \leq d[v_j]$ at the time that v_j is enqueued.*
Proof. By Lemma 3 and the property that all node receives a finite d values during the course of MBFS. □

With the lemmas discussed above, we now prove that the MBFS correctly finds shortest-path distances.

Theorem 1. *Given graph $G(V, E)$ is represented with CESDAM scheme, and suppose MBFS is run on CESDAMS of G from any arbitrary start node $s \in V$. Then, during its execution, MBFS discovers every node $v \in V$ that is reachable from the start node s, and after termination, $d[v] = \delta(s, v)$ for all $v \in V$.*
Proof. Proof by contradiction. Let us consider that some node that are reachable from s gets wrong d value, i.e., d value not equal to shortest-path distance. Say, v be the node that gets incorrect d value. As shown in Lemma 2, the node v must follow that $d[v] \geq \delta(s, v)$. Since we assume $d[v] \neq \delta(s, v)$, the node v must satisfy $d[v] > \delta(s, v)$. Let u be the node predecessor of v on a shortest path from s to v so $\delta(s, v) = \delta(s, u) + 1$ and $d[u] = \delta(s, u)$. Thus, we have

$$d[v] > \delta(s, v) = \delta(s, u) + 1 = d[u] + 1 \tag{5}$$

When the node u is dequeued from the Q in line 12, the node v is either not discovered, partially visited or visited. We have to show that Eq. (5) contradicts in all these cases. If v is not discovered, then line 38 implies $d[v] = d[u] + 1$, contradicting Eq. (5), where $v = nnode$ and $u = cnode$. If v is visited, then it was already removed from the queue and, by Lemma 4, we have $d[v] \leq d[u]$, it contradicts Eq. (5). If v is partially visited, then it must be discovered during scanning of neighbors any node w that was dequeued from Q before u for which $d[v] = d[w] + 1$. By Lemma 4 we have $d[w] \leq d[u]$, which implies $d[v] \leq d[u] + 1$, again contradicts Eq. (5). Thus, we conclude that $d[v] = \delta(s, v)$ for all nodes $v \in V$ that are reachable from start node s. $\square$

5. Efficiency of CESDAM scheme in terms of space requirement

In CESDAM representation, the graph with n nodes is divided into subgraphs that are presented in the CESDAMs. If there are k CESDAMs that represent the graph then we have k subgraphs each having $m_1, m_2, ..., m_k$ number of nodes. The sizes of CESDAMs are not dependent on the number of nodes in the subgraphs, instead it is dependent on connectivity pattern among the nodes. Therefore, number of rows and columns in the matrix representing CESDAM is not predictable in prior by simply considering nodes. Thus, we assume that for k matrices of sizes are $m_1^r \times m_1^c, m_2^r \times m_2^c, ..., m_k^r \times m_k^c$ that are the CESDAMs for subgraphs. For the simplicity of the analysis, we consider $m_i = m_i^r = m_i^c$. Thus, we have k CESDAMs of sizes $m_1 \times m_1, m_2 \times m_2, ..., m_k \times m_k$. Note that subgraphs that are represented in the CESDAMs are overlapped. Thus, $m_1 + m_2 + \cdots + m_k > n$, if at least one overlapping node exists. In case of adjacency matrix, all the n nodes are represented in a single matrix of size $n \times n$. If k nonoverlapped subgraphs are represented with adjacency matrix then we have $m_1 \times m_1, m_k \times m_2, ..., m_k \times m_k$ sized matrices. Therefore, we have following theorems.

Theorem 2. *Individual adjacency matrix representation of nonoverlapping subgraphs in total requires less space than the adjacency matrix for entire graph.*

Proof. Given a graph $G(V, E)$, the size of adjacency matrix representation of G is $n \times n = n^2$, where $n = |V|$ is number of nodes present in G. Suppose, the graph G is divided into k nonoverlapping subgraphs. Each subgraph

$g_i(V_i, E_i)$ contains $m_i = |V_i|$ number of nodes. Hence, size of adjacency matrix representation of each subgraph g_i is $m_i \times m_i$. Since, all k subgraphs are nonoverlapping so $\sum_{i=1}^{k} m_i$, i.e., $m_1 + m_2 + \cdots + m_k = n$.

Now, if G is divided into two nonoverlapping subgraphs of sizes 1 and $n - 1$,

$$1 \times 1 + (n - 1) \times (n - 1) < n \times n$$

Similarly, for sizes 2 and $n - 2$,

$$2 \times 2 + (n - 2) \times (n - 2) < n \times n$$
$$\ldots \text{ and so on}$$

for sizes k and $n - k$,

$$k \times k + (n - k) \times (n - k) < n \times n \ldots \tag{6}$$

Again, if G is divided into two nonoverlapping subgraphs of sizes 1 and $n - 1$,

$$1 \times 1 + (n - 1) \times (n - 1) < n \times n$$

If G is divided into three nonoverlapping subgraphs of sizes 1, 1 and $n - 2$,

$$1 \times 1 + 1 \times 1 + (n - 2) \times (n - 2) < n \times n$$
$$\ldots \text{and so on}$$

for $k - 1$ subgraph of size 1 and one subgraph of size $n - k$,

$$1 \times 1 + 1 \times 1 + \cdots + (k - 1)^{th} 1 \times 1 + (n - k) \times (n - k) < n \times n \ldots \tag{7}$$

Therefore, with (6) and (7), for k nonoverlapping subgraphs of sizes $m_1, m_2, \ldots, m_k$,

$$m_1 \times m_1 + m_2 \times m_2 + \cdots + m_k \times m_k < n \times n$$

$\square$

Corollary 1. *If k nonoverlapping subgraphs contain equal number of nodes m then for n total number of nodes in the graph $k \times m^2 < n^2$.*

Proof. From Theorem 2, for k nonoverlapping subgraphs containing $m_1, m_2, \ldots, m_k$ nodes,

$$m_1 \times m_1 + m_2 \times m_2 + \cdots + m_k \times m_k < n \times n$$

if $m_1 = m_2, \dots, m_{k-1} = m_k = m$ then,

$$k \times m \times m < n \times n \Rightarrow k \times m^2 < n^2$$

$\square$

Lemma 5. *Level 2 external nodes in CESDAM are the overlapping nodes and shared by only two CESDAMs.*

Proof. Level 2 external nodes are the outermost nodes of the CESDAM representing ego network level 2.5. Thus, these nodes will be shared by other CESDAMs. In proposed approach, a node is allowed to be shared with at most two CESDAMs. However, a CESDAM can have multiple nodes shared with multiple CESDAMs. $\square$

Theorem 3. *CESDAM representation of graph requires less space than adjacency matrix representation.*

Proof. Given a graph $G(V, E)$, $n = |V|$ and k CESDAMs that represent the graph G. With Lemma 5, k CESDAMs can have multiple combinations of overlapping pairs. Suppose, pairing of two CESDAMs follow the pattern as given bellow:

m_1 with m_2, m_2 with m_3, m_3 with m_4 and so on $\dots m_{k-2}$ with m_{k-1}, m_{k-1} with m_k, m_k with m_1.

All CESDAMs are paired with other two CESDAMs. All CESDAMs pairs have a level 2 external nodes, i.e., a overlapped nodes between two CESDAMs. Therefore, all CESDAMs will have $2a$ number of overlapped nodes with their predecessor and successor CESDAMs each contain a number of overlapped nodes. Suppose, we covert subgraphs representing CESDAMs to nonoverlapping subgraphs by assigning overlapped nodes to predecessors as follows.

$$m_1 - a, m_2 - a, m_3 - a \text{ and so on } \dots m_{k-1} - a, m_k - a,$$

such that,

$$(m_1 - a) + (m_2 - a) + \cdots + (m_k - a) = n$$

$$m_1 + m_2 + \cdots + m_k = n + ka.$$

Space requirement will be,

$$
\begin{aligned}
&= (m_1 - a)^2 + (m_2 - a)^2 + \cdots + (m_k - a)^2 \\
&= m_1^2 + m_2^2 + \cdots + m_k^2 - 2a(m_1 + m_2 + \cdots + m_k) + ka^2 \\
&= m_1^2 + m_2^2 + \cdots + m_k^2 - 2a(n + ka) + ka^2
\end{aligned}
$$

With Theorem 2, it is clear that the portion $(-2a(n + ka) + ka^2)$ in above expression is overhead due to overlapped nodes. In order to have less space requirement than adjacency matrix representation of entire graph $- 2a(n + ka) + ka^2 \leq 0$ has to be satisfied. Now,

$$= -2a(n + ka) + ka^2 = -2an - 2ka^2 + ka^2$$

$$= -2an - ka^2 < 0, \forall a, n, k > 0.$$

$\square$

Theorem 4. *Given a graph G(V, E), $n = |V|$, k CESDAMs and each pair of CESDAMs existed as per G have a level 2 external nodes then space requirement will be less than adjacency matrix representation of entire graph if and only if $k \geq \frac{-n^2 - 2an + \sum_{i=1}^{k} m_i^2}{a^2}$.*

Proof. With Theorem 3,

$$= m_1^2 + m_2^2 + \cdots + m_k^2 - 2a(n + ka) + ka^2 \leq n^2$$

$$= -2a(n + ka) + ka^2 \leq n^2 - \sum_{i=1}^{k} m_i^2$$

$$= -2an - 2ka^2 + ka^2 \leq n^2 - \sum_{i=1}^{k} m_i^2$$

$$= -ka^2 \leq n^2 + 2an - \sum_{i=1}^{k} m_i^2$$

$$= ka^2 \geq -n^2 - 2an + \sum_{i=1}^{k} m_i^2$$

$$= k \geq \frac{-n^2 - 2an + \sum_{i=1}^{k} m_i^2}{a^2}$$

$\square$

Corollary 2. *Given a graph G(V, E), $n = |V|$, k CESDAMs are of equal size m and each pair of CESDAMs existed as per G have a external level-2 nodes then space requirement will be less than adjacency matrix representation of entire graph if and only if $k \leq \frac{n^2 + 2na}{m^2 - a^2}$.*

Proof. With Theorem 4,

$$= k \geq \frac{-n^2 - 2an + \sum_{i=1}^{k} m_i^2}{a^2}$$

$$= ka^2 \geq -n^2 - 2an + \sum_{i=1}^{k} m_i^2$$

Since, $m_1 = m_2, \ldots, m_{k-1} = m_k = m$ so,

$$= ka^2 \geq -n^2 - 2an + km^2$$
$$= ka^2 - km^2 \geq -n^2 - 2an$$
$$= km^2 - ka^2 \leq n^2 + 2an$$
$$= k \leq \frac{n^2 + 2na}{m^2 - a^2}.$$

□

Theorem 5. *For requiring less space than adjacency matrix, equal sized CESDAMs of size m can have maximum $\frac{n^2 - mn + 2an}{a + m}$ number of level 2 external nodes of n if CESDAM pairs contain equal overlapped nodes a.*
Proof. With Theorem 3,

$$\Rightarrow m_1^2 + m_2^2 + \cdots + m_k^2 - 2a(n + ka) + ka^2 \leq n^2$$

Since equal sized CESDAMs so,

$$m_1 = m_2, \ldots, m_{k-1} = m_k = m,$$

$$km = n + ka \text{ and}$$

$$ka = y, \text{ total number of external level} - 2 \text{ nodes.}$$

Thus,

$$\Rightarrow km^2 - 2an + ka^2 \leq n^2$$
$$\Rightarrow (n + ka)m - 2an + ka^2 \leq n^2$$
$$\Rightarrow (n + y)m - 2an + ya \leq n^2$$
$$\Rightarrow mn + ym - 2an + ya \leq n^2$$
$$\Rightarrow ym + ya \leq n^2 - mn + 2an$$
$$\Rightarrow y \leq \frac{n^2 - mn + 2an}{a + m}.$$

□

Theorem 3 proves that the CESDAM scheme requires less space than the adjacency matrix. It indirectly implies that the CESDAM is also space efficient than the another matrix-based scheme, i.e., the incidence matrix. Incidence matrix requires space $O(e \times n)$ and adjacency matrix requires $O(n^2)$, where $e = |E|$ is the total number of edges and n is the total number of nodes. Since most of the connected graphs have $e > n$, adjacency matrix requires less space than incidence matrix. Thus, with Theorem 3 we can show that the CESDAM also requires less space than incidence matrix. On the other hand, sparse matrix format such as CSR requires less space than CESDAM, but CSR has its own limitations [18, 19]. Similarly, the list-based schemes such as adjacency list or incidence list also require a little less space than CESDAM due to the blank spaces in the matrix.

6. Conclusions

Graph representation is very important for efficient processing of graphs both in terms of memory requirement and time requirement. The processing of large graphs with CESDAM scheme has been investigated, from the perspective of both memory requirement and time requirement. The CESDAM scheme is inspired by the notion of data partitioning. The graph is represented in multiple CESDAMs, each containing subgraphs of the graph. Representation of graph in parts with CESDAM reduces the memory requirement significantly, which requires space $O(k \times m^2)$, where k is the number of CESDAMs and m is the size of each CESDAM. Theoretical analysis shows that CESDAM experiences less space requirement than adjacency matrix that requires space $O(n^2)$ to represent a graph, where n is number of nodes in the graph.

The processing time required for CESDAM scheme is analyzed in perspective of breadth-first searching. In CESDAM scheme, the graph is divided into multiple subgraphs; each subgraph is then represented in one CESDAM. Therefore, to analyze time requirement, all the CESDAMs representing the graph has to be taken into account. A meta-graph is prepared on top of CESDAM ids to accumulate all the CESDAMs during processing of graph. The analysis shows, breadth-first searching on CESDAM scheme requires time $O(n \times B \times m^2)$, where B is the branching factor of meta-graph. The parts of a graph are represented in multiple CESDAMs, which require assurance of correctness during processing and accumulating the subgraphs. We have proved that with the CESDAM scheme, parts of any graph represented as CESDAM can be processed correctly without compromising its meaning when the graph was undivided.

References

[1] S. Eubank, H. Guclu, V.S.A. Kumar, M.V. Marathe, A. Srinivasan, Z. Toroczkai, N. Wang, Modelling disease outbreaks in realistic urban social networks, Nature 429 (6988) (2004) 180–184.

[2] T. Lin, Mausam, O. Etzioni, Entity linking at web scale, in: Proceedings of the Joint Workshop on Automatic Knowledge Base Construction and Web-scale Knowledge Extraction, Association for Computational Linguistics, Stroudsburg, PA, USA, 2012, pp. 84–88. URL http://dl.acm.org/citation.cfm?id=2391200.2391216.

[3] X. Wu, X. Zhu, G.-Q. Wu, W. Ding, Data mining with big data, IEEE TKDE 26 (1) (2014) 97–107, https://doi.org/10.1109/TKDE.2013.109.

[4] A. Anand, A. Grover, Mausam, P. Singla, A novel abstraction framework for online planning: extended abstract, in: Proceedings of the 2015 International Conference on Autonomous Agents and Multiagent Systems, International Foundation for Autonomous Agents and Multiagent Systems, Richland, SC, 2015, pp. 1901–1902, ISBN: 978-1-4503-3413-6. URL http://dl.acm.org/citation.cfm?id=2772879.2773494.

[5] J. Zhang, Z. Fang, W. Chen, J. Tang, Diffusion of "following" links in microblogging networks, IEEE Trans. Knowl. Data Eng. 27 (8) (2015) 2093–2106, https://doi.org/10.1109/TKDE.2015.2407351.

[6] K.S.M.T. Hossain, D. Patnaik, S. Laxman, P. Jain, C. Bailey-Kellogg, N. Ramakrishnan, Improved multiple sequence alignments using coupled pattern mining, IEEE/ACM Trans. Comput. Biol. Bioinform. 10 (5) (2013) 1098–1112, https://doi.org/10.1109/TCBB.2013.36.

[7] A. Birlutiu, F. d'Alche Buc, T. Heskes, A Bayesian framework for combining protein and network topology information for predicting protein-protein interactions, IEEE/ACM Trans. Comput. Biol. Bioinform. 12 (3) (2015) 538–550, https://doi.org/10.1109/TCBB.2014.2359441.

[8] C.A. Hill, G.P. Zhang, G.D. Scudder, An empirical investigation of EDI usage and performance improvement in food supply chains, IEEE Trans. Eng. Manag. 56 (1) (2009) 61–75, https://doi.org/10.1109/TEM.2008.922640.

[9] M. Eskandarpour, P. Dejax, J. Miemczyk, O. Péton, Sustainable supply chain network design: an optimization-oriented review, Omega 54 (2015) 11–32.

[10] Y. Zhu, C. Yan, Graph methods for predicting the function of chemical compounds, in: 2014 IEEE International Conference on Granular Computing (GrC), October, 2014, pp. 386–390, https://doi.org/10.1109/GRC.2014.6982869.

[11] W. Zheng, L. Zou, X. Lian, D. Wang, D. Zhao, Efficient graph similarity search over large graph databases, IEEE Trans. Knowl. Data Eng. 27 (4) (2015) 964–978, https://doi.org/10.1109/TKDE.2014.2349924.

[12] J. Cai, Z.-J. Zha, M. Wang, S. Zhang, Q. Tian, An Attribute-Assisted Reranking Model for Web Image Search, IEEE Trans. Image Process. 24 (1) (2015) 261–272.

[13] D.B. West, et al., Introduction to Graph Theory, vol. 2, Prentice Hall Upper Saddle River, 2001.

[14] J.L. Gross, J. Yellen, Graph Theory and Its Applications, CRC press, 2005.

[15] Y. Saad, SPARSKIT: A Basic Tool Kit for Sparse Matrix Computations, Research Institute for Advanced Computer Science (RIACS), 1990.

[16] J.E. Gonzalez, R.S. Xin, A. Dave, D. Crankshaw, M.J. Franklin, I. Stoica, Graphx: graph processing in a distributed dataflow framework, in: 11th USENIX Symposium on Operating Systems Design and Implementation (OSDI 14), 2014, pp. 599–613.

[17] D. Merrill, M. Garland, Merge-based sparse matrix-vector multiplication (SpMV) using the CSR storage format, in: Proceedings of the 21st ACM SIGPLAN Symposium on Principles and Practice of Parallel Programming, ACM, 2016, p. 43.

[18] T. Oberhuber, A. Suzuki, J. Vacata, New row-grouped CSR format for storing the sparse matrices on GPU with implementation in CUDA, Acta. Tech. 56 (4) (2010) 447–466.
[19] F. Khorasani, K. Vora, R. Gupta, L.N. Bhuyan, CuSha: vertex-centric graph processing on GPUs, in: Proceedings of the 23rd International Symposium on High-Performance Parallel and Distributed Computing, ACM, New York, NY, USA, 2014, pp. 239–252, ISBN: 978-1-4503-2749-7, https://doi.org/10.1145/2600212.2600227.
[20] G. Malewicz, M.H. Austern, A.J.C. Bik, J.C. Dehnert, I. Horn, N. Leiser, G. Czajkowski, Pregel: a system for large-scale graph processing, in: Proceedings of the 2010 ACM SIGMOD International Conference on Management of Data, ACM, 2010, pp. 135–146, https://doi.org/10.1145/1807167.1807184.
[21] J.S. Vitter, External memory algorithms and data structures: dealing with massive data, ACM Comput. Surv. 33 (2) (2001) 209–271, https://doi.org/10.1145/384192.384193.
[22] A. Kyrola, G. Blelloch, C. Guestrin, Graphchi: large-scale graph computation on just a pc, in: Presented as Part of the 10th USENIX Symposium on Operating Systems Design and Implementation (OSDI 12), 2012, pp. 31–46.
[23] D. Ajwani, R. Dementiev, U. Meyer, A computational study of external-memory BFS algorithms, in: Proceedings of the 17th Annual ACM-SIAM Symposium on Discrete Algorithm, SIAM, 2006, pp. 601–610.
[24] M. Bernaschi, G. Carbone, E. Mastrostefano, M. Bisson, M. Fatica, Enhanced GPU-based distributed breadth first search, in: Proceedings of the 12th ACM International Conference on Computing Frontiers, ACM, 2015, pp. 10:1–10:8, ISBN: 978-1-4503-3358-0, https://doi.org/10.1145/2742854.2742887.
[25] J. Zhong, B. He, Medusa: a parallel graph processing system on graphics processors, SIGMOD Rec. 43 (2) (2014) 35–40, https://doi.org/10.1145/2694413.2694421.
[26] J. Xue, Z. Yang, Z. Qu, S. Hou, Y. Dai, Seraph: an efficient, low-cost system for con-current graph processing, in: Proceedings of the 23rd International Symposium on High-Performance Parallel and Distributed Computing, ACM, 2014, pp. 227–238, https://doi.org/10.1145/2600212.2600222.
[27] V. Arnaboldi, M. Conti, A. Passarella, F. Pezzoni, Analysis of ego network structure in online social networks, in: 2012 International Confernece on Social Computing (SocialCom) Privacy, Security, Risk and Trust (PASSAT), September, 2012, pp. 31–40, https://doi.org/10.1109/SocialCom-PASSAT.2012.41.
[28] A. Biswas, B. Biswas, Investigating community structure in perspective of ego network, Expert Syst. Appl. 42 (20) (2015) 6913–6934, https://doi.org/10.1016/j.eswa.2015.05.009.

About the authors

Ravi Kishore Devarapalli received the B.Tech. degree in Computer Science and Engineering from the JNTU Hyderabad, Telangana, India, in 2012, M.Tech. degree in Computer Science and Engineering from the JNTU Hyderabad, Telangana, India, in 2015, and he is currently pursuing the Ph.D. degree with the Department of Computer Science and Engineering, National Institute of Technology Silchar (NITs), Silchar, India. His research interests include social network analysis.

Dr. Anupam Biswas received the B.E. degree in Computer Science and Engineering from the Jorhat Engineering College, Jorhat, India, in 2011, M.Tech. degree in Computer Science and Engineering from the Nehru National Institute of Technology Allahabad, Prayagraj, India, in 2013, and the Ph.D. degree in Computer Science and Engineering from IIT (BHU) Varanasi, Varanasi, India, in 2017. He is currently working as an Assistant Professor with the Department of Computer Science and Engineering, National Institute of Technology Silchar, Silchar, India. He has authored or coauthored several research articles in reputed international journals, conference, and book chapters. His research interests include machine learning, deep learning, computational music, information retrieval, social networks, and evolutionary computation. He has served as a Program Chair for the International Conference on Big Data, Machine Learning and Applications (BigDML 2019). He has served as a General Chair for the 25th International Symposium on Frontiers of Research in Speech and Music (FRSM 2020) and co-edited proceedings of FRSM 2020 published as book volume in Springer AISC Series. He has edited three books titled Health Informatics: A Computational Perspective in Healthcare, Principles of Social Networking: The New Horizon and Emerging Challenges in different Springer book series.

## CHAPTER SEVEN

MapReduce based convolutional graph neural networks: A comprehensive review

U. Kartheek Chandra Patnaik and Ripon Patgiri
Department of Computer Science and Engineering, National Institute of Technology Silchar, Silchar, Cachar, Assam, India

Contents

Abstract

Deep learning has brought a lot of changes in the digital era. Big data and deep learning models have broad research scope for future researchers since data is represented in different forms like audio, video, image, text and graph, etc. Out of which, handling Graph data is the cumbersome approach, as the hidden patterns are in plenty. Capturing hidden patterns in graph data is a challenging issue these days. This review provides a global view of convolutional graph neural networks using different machine learning models, and map reduce based neural graph networks. We discuss different state-of-art learning approaches for handling graph data. We further discuss the limitations of few existing models in handling massive data called BigGraph. We also identify the applications and challenges for Convolutional Graph neural Networks (ConvGNN).

Advances in Computers, Volume 128
ISSN 0065-2458
https://doi.org/10.1016/bs.adcom.2021.10.002
213

1. Introduction

The Neural Networks brought a significant change in different success stories in Data Analytics as a Service (DaaS). Various applications and models using machine learning algorithms have been developed with less computational cost. Some real-time technologies like Machine Intelligence, Language Translation, Speech-Text Processing used these deep learning models for their existence in this real world. There is still a lot of research scope for Deep Neural Networks for handling Graph Data on another side of the coin. Underwater communication, positioning, target identification and information processing is some of the best graph-based data analytics services which need a lot of attention to extend the model for better accuracy in identifying hidden patterns. Detecting and identifying different objects in the ocean like sea urchins live on the sea bed, target weapons in war base, various fish species, etc., is one of the most promising applications for the near future, which needs graph analytics for finding the object from different directions.

Graphs are simple but effective data structures for modeling and analyzing real-world data relationships. Graph analytics is appealing in a number of contexts, such as web information systems, social networks, business intelligence, and life sciences, due to the simplicity of graph network models and its various algorithms. Web pages, users, and proteins can all be represented as vertices in a graph, with their links represented as edges. Graph algorithms [1] use this abstraction to help rank websites, detect groups in social graphs, and classify pathways in biological networks, among other things. These domains' graphs are often very wide, with a huge amount of vertices and edges, making efficient data management and graph algorithm execution difficult. Graphs should be able to accurately represent entities and relationships of various types for graph-oriented analysis of heterogeneous data, likely integrated from various sources.

Big graph is popular in real-world applications today, for instance, with billions of users of social online and mobile communication networks and Web graphics and weekly websites. The treatment of such large graphs usually requires a special infrastructure, the most frequently used being systems like Pregel [2]. A Pregel-like system programmer thinks like a vertex, and only needs to define the action of one vertex, and the system automatically plans the execution of the computer logic at all vertices. The framework also manages fault tolerance and scales without the need for programmers to make additional programming effort.

At the present days they are no such good models which give good accuracy for identifying the sea collection. Especially target detection algorithms are the one which is mostly used in identifying those objects in the very Deep Ocean. Machine learning models Scale-Invariant Feature Transform (SIFT) [3] and Histogram of Oriented Gradients (HOG) to define features, and finally, use the Support Vector Machine (SVM), Adaptive Boosting (AdaBoost) and other technologies for classification. Moreover, the earlier algorithms have many issues in identifying and detecting objects accurately due to uneven surfaces, indeterminate light and resonance in deep water. So neural network-based analysis is required for analyzing such things very deeply by choosing a more number of hidden patterns. Since the underwater designs are mostly 2D, 3D and n-dimensional Convolutions.

2. Motivation

Motivated by deep learning CNNs, RNNs and encoders, different operations and techniques are rapidly developed for handling such complex graph data. Graphs are the data structure which is the collection of nodes and their relationship. Graphs are used to denote the large number of nodes linked together to operate for high performance and more interpretability. Due to different advantages in graphs, machine learning was drawn a lot of attention to the current working models. The initial scope of Graph Neural networks is Convolution Neural Networks (CNNs), which is used for introspecting multidimensional features and data representations, as shown in Fig. 1. Since CNN's found a lot of advantages like less computational cost, more efficiency and less complex structure. However, the number of hidden layers in CNN brought more convolutions to find perfect patterns. Understanding graph Convolutions is generalized by considering adjacent pixels of an image in the form of a graph. A graph convolution is the average weighted node neighborhood information.

The concept of Graph Neural Networks is required for the real time data analytics, especially for handling unstructured data. Unstructured data is the data that does not have predefined schema. Hence, a lot of data in the current world like social media, sensor data, geo-spatial data which are very difficult to handle and analyze, and needs complex algorithms to find the relationship between the aggregated data. The current existing machine learning models and techniques haven't proved to be good for the faster way of analyzing data. Hence Convolutional graph neural networks show good response in terms of efficiency and accuracy. The contribution of this review is explained as follows:

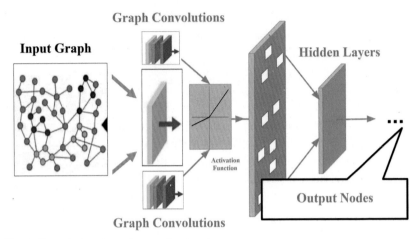

Fig. 1 Convolutional graph neural network.

- We identify the different machine learning models for examining graph data and how can we resolve them.
- Understanding different Deep Neural Network models are analyzed and their operations are specified. The types of neural networks are RNNs, GNNs and CNNs.
- A contrast between these models is done for understanding the well usage of map reduce based convolutional neural networks.
- This survey also focuses the applications and challenges that are currently going on and also identifies the accuracy of each model.

3. Related work

GCOMB [4] analyzes the large graphs to improve the quality of the node to attain efficiency. It is generally used for models that combines both supervised and reinforcement learning. GCOMB, along with Q-learning lightweight architecture for a set of nodes, is a combinatorial solution on BigGraphs. GCOMB is deficient with combinatorial graph problems.

Novel Graph LSTM [5] parses semantic nature of an object. It converts LSTM into vertices and edges. The node's state is altered right from the starting node to all neighboring nodes with a confidence-driven approach. Each image pixel is taken as a neighborhood connection in a structured graph. Each node is updated statically in the graph, thereby creates an arbitrary shape. It lacks in creating a dynamic structure of the graph for every neighboring node.

RN-MCTS [6] is a combined form of recurrent neural network and graph-based approach that states the nodes into the upper and lower tree. Updating and searching of nodes in the tree either from the upper bound or lower bound is faster when compared to earlier techniques. The optimization techniques give more speed in searching the nodes. Monte Carlo Tree Structure (MCT) achieves perfect balance with both exploration and exploitation in terms of tree breadth than depth. Optimization can be achieved even after revisiting the same nodes from root to leaf. RV-MCTS lacks in the run-time matching of nodes behavior that leads to less efficiency.

GraphRNN [7] generates high-quality graphs that match the features of graphs in a variety of metrics. Based on degrees, cluster metrics, and bit-counts, graphRNN can capture graph statistics with average statistics that match the distribution of the data set. GraphRNN is highly robust since it has a strong performance in generating graph node distributions. GraphRNN achieves good efficiency than previous models with high scalability and high robustness. When scaling larger graphs, GraphRNN lacks in developing graph efficiency.

GRAN [8] is very efficient in the generation of grid graphs with adjacency matrix in O(N) time complexity. Block size and stride can be varied with edge to edge graph. It generates the block even in the existing chart with conditions correlation. The performance is the benchmark and shown the best results even for extensive graph data. It is lazy when the graphs are latent, and graphs are partial.

GNDP [9] a novel diagnosis prediction method is used to predict the patient future health condition. It is used to find the spatial and temporal graph pattern data for the given knowledge. Irrespective of sequential or random graph patterns, the prediction accuracy is very high in GNDP. The patient data is converted into an adjacency matrix for further normalization. The temporal dependency of the information is the extraction of the features from the given graph pattern. It performs both 2D and 3D convolution operations on temporal data. More accuracy in prediction is required in this model when data is so large.

RGNN [10] is the next-period prescription prediction method used to represent hospital patient event graphs. Different types of patient medical events will be categorized into graphs with edges and nodes. Nodes represent patient medical events, and edges represent relations between any two events. It achieved good performance in identifying the patient event co-occurrence relations. Temporal graph representation proved its efficiency in prescriptive prediction. It lacks for predicting massive graphs, i.e., BigGraphs.

CNNG [11] is the combination of CNN and GNN, which segments the images with its local features and identifies semantic comparison between structures. The graph for the semantic segments is constructed to extract the quality and thereby learn and extract the feature. Segmentation is a way of transforming pixels into node classifications, i.e., graphs. It also compares the local feature information and following receptive features. Without the complexity of the network, the efficiency of segmentation is improved. Hence several hidden layers are to be enhanced for further segmentation.

GCNN [12] is a graph convolutional neural network operation that is used to convert vertex domain into Fourier domain for classifying patterns. Since the data can be viewed only in the graph structure, each knowledge is transformed into edges and nodes to meet the regularity. GCNN is very good at analyzing spatial data like traffic patterns, cloud structure identification.

4. MapReduce and CGNN

In hierarchical computing, MapReduce framework [13] is distributed data processing technique used to process data in parallel using a map and reduce tasks. When MapReduce framework is integrated with Graph, it requires multiple processing levels with edges and vertices. Especially Join and Combine operations are widespread in leveraging graph optimization. Since Graph is the iterative kind of approach, it needs a lot of aggregations and optimizations. The time is taken to process large graphs even more complex using the MapReduce framework. So, it requires many hidden layers to be constructed for analyzing such large graphs, i.e., BigGraph.

In contrast, the CNN (convolutional neural network), known as ConvNet, generalizes better than the latter type, known as Feed-Forward Neural Networks (F CNN). The design of CNN is an example of hierarchical feature detectors that use biologically inspired neural networks since it can learn highly complex features and objects it can identify individuals. These are just some of the compelling reasons why CNNs are commonly used instead of traditional models. Researchers have thus far been intrigued by using the principle of weight sharing to reduce the number of training parameters, which could also improve prediction accuracy. Fewer parameters mean fewer degrees of freedom; therefore, we can effectively program CNNs without losing accuracy. The classification stage also involves the feature extraction stage, which uses both common knowledge and understanding to compile the results. Furthermore, using enormous neural networks in large networks is more complex than using ANNs for other models. CNN's [14] perform well and are frequently used in areas such as image analysis, entity detection, face detection, speech-to-enabled text messaging, and vehicle identification for people with diabetes.

4.1 ConVol model

It is customary in this type of neural network for a model to have a single input and output layer and many hidden layers. A neuron applies an algorithm to an algorithm with input vector X to produce an output vector as in the general formula: $F(P,R) = Q$ where R is the quantity vector indicating the degree of intercommunication between neurons in different adjacent layers. The quantity vector that was obtained can now be used to perform image classification. There is a large amount of literature on image classification based on pixels. Contextual details, such as the image's shape, produce good outcome. CNN is a best model that is grabbing popularity due to its ability to identify objects based on contextual details. In Fig. 1, the general CNN model is depicted. A convolution layer, a pooling layer, an activation feature, and a completely connected layer are the four components of a general CNN model. The functionality of each part is shown in below Fig. 2.

4.1.1 Convolution layer

The input layer gets a classified image, and the class mark expected is calculated with extracted image features. The local connection between an individual neuron in the next layer and a few neurons is known as the receptive field in the previous layer. The field of reception is used for removing local characteristics from the image. The receptive field of a neuron associated with a specific region of the previous layer generates a weight vector that at all points of the plane remains the same as the plane where the neurons in the next layer are referred to. As neurons on the aircraft share the exact weights, the associated features at different locations in the data can be found. This is illustrated in Fig. 2. A weight vector, also known as the filter or kernel, slides through the input vector to create the feature chart.

Fig. 2 CNN model structure.

Convolution refers both horizontally and vertically to the shape of the filter to slide. It extracts N number of functionality in one layer from the input image, resulting in N filters and N maps. Due to the local receptive field phenomenon, the number of trainable parameters is significantly reduced.

4.1.2 Pooling layer
The exact location becomes less important once a feature is detected. Thus, in the convolution layer a pooling or subsampling layer follows. It reduces considerable the number of parameters to be trained and introduces invariance in the translation system. A pooling window is chosen and the window inserts are passed through a pooling function, as shown in Fig. 3. The pooling function produces another vector. There is little pooling technique such as an average pooling and max pooling, the most used of which is max pooling which reduces map size considerably. During calculation errors, the error will not be propagated back to the winning device because the flow is not involved.

4.1.3 Fully connected layer
The fully connected layer is identical to the full connected network in conventional models. In the fully connected stratum is entered the first stage output (including convolution and multiple clustering) and the weight vector and input point product will be calculated for the ultimate output. Gradient decay, also known as batch mode learning, decreases the cost function by estimating costs over a complete data set, and only after a time when the entire data set has passed through an era, it updates its parameters. The result is a global minimum, but if the training data is large, the time required to train the network increases considerably. This cost reduction approach was replaced by stochastic gradient descent.

4.1.4 Activation function
In conventional machine learning algorithms, there is a large literature that uses the sigmoid activator function. Rectified Linear Unit (ReLU) has proven better than the first due to two main factors to introduce nonlinearity. First, it is easy to calculate the ReLU partial derivative. Second during the training period nonlinearity is foremost factor, in which ReLU function $Relu(x) = max(0,x)$ is faster when compared to Sigmoid function

$$\sigma\left(w^T x + b\right) = \frac{1}{1 + e^{-(w^T x + b)}}.$$

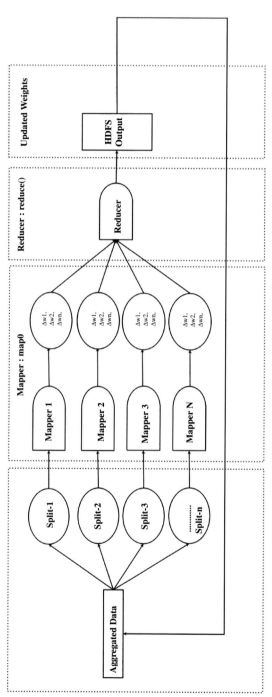

Fig. 3 MapReduce-based CNN.

5. Convolutional graph neural network: Evaluation

Convolutional can take advantage of MapReduce data processing. After each set of training data has been distributed, CNNs will update their gradient calculations each time. When the data is divided into many parts, parallelization can be executed during training because of many samples distributed. Next, multiple computers use different pieces of data to train its classifier. At this point, the results can be aggregated to yield new results, which are used to refine the parameters. Since fully trained GCNN is still ambiguous, because of the different combination of nodes is seen during the training phase. This may lead to a maximum number of nearest neighbor nodes with multiple computations. To overcome this issue and to decrease computational cost, we introduce mapreduce-based CGNN, which decomposes the data aggregation into the map, join and reduce stages. Mapper job is to map all the nodes of the graph with key-value pairs in which key is represented as Node Id and Value is the Node object. Join function will join subsequent nodes having higher priority and Reduce method is used to aggregate all the nodes which are with more unique while joining.

PinSage [15] an extremely scalable and random Graph Convolution Network, is able to learn embedding nodes in Web graphs with trillions of objects. This work overloads the background of product discovery and promotes the shopping future with Pinterest. The quality of learning embedding products on a variety of recommendation tasks is thoroughly assessed, with offline metrics, user studies and A/B tests showing a significant increase in recommended performance with a good improvement of impressions of our Shop the Look product.

By removing regions locally connected from graphs, PATCHY-SAN [16] is addressing these challenges. The model first determines the nodes ordered by a particular graphical approach and selects a fixed-length node sequence. Second, the problem of the arbitrary node size is being solved in a set-size neighborhood for each node. Finally, the neighborhood diagram is normalized according to the graphic labeling process, to which nodes with similar structural roles have a relatively similar position and the representational lessons with classic CNNs are followed. But PATCHY-SAN lacks flexibility and a wide range of applications because the graphic labelling approach often based only on the graphic structure determines the spatial order of the knots.

Hamilton et al. [17] propose the inductive aggregation-based model of representation learning called graphSAGE. A full version of the algorithm is simple: The node layer of graphSAGE (1) is added to a learning aggregator with vectors of all its immediate neighbors, (2) the representative vector u is combined to feed into one completely connected layer.

Xu et al. [18] suggests skip connection architecture to the Jumping Knowledge Network. The Jumping Knowledge Network can therefore choose the aggregates of the different layers of convolution. The intermediate images for each node can be selectively aggregated in the last model layer. The layer wide range includes the underlying aggregator, the max pool add-on and the LSTM attention aggregator. In addition, the Jumping Knowledge Network model recognizes the combination of the other current neural graph network models.

The above algorithms have a gap between spectral and spatial-based methods that is often considered to be bridging. However, the training process for large-scale graphs can be expensive in memory. In addition to this, GCN translation interferes with generalization, making it harder to learn the representations of the invisible nodes in the same graph and the nodes in a completely different graph.

To address the above issue, MapReduce-based Convolutional Graph Neural Networks (MapRed-CGNN) is the best approach to scale large graphs by generalizing and representing the nodes that are invisible and, thereby decrease the computational cost.

5.1 MapReduce-based convolutional graph neural networks

As a result of its use in a commodity cluster, MapReduce has been regarded as the de facto standard for large-scale data processing. Since it is an open-source, the Hops framework has gained much popularity with the online community. It is the Hadoop Distributed File System (HDFS). The Hadoop namespace has a name node and workflow nodes (data nodes), the central processing node is responsible for data, and the node manages the metadata. Data-mapper and data-reducer run on the nodes. When work is applied to a Hadoop machine, the data is broken down into equal-sized fragments. Each piece of data has one or multiple copies depending on the integrity configuration. Hadoop mappers copy and retrieve data by accessing nodes that are local or across the network. Finally, the final result is examined by the HDFS reducers, which combine, sort, and produce it.

5.1.1 Mapper of MapRedGCNN

Mappers load the input data (x, y, w1, w2, …, w$_n$) and a uniform distribution (u1, u, …, n) into the mapping, and random values (u, n) are assigned to the output data) The weights used to obtain an "I" results contain a night weight, such that the consequences used to get "I" layer results are the same. Y is just a representative of the truth for each instance in the training data. To make a network for training, you need to initialize the web with the given weights and input vectors and then train the network with the samples. The newly qualified group [W1, W2, W3…] is fed to the reducer, which reduces the new values [W$_n$]. After each step in the iterative process, new weight values are calculated and are fed into the mapping system.

5.1.2 Reducer of MapRedGCNN

The reducer takes the input from the mapper and applies the weighting function to produce the final output. Since the mapper output layer is cached, the reducer can index them by output weight. The calculation finds the total of all instances in the batch and divides it by the number of times you've trained. Weight calculation is used to determine the network results, which are then sent to the mapper.

5.1.3 Join of MapRedGCNN

The jobs in MapReduce use a driver to be the task planner. The driver produces a guided acyclic graph (DAG) or program implementation plan divided into smaller tasks. The driver communicates with Hadoop for maps and reducing classes and for job configuration to be specified. The user sets and includes the training path to the data, the output path, the training parameters, and the network settings. Training parameters include the number of exercise samples, number of validation samples, the maximum number of iterations, maximum times, and batch size. Network configurations comprise reception field size, stride, number and layer type, learning rate, and optimization method.

5.2 Benchmarking GCNN through public data sets

In every GNN model included currently in our benchmarking framework, we perform extensive experimentation on all data sets. The experiments help us to gain many insights, of which few are discussed:

Rasool et al. [19] have introduced a new graphics semi-supervised learning optimization framework. After clearly defining half-controlled problems with the adjacent distribution, we provided a comprehensive examination of subjects such as semi-controlled learning, graph neural network, and optimization of preconditioning (and NGD as its particular case). It commonly used a probabilistic framework for lower-squared regression and classification of cross-entropy.

In order to solve node classification tasks of quotation graphs, Dabhi et al. [20] propose a model using the NodeNet NGL. It surpassed different data sets with good accuracy.

SplineCNN [21], a spline-based neural network that features a new trainable convolution operator, learns about irregularly structured geometric data input. Convolution filter works in the space domain, adding local features and using a continuing, trainable kernel function with B-spline controller values parametrized. It showed that in many benchmark tasks, including the graph classification, graph node classification and the form correspondence on meshes, SplineCNN can improve the state of the art results while enabling extremely fast workout and inference calculations.

Hongwei Wang et al. [22] proposed an end-to-end model to unify the node classification GCN and LPA. The LPA is used as a standardization to help GCN learn proper edge weighs and to improve the classification performance in this model. The model can also be considered as learning weights based on node labels that are task-oriented rather than existing function-based attention models. GCN-based methods for the node classification are very accurate on different data sets in a number of experiments with real world graphs.

AdaGCN (Adaboosting Graph Convolutional Network) has the ability to extract knowledge efficiently from the top-notch neighbors of current nodes, and integrates knowledge from different neighbors' hops into the network into a new RNN, a newer RNN-style neural network architecture that integrates AdaBoost to network computation. AvaGCN [23] has the same basic neural network architecture among all "layers," different from other graph neural networks that stack numerous graph convolution layers directly and are recursively optimized similar to RNN. The links between AdaGCN and existing graphic convolutionary methods, which present the benefits of the proposal, are also theoretically established. Extensive experiments show the consistent state-of-the-art graphical prediction with good performance on multiple data sets.

The quantitative comparison of the above techniques on different data sets like Citeseer, Cora and PubMed has shown with good accuracy on Node Classifications, which is shown in Fig. 4. Still the above models are less accurate with multidimensional and huge graphs, i.e., BigGraphs. So

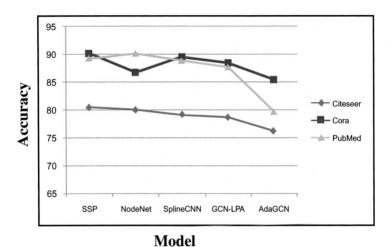

Model

Fig. 4 Accuracy of different models.

to overcome this, an integration of MapReduce and GCNN-based approach is expected to perform with more accuracy on all different benchmark data sets.

6. Applications and challanges of CGNN

6.1 Applications of CGNN

6.1.1 Web recommender systems

Graphs obviously emerge as part of the user interactions in e-commerce platforms with the products, resulting in many companies that use GNNs to recommend their products. A standard case of use is for modeling user and product graphic interactions, learning node embedding with certain negative sampling losses and using the kNN index to retrieve users' similar information in real time.

6.1.2 Combinatorial optimization

Combined optimization (CO) solutions are the workplace for a number of significant applications in finance, logistics, energy, the life sciences and hardware design. Graphs formulate most of these problems. Throughout the past century, many inks have been used to resolve CO problems more effectively through algorithmic approaches; however, the ML–driven revolution of modern computing has provided a new and compelling way to learn solutions. In Gasse et al. [24] a chart network was proposed that would be able to learn branch–and–bound variable selection policies: a critical step in MILP. The learned representations therefore try to minimize the solver's time and have shown that the inference time and the quality of the decisions are a good compromise.

6.1.3 Computer vision

Since objects in the world are deeply connected, GNNs can also help images containing these objects. One way to perceive the image is through the scene graphs, a set of objects in the image and their connections. Applications for image collection, understanding and argumentation, subtitling, visually answering questions and image generation have been found in scene graphs showing that it can improve performance greatly. Not only can GNNs simulate the dynamics during the transition, but also provide a better understanding of the influence of particles over time and distance.

6.2 Pharma-drug discovery

Pharmaceutical companies are intensively seeking a new paradigm of drug discovery with billions of dollars in research and development and tough competition. Graphs can represent interactions on different scales in biology. At a molecule, the edges of an amino acid residue can be the connections of atoms in a molecule or the interactions between them. Graphs can represent interactions in a larger scale between more complex structures, such as proteins, mRNA or metabolites. The charts can be used for target identification, molecular property prediction, high-performance screening, novel drug design, protein engineering, and re-purposing of drugs depending on the specific abstraction level.

6.3 Challenges of CGNN

6.3.1 Scalability

One of the challenging vital factors is the scalability that limits industrial applications, which often require large diagrams and low latency constraints (consider a Twitter social network with hundreds of millions of nodes and trillions of edges). This aspect has almost been overlooked by the academic research community until recently, with numerous models described in literature insufficient for large sizes. In addition, graphics hardware (GPU), which has been gladly married to traditional deep learning architects, is not necessarily the best for graphs. Specialized graphics hardware is needed.

6.3.2 Dynamic graphs

Dynamic graphs are another foremost challenge now-a-days where charts are a common way in which systems are modeled, such an abstraction is often too simplistic as real-world systems evolve and are dynamic. It is the temporal behavior that sometimes gives the system crucial insights. Despite some recent developments, it remained open to research to design graphical neural network models that can efficiently handle continuous-time charts as a stream of nodes or edge-specific events.

6.3.3 Complex structures

In complex networks, higher-order structures such as motifs, graphs, or simplified complexes are known to be relevant, such as the description of biological protein–protein interactions. However, the most neural graph networks have only nodes and edges. By adding such structures to the communication mechanism, graphic models could be more expressive.

7. Conclusion

In connected graph analysis, we found few problems, such as graph mining, graph coordination, and graph queries for big graphs. We looked at the challenges of scaling current methods on a large graph. The critical analysis shows us difficulty working on the existing techniques for massive graph data sets and sets the platform for measuring them using the map/reduce paradigm. In future work, we shall consider whether MapReduce is the best analysis model for charts or problems of the same class or whether the highest number of hidden layers of convergence function can better solve other paradigms. The performance and scale-up of graph analysis procedures in massive data sets on multiple machines would be analyzed thoroughly.

References

[1] D. Yan, J. Cheng, M.T. Özsu, F. Yang, Y. Lu, J.C. Lui, Q. Zhang, W. Ng, A general-purpose query-centric framework for querying big graphs, Proceedings of the VLDB Endowment 9 (7) (2016) 564–575.

[2] M. Han, K. Daudjee, K. Ammar, M.T. Özsu, X. Wang, T. Jin, An experimental comparison of pregel-like graph processing systems, Proceedings of the VLDB Endowment 7 (12) (2014) 1047–1058.

[3] Y. Tao, Y. Xia, T. Xu, X. Chi, Research progress of the scale invariant feature transform (sift) descriptors, J. Convergence Inf. Technol. 5 (1) (2010) 116–121.

[4] A. Mittal, A. Dhawan, S. Manchanda, S. Medya, S. Ranu, A. Singh, Learning heuristics over large graphs via deep reinforcement learning, arXiv (2019) (preprint arXiv:1903.03332).

[5] X. Liang, X. Shen, J. Feng, L. Lin, S. Yan, Semantic object parsing with graph lstm, in, European Conference on Computer Vision, Springer (2016) 125–143.

[6] K. Wu, G. Liu, J. Lu, Graph-based node finding in big complex contextual social graphs, Complexity (2020).

[7] J. You, R. Ying, X. Ren, W. Hamilton, J. Leskovec, Graphrnn: generating realistic graphs with deep auto-regressive models, in: International Conference on Machine Learning, PMLR, 2018, pp. 5708–5717.

[8] R. Liao, Y. Li, Y. Song, S. Wang, C. Nash, W.L. Hamilton, D. Duvenaud, R. Urtasun, R.S. Zemel, Efficient graph generation with graph recurrent attention networks, arXiv (2019) (preprint arXiv:1910.00760).

[9] Y. Li, B. Qian, X. Zhang, H. Liu, Graph neural network-based diagnosis prediction, Big Data 8 (5) (2020) 379–390.

[10] S. Liu, T. Li, H. Ding, B. Tang, X. Wang, Q. Chen, J. Yan, Y. Zhou, A hybrid method of recurrent neural network and graph neural network for next-period prescription prediction, International Journal of Machine Learning and Cybernetics 11 (12) (2020) 2849–2856.

[11] Y. Lu, Y. Chen, D. Zhao, B. Liu, Z. Lai, J. Chen, Cnn-g: convolutional neural network combined with graph for image segmentation with theoretical analysis, IEEE Transactions on Cognitive and Developmental Systems. (2020).

[12] X. Yan, T. Ai, M. Yang, H. Yin, A graph convolutional neural network for classification of building patterns using spatial vector data, ISPRS journal of photogrammetry and remote sensing 150 (2019) 259–273.

[13] R. Elshawi, O. Batarfi, A. Fayoumi, A. Barnawi, S. Sakr, Big graph processing systems: state-of-the-art and open challenges, in: 2015 IEEE First International Conference on Big Data Computing Service and Applications, IEEE, 2015, pp. 24–33.

[14] S. Indolia, A.K. Goswami, S. Mishra, P. Asopa, Conceptual understanding of convolutional neural network-a deep learning approach, Procedia computer science 132 (2018) 679–688.

[15] R. Ying, R. He, K. Chen, P. Eksombatchai, W.L. Hamilton, J. Leskovec, Graph convolutional neural networks for web-scale recommender systems, in: Proceedings of the 24th ACM SIGKDD International Conference on Knowledge Discovery & Data Mining, 2018, pp. 974–983.

[16] M. Niepert, M. Ahmed, K. Kutzkov, Learning convolutional neural networks for graphs, in: International conference on machine learning, PMLR, 2016, pp. 2014–2023.

[17] W.L. Hamilton, R. Ying, J. Leskovec, Inductive representation learning on large graphs, arXiv (2017) (preprint arXiv:1706.02216).

[18] K. Xu, C. Li, Y. Tian, T. Sonobe, K.-I. Kawarabayashi, S. Jegelka, Representation learning on graphs with jumping knowledge networks, in: International Conference on Machine Learning, PMLR, 2018, pp. 5453–5462.

[19] M. Rasool Izadi, Y. Fang, R. Stevenson, L. Lin, Optimization of graph neural networks with natural gradient descent, arXiv (2020) (e-prints (2020) arXiv–2008).

[20] S. Dabhi, M. Parmar, Nodenet: a graph regularised neural network for node classification, arXiv (2020) (preprint arXiv:2006.09022).

[21] M. Fey, J.E. Lenssen, F. Weichert, H. Müller, Splinecnn: fast geometric deep learning with continuous b-spline kernels, in: Proceedings of the IEEE Conference on Computer Vision and Pattern Recognition, 2018, pp. 869–877.

[22] H. Wang, J. Leskovec, Unifying graph convolutional neural networks and label propagation, arXiv (2020) (preprint arXiv:2002.06755).

[23] K. Sun, Z. Lin, Z. Zhu, Adagcn: Adaboosting graph convolutional networks into deep models, arXiv (2019) (preprint arXiv:1908.05081).
[24] M. Gasse, D. Chételat, N. Ferroni, L. Charlin, A. Lodi, Exact combinatorial optimization with graph convolutional neural networks, arXiv (2019) (preprint arXiv:1906.01629).

About the authors

U. Kartheek Chandra Patnaik Uriti is Research Scholar in NIT, Silchar and Faculty at Lendi Institute of Engineering and Technolgy. He is specialized in Medical Image Processing and Big Data. His interest towards research and technology is the positive strength of his success right from the begining of his career.

Dr. Ripon Patgiri has received his Bachelor Degree from Institution of Electronics and Telecommunication Engineers, New Delhi in 2009. He has received his M. Tech. degree from Indian Institute of Technology Guwahati in 2012. He has received his Doctor of Philosophy from National Institute of Technology Silchar in 2019. After M.Tech. degree, he has joined as Assistant Professor at the Department of Computer Science & Engineering, National Institute of Technology Silchar in 2013. He has published numerous papers in reputed journals, conferences, and books. His research interests include distributed systems, file systems, Hadoop and MapReduce, big data, bloom filter, storage systems, and data–intensive computing. He is a senior member of IEEE. He is a member of ACM and EAI. He is a lifetime member of ACCS, India. Also, he is an associate member of IETE. He was General Chair of 6th International Conference on Advanced Computing, Networking, and Informatics (ICACNI 2018, http://www.icacni.com) and

International Conference on Big Data, Machine Learning and Applications (BigDML 2019, http://bigdml.nits.ac.in). He is Organizing Chair of 25th International Symposium on Frontiers of Research in Speech and Music (*FRSM 2020*, http://frsm2020.nits.ac.in) and International Conference on Modeling, Simulations and Applications (CoMSO 2020, http://comso.nits. ac.in). He is convenor, Organizing Chair and Program Chair of 26th annual International Conference on Advanced Computing and Communications (ADCOM 2020). He is guest editor in special issue "Big Data: Exascale computation and beyond" of EAI Endorsed Transactions on Scalable Information Systems. He is also an editor in a multi-authored book, title *"Health Informatics: A Computational Perspective in Healthcare"*, in the book series of *Studies in Computational Intelligence*, Springer. Also, he is writing a monograph book, titled *"Bloom Filter: A Data Structure for Computer Networking, Big Data, Cloud Computing, Internet of Things, Bioinformatics and Beyond"*, Elsevier.

## CHAPTER EIGHT

Fast exact triangle counting in large graphs using SIMD acceleration

Kaushik Ravichandran, Akshara Subramaniasivam, P.S. Aishwarya, and N.S. Kumar
Department of Computer Science and Engineering, PES University, Bangalore, India

Contents

Abstract

Triangle counting is a cornerstone operation used in large graph analytics. Hash-based algorithms for triangle counting fail to take advantage of available vectorization in modern processors. Linear algebraic-based methods often involve sparse matrix multiplication which is inherently expensive. Nonlinear algebraic methods have a slow implementation and count each triangle multiple times which is then rescaled to obtain the exact triangle count. We propose a fast vector instruction implementation of a set operation-based triangle counting algorithm, which avoids matrix multiplication and finds the exact triangle count directly. Our implementation outperforms reference implementations proposed by the MIT graph challenge and miniTri when tried on about 40 graphs from the SNAP large network dataset, giving a speedup ranging from 41× to more than 1500×. A comparison against existing state-of-the-art techniques gave a speedup of 3× on average. We additionally show that this algorithm can easily be plugged into graph frameworks that either use or can be modified to use

Advances in Computers, Volume 128
ISSN 0065-2458
https://doi.org/10.1016/bs.adcom.2021.10.003

compressed sparse row like representation, to give faster compute times. We also propose an optimization to the k truss decomposition algorithm that can be used with the
optimized triangle counting algorithm to give better performance.

1. Introduction

Advancements in technology have led to the real world becoming
richly interconnected which graphs aim to mimic accurately. Hence,
research on graph structures and algorithms has accelerated and is critical
to a variety of domains. Triangle counting serves as a key building block
for many important graph algorithms. A triangle can be defined as a set of
three mutually adjacent vertices in a graph, as shown in Fig. 1. A triangle
is a subgraph that conveys information about the cohesiveness of a graph,
and is commonly used in computing metrics such as clustering coefficients
and transitivity for the graph. The number of triangles in a graph or a subgraph is important information used in applications such as social network
mining, link classification, subgraph isomorphism [1], spam detection and
the like [2].

A large amount of work exists on improving algorithms that compute the
exact number of triangles in a graph. A simple brute-force approach to triangle counting involves enumerating all triplets of nodes, and counting
those which form triangles. The triangle counting problem was initially
made more efficient by translating it to a matrix multiplication problem
[3,4] which has been solved using linear and nonlinear algebraic methods
over the years. Fast and simple shared memory parallel algorithms using
techniques such as merging and hashing for intersecting adjacency lists [5]
have been used to approach this problem. Various other specialized algorithms based on the MapReduce framework [6], the wedge sampling idea
[7], have also been proposed.

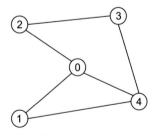

Fig. 1 Sample graph with triangle {0,1,4}.

Various communities such as Minitri [8], GraphBLAS [9] tackle optimizing graph algorithms from various aspects. The MIT/Amazon/IEEE Graph Challenge [1] has been pivotal in pushing the graph algorithms community forward which has led to much advancement to the performance of the triangle counting algorithm with the use of innovative hardware, software, GPUs. This has reflected in improvements on power efficiency, computational efficiency, and scale [10,11]. The authors have presented benchmarks and metrics for standardizing evaluation of such algorithms, which have been the most commonly used baseline since its inception.

We propose an exact triangle counting algorithm using a modified compressed sparse row (CSR) representation and set operations. Many algorithms perform additional computations to count the triangles more than once. Our algorithm gives the exact triangle count directly, without the need for rescaling. Our algorithm also does not perform any matrix multiplications, thus avoiding the shortcomings associated with sparse matrix multiplications entirely. The set operations use vector processors to allow faster processing of list elements. Our goal is to obtain an exact count of triangles, not an approximate one, and also does not involve triangle enumeration. We present a comparison of our performance against the Graph Challenge Baseline [1], Minitri [8], a linear algebraic approach (GraphBLAS) [9], and Ligra [12].

The structure of the rest of this chapter is as follows. In Section 2, we give a brief idea on the base algorithm for triangle counting using set operations. Section 3 discusses the data structure used and the optimizations made, in detail followed by a discussion of the results in Section 4. We further cover the applicability of the research in two parts, in Section 5. In the first part, we propose an optimization for k truss decomposition algorithm which can be coupled with triangle counting to tackle the subgraph isomorphism problem efficiently. In the second part, we compare performance gains made by our algorithm when plugged into graph processing frameworks. We finally conclude in Section 6 with a discussion on the next steps of this research.

2. Triangle counting algorithm

There are several techniques for triangle counting in graphs that vary depending on the data structure used. We will discuss an approach that uses set operations to count the number of triangles, as shown in Algorithm 1. The input graph is a simple, undirected graph G, represented by a set of edges E and a set of vertices V. A data structure such as the adjacency list

ALGORITHM 1 Triangle counting using set operations.

Result: The number of triangles in graph G

1 numTriangles = 0;

2 G = simple, undirected graph (V, E);

3 V = {1, 2, 3 ... n};

4 E = {$(u_1, v_1), (u_2, v_2)...(u_m, v_m)$} where (u_1, v_1) represents an edge;

5 **for** *i in V* **do**

6   nb(i) = neighbours of i in G;

7   **for** *j in nb(i)* **do**

8   nb(j) = neighbours of j in G;

9   numTriangles += $nb(i) \cap nb(j)$;

10   **end**

11 **end**

12 numTriangles = numTriangles/6;

13 return numTriangles;

is used to optimize the retrieval of neighbors for a node. The underlying principle of the algorithm is that the number of triangles that an edge, (u_1, v_1), is a part of can be obtained from the set intersection of the neighbors of the two vertices, u_1 and v_1, that form the edge. This holds true as the common neighbors of the two vertices that form an edge represents the third vertex of the triangle. In this approach, each triangle is counted six times, twice for each edge. This count is then rescaled, to get the exact triangle count.

3. Optimizations

In this section, we discuss the optimizations made to the baseline algorithm for triangle counting using set operations. The data structure used to store the graph for triangle counting is similar to the CSR format with an additional list maintained to facilitate edge removal by forward referencing. The representation of the sample graph in Fig. 1 is given in Fig. 2. Two offset lists and one edge list are maintained and initialized during construction of the graph. The neighbors for each vertex in the edge list, is stored in

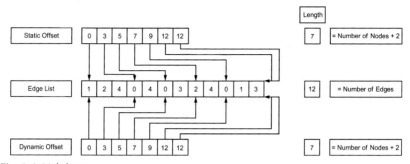

Fig. 2 Initial data structure.

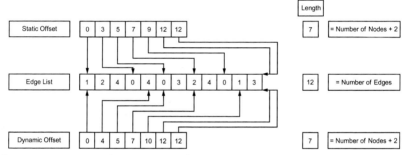

Fig. 3 Data structure after removing edges (1, 0) and (4, 0).

increasing order to ensure ordered traversal which aids us in correctly identifying the edges to be deleted, as shown in Fig. 3. Since the graph is undirected, the edges form an ordered set, which implies that each edge will be stored twice, as (i, j) and (j, i).

The graph is serialized to ensure that the index of the offset lists represents the vertex id. The values in the offset list stores the starting index of the list of neighbors to that vertex, in the edge list. The offset lists are first initialized for the entire graph. The static offset list is unaltered, whereas the dynamic offset list is modified to allow forward referencing. Index "i" in the dynamic offset list stores a reference to the beginning of the list of neighbors for vertex i. Similarly, index "$i + 1$," in the static offset list stores a reference to the end of the list of neighbors for a vertex i. Thus, both offset lists are required to extract the neighbors of a vertex. The optimized algorithm using the CSR format is given in Algorithm 2.

The base algorithm counts a triangle formed by vertices i, j, k six times, twice for each edge. These additional computations are redundant, and can

ALGORITHM 2 Optimized triangle counting using set operations.

Result: The number of triangles in graph G

1 numTriangles = 0;

2 staticOffset = $[x_1, x_2...x_n]$;

3 edgeList = $[v_1, v_2, v_3...v_m]$;

4 dynamicOffset = $[x_1, x_2...x_n]$;

5 **for** i *in len(staticOffset)* **do**

6   nb(i) = edgeList[dynamicOffset[i]] to edgeList[staticOffset[i+1]];

7   **for** j *in nb(i)* **do**

8   **if** $j > i$ **then**

9   break;

10   **end**

11   nb(j) = edgeList[dynamicOffset[j]] to edgeList[staticOffset[j+1]];

12   numTriangles += $nb(i) \cap nb(j)$;

13   dynamicOffset[j]++;

14   **end**

15 **end**

16 return numTriangles;

be avoided by taking advantage of the vertex-based ordering of the elements in the data structure. We do so in two steps, by first reducing the number of duplicate counts from 6 to 3 and subsequently incorporating edge deletion to count only unique triangles.

3.1 Exact counting

Given two vertices of a triangle i and j, the triangle is counted twice for that edge, once for the pair of vertices (i, j) and once for the pair (j, i). We can avoid one of these counts for every edge, by additionally adding a constraint to count the triangle only if $i < j$, as shown in lines 8–9 in Algorithm 2. This ensures that all the triangles for each edge are counted only once, which brings down the multiplicity of count of each triangle down to 3, as one count per edge per triangle is reduced.

Given a triangle with vertices i, j, and k where $i < j < k$, we now count a triangle thrice, for the edges (i, j), (j, k), and (i, k) of the triangle. After counting the triangle for the first pair, which is always edge (i, j) since $i < j < k$, we can simply remove the edge (i, j) from the graph which will no longer be needed. We will not count the same triangle for edges (j, k) and (i, k), since edge (i, j) has been removed. This solution is complete because our CSR is ordered, as well as the constraint we added earlier, that $i < j$. This edge removal is done by simply incrementing the start index value corresponding to the node id in the dynamic offset list. By doing so, we remove the edges in a nondecreasing order which is in harmony with the rest of the algorithm, which requires traversal of the vertices in the same order.

3.2 Vectorized intersection

The algorithm for vectorized sorted set intersection was adopted from the work done by Schlegel et al. [13]. This algorithm allows us to process multiple elements of a list of elements at once, depending on the register size used. The algorithm mentioned in this section uses streaming SIMD extensions (SSE) [14], which can process four elements at once by taking advantage of the full 128 bits in a register. This can be extended to instruction set extensions such as advanced vector extensions (AVX) and AVX-512 [15] which allows use of 256 bit and 512 bit register sizes, respectively.

The building block of the intersection algorithm requires finding the intersection of two short sorted arrays referred to as segments, each with four 32 bit elements stored in 128 bit registers. We compare one segment with all cyclic shifts of the elements in the other segment and obtain one bit mask as a result of each comparison, thus resulting in 4 bit masks. A logical disjunction of these bit masks gives the common elements between the two segments.

Fig. 4 shows how this idea can be used to find the intersection of two sorted lists. The example shows two lists, A and B, with 12 elements each. Each list is divided into three segments. Each comparison operation takes two segments, one from each list, and produces two outputs, the first is a list of common elements between the segments compared, and the second is the segment with the larger last element. The first output forms a part of the final output and the second output is compared with the next segment from the other list, to obtain two more outputs. We successively perform this comparison until there are no more segments in one of the lists, and copy over the remaining distinct elements from the other list. Using this technique, we can efficiently compute the intersection of two 12 element lists using 5 comparisons, which can be extended to larger list sizes.

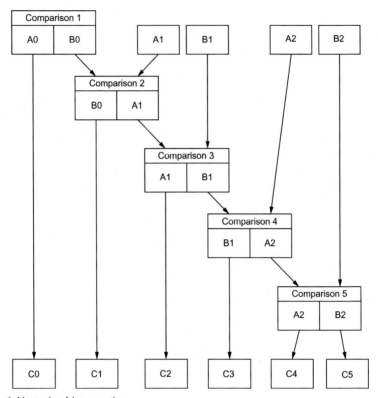

Fig. 4 Vectorized intersection.

4. Evaluation

In this section, we evaluate our performance with the reference implementations for triangle counting, on 40 graphs from the Stanford Large Network Dataset collection (SNAP) [16].

The experiments were carried out on a 64-bit Intel(R) Xeon(R) CPU E5-2676 v3 @ 2.40 GHz on a Haswell architecture. In all experiments, the input data is stored in the main memory and uses the optimized CSR format for our implementation. Our implementation was compiled using g++ for c++17 with the -flto and -O3 flags enabled. All the benchmarks were compiled using the default flags provided for best performance. Every number quoted is the execution time to count the number of triangles in the graph, for an average of 20 test runs. For the task, correctness is evaluated by comparing the reported triangle count with the actual triangle count for the input graph, obtained from the SNAP dataset. As correctness is a necessary metric for our task, 100% correctness is ensured for all runs.

We first evaluated our performance against a serial C++ implementation of miniTri [8] and the benchmark algorithm provided by MIT/Amazon/IEEE Graph Challenge [1] in Julia (fastest among the provided implementations of C++, Julia, and Python). We compared against the above mentioned triangle counting algorithms as they are widely used reference implementations in the graph algorithm benchmarking community. We also provide a comparison with two state of the art triangle counting algorithms, one that uses GraphBLAS [9]—one of the fastest implementations of the Triangle Counting algorithm in the recent times and Ligra [12]—a lightweight graph processing framework that optimizes for shared memory and multicore machines. The performance of our algorithm is evaluated using the following metrics:

- *Execution time*: Time required to count the triangles in milliseconds (preprocessing time is not included in the runs for any of the algorithms).
- *Rate*: The ratio of the number of edges in the graph to the execution time.
- *Speedup*: The ratio of the execution times of a given benchmark algorithm to that of our implementation.
- *Rate v/s no. of edges*: A plot of the rate v/s number of edges shows how well the algorithm scales for larger graph sizes.

4.1 Results

The time taken and the speedup for our algorithm, the graph challenge baseline and the miniTri algorithm are shown in Table 1. Our algorithm outperforms miniTri and Graph Challenge Baseline with a significant speedup, averaging about 796 and 1080, respectively, which shows large performance benefits against standard benchmarking references. A few of the runs for the benchmark implementations, indicated by a hyphen in the table, were not done as the implementation was too memory/computation intensive. Additionally, we compare our speedups obtained against miniTri with the speedups quoted by the serial GraphBLAS-based approach [9] on the Haswell architecture. The comparable speedups obtained is only indicative that our algorithm is at par with the approach. We outperform Ligra [12] in all datasets to obtain a speedup of at least 2.5, as shown in Table 1.

Table 1 Results.

Dataset	Ours Time taken (ms)	Graph challenge baseline Time taken (ms)	Graph challenge baseline Speedup (Ours)	miniTri Time taken (ms)	miniTri Speedup (Ours)	miniTri Speedup (graphBLAS)	Ligra Time taken (ms)	Ligra Speedup (Ours)
amazon0302	49.56	1999.94	41	5916.78	120	**135**	130	2.62
ca–AstroPh	24.98	3760.71	151	14071.9	**564**	346	78.40	3.14
ca–CondMat	7.06	2885.62	409	1505.71	**214**	187	20.90	2.96
ca–GrQc	0.76	2408.80	3162	117.48	155	**195**	2.60	3.41
ca–HepPh	18.98	3065.73	162	17224.30	**908**	427	58.20	3.07
ca–HepTh	1.38	2464.10	1790	154.67	113	**150**	5.30	3.85
cit–HepPh	53.91	5341.22	100	29998.86	**557**	358	442	8.20
cit–HepTh	54.72	–	–	44638.70	**816**	542	443	8.10
email–Enron	24.81	–	–	27535.59	**1111**	591	62.40	2.52
email–EuAll	27.51	–	–	123308.5	**4483**	2367	87.1	3.17
facebook_combined	12.50	2873.61	230	9252.46	**741**	350	38.80	3.10
loc–brightkite_dges	20.01	4151.14	208	14249.04	**713**	412	50	2.50
oregon1_10331	2.01	3123.92	1555	2487.22	1239	**1373**	5.80	2.89
oregon1_10407	2.09	3115.19	1494	2466.19	1183	**1316**	5.80	2.78
oregon1_10414	2.16	3138.12	1454	2587.07	1199	**1359**	6.10	2.83

oregon1_10421	2.22	3143.17	1418	2661.63	1201	**1273**	6.20	2.80
oregon1_10428	2.26	3155.56	1425	2651.70	1198	**1357**	6.10	2.75
oregon1_10505	2.18	3159.87	1449	2691.20	1234	**1371**	6.10	2.80
oregon1_10512	2.20	3163.65	1438	2673.38	1215	**1411**	6.20	2.82
oregon1_10519	2.30	3119.86	1355	2703.83	1175	**1330**	6.20	2.69
oregon1_10526	2.22	3181.03	1434	2762.77	1246	**1330**	6.30	2.84
oregon2_10331	3.36	3222.96	959	3529.72	**1050**	865	9.40	2.80
oregon2_10407	3.36	3212.35	955	3591.83	**1068**	911	9.36	2.78
oregon2_10414	3.51	3233.01	922	3723.59	**1061**	866	9.70	2.76
oregon2_10421	3.48	3245.66	932	3632.83	**1043**	908	9.50	2.73
oregon2_10428	3.44	3259.15	948	3838.67	**1117**	886	9.50	2.76
oregon2_10505	3.38	3255.38	965	3829.82	**1135**	932	9.20	2.73
oregon2_10512	3.45	3257.13	945	3860.71	**1120**	936	9.40	2.73
oregon2_10519	3.61	3271.81	908	3888.42	**1079**	926	9.90	2.75
oregon2_10526	3.66	3350.03	917	4024.00	**1101**	918	10.10	2.76
p2p-Gnutella04	2.15	2561.64	1194	228.64	107	**164**	5.50	2.56
p2p-Gnutella05	1.73	2531.90	1462	196.46	114	**166**	4.40	2.54
p2p-Gnutella06	1.72	2528.83	1472	186.49	109	**167**	4.30	2.50

Continued

Table 1 Results.—cont'd

Dataset	Ours	Graph challenge baseline		miniTri			Ligra	
	Time taken (ms)	Time taken (ms)	Speedup (Ours)	Time taken (ms)	Speedup (Ours)	Speedup (graphBLAS)	Time taken (ms)	Speedup (Ours)
p2p-Gnutella08	1.08	2466.59	2278	160.29	149	**197**	2.80	2.58
p2p-Gnutella09	1.37	2466.18	1803	188.45	138	**192**	3.60	2.63
p2p-Gnutella24	3.26	2753.45	845	306.60	95	**144**	8.50	2.61
p2p-Gnutella25	2.58	2704.23	1048	228.58	89	**137**	6.90	2.67
p2p-Gnutella30	4.40	2807.70	638	399.97	91	**139**	11.60	2.63
p2p-Gnutella31	7.41	3006.43	406	684.55	94	**134**	20.00	2.70
soc-Epinions1	82.57	–	–	–	–	–	218.00	2.64
soc-Slashdot0811	83.21	–	–	–	–	–	220.00	2.64
soc-Slashdot0902	91.61	–	–	–	–	–	239.00	2.62

Bold values indicate the faster of the two speedups being compared.

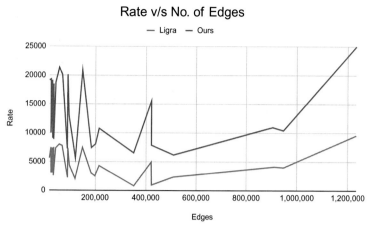

Fig. 5 Rate v/s no. of edges.

A comparison of the rate (Edges/s) vs number of edges for our algorithm against Ligra is shown in Fig. 5. This graph depicts the number of edges processed per second with an increase in the graph size, which indicates the order of compute time. As the number of edges increases, we notice that Ligra shows a small increase whereas our implementation shows a steep rise in the number of edges processed per second. From this we can see that we perform notably better with an increase in size of the graph, which would be significant given an exponential increase. It further indicates that the data level parallelism exploited in our algorithm helps us scale better.

5. Applicability

The triangle counting algorithm is integral to several applications, including community detection, spam detection and link recommendation. Apart from this, triangle counting plays a significant role in graph operations such as clique detection and k truss decomposition. Below, we propose an idea to optimize k truss decomposition that can be used for graph analytics, together with the optimized triangle counting.

5.1 Truss decomposition

In massive networks, it is more feasible to work on the smaller and important core areas in the graph, as we can get a good approximation of the results quickly. The k truss decomposition algorithm is one such method that can be used to obtain a smaller subset of the huge network. A k-truss for a simple, undirected graph G is defined as the largest possible subgraph of G with k nodes, such that each edge in the subgraph is incident to at least k-2 triangles. The task of k truss decomposition involves finding all such subgraphs for all possible values of k for the graph, where $k \in [2, k\text{-max}]$ where k-max is a constant derived from the input graph. There are two approaches to truss decomposition which were explored in detail by Jia Wang et al. [17], both of which require triangle counting:

- *Bottom up approach*: This approach is used to find the subgraphs for all values of k, starting from the smallest possible value of $k(= 2)$, incrementally building on the initial solution for larger values of k. This is the faster approach for a general case and is used when a list of subgraphs for all values of k is required.
- *Top down approach*: In this approach, we start by finding the k truss decomposition for the largest possible value of $k(= k -\text{max})$, further working our way down, iteratively till $k = 2$. It is slower than the bottom up approach for the general case, but faster when only the subgraphs for the larger values of k are required.

Since the top down and bottom up approaches can work independently using the same representation of the graph, we can formulate the concept of obtaining the truss decomposition for all values of k, by using both techniques in parallel. The bottom up approach is used to find the required subgraphs for all $k = 2$ up to k equal to a certain value "C," and at the same time, the top down approach would find the required subgraph from the maximum value of $k(= k -\text{max})$ till k equal to "C." This improves the performance due to the existing parallelism between the two approaches. For example, if k-max $= 50$ and $C = 40$, the top down approach can find the decomposition for the top 10 values of k in the same duration that it takes the bottom up approach to find the decomposition for the first 40 values of k. We can now find the k truss decompositions of the graph for all k, in the time it takes to find $C(= 40)k$ truss decompositions. This optimization when implemented with our optimized triangle counting algorithm can be used to obtain better results for the k truss decomposition task as a whole.

5.2 Frameworks

Several graph frameworks [12,18–20] contain common graph operations and use similar underlying principles to perform important graph functionalities in an optimized manner. This can be achieved by mapping the program into high performing sparse matrix operations [19] or based on domain specific languages [18]. The CSR format used by our algorithm is simple to generate, thus making the algorithm easy to use within various frameworks. We were successfully able to plug in our algorithm to two frameworks with ease. They evaluate the performance of our algorithm when used within it, against the one currently used.

GAP benchmark suite: The GAP benchmark suite (GapBS) [20] is intended to help graph processing research by standardizing evaluations. This benchmarking suite also provides optimized reference implementations for a range of graph analytic algorithms. The reference implementation provided for the triangle counting algorithm is representative of state-of-the-art performance. Our algorithm outperformed the triangle counting algorithm provided by gapbs on most of the real world graphs from SNAP, giving speedups ranging from 1× to 10×, with an average speedup of about **3.26 ×**.

Ligra: Ligra is a lightweight graph processing framework developed by Shun et al. [12] that optimizes for shared memory and multicore machines. A comparison of our performance against the current algorithm in the framework is captured in Table 1, showing an average speedup of about **3.04 ×** when tried on several real world graphs from SNAP. This shows that our easy to use algorithm performs at least 2.5× better than the current algorithm used within Ligra.

6. Conclusion

We propose an exact triangle counting algorithm using a modified CSR representation and set operations. The algorithm gives the exact triangle count directly, without the need for rescaling. The set operations implemented uses vector instructions which allows faster processing of the list elements. Our algorithm when compared against the reference implementations of MIT Graph Challenge, miniTri, GraphBLAS, and Ligra showed encouraging results. We also show how the algorithm's simple CSR format makes it easy to use and can be easily plugged into graph frameworks such as the GapBS and Ligra.

This chapter is a first look into exact triangle counting using set operations implemented with vector instructions. There is scope for exploration with other vector extensions such as AVX-512 and more efficient set intersection algorithms which further exploit data level parallelism. An efficient implementation of the optimized *k* truss decomposition algorithm we propose, along with the optimized triangle counting algorithm can be used to get optimal performance for the former task.

References

[1] S. Samsi, V. Gadepally, M. Hurley, M. Jones, E. Kao, S. Mohindra, P. Monticciolo, A. Reuther, S. Smith, W. Song, Static graph challenge: subgraph isomorphism, in: 2017 IEEE High Performance Extreme Computing Conference (HPEC), IEEE, 2017, pp. 1–6.
[2] C.-Y. Kuo, C.N. Hang, P.-D. Yu, C.W. Tan, Parallel counting of triangles in large graphs: pruning and hierarchical clustering algorithms, in: 2018 IEEE High Performance extreme Computing Conference (HPEC), IEEE, 2018, pp. 1–6.
[3] J.R. Bunch, J.E. Hopcroft, Triangular factorization and inversion by fast matrix multiplication, Math. Comput. 28 (125) (1974) 231–236.
[4] A. Azad, A. Buluç, J. Gilbert, Parallel triangle counting and enumeration using matrix algebra, in: 2015 IEEE International Parallel and Distributed Processing Symposium Workshop, IEEE, 2015, pp. 804–811.
[5] J. Shun, K. Tangwongsan, Multicore triangle computations without tuning, in: 2015 IEEE 31st International Conference on Data Engineering, IEEE, 2015, pp. 149–160.
[6] T.G. Kolda, A. Pinar, T. Plantenga, C. Seshadhri, C. Task, Counting triangles in massive graphs with MapReduce, SIAM J. Sci. Comput. 36 (5) (2014) S48–S77.
[7] C. Seshadhri, A. Pinar, T.G. Kolda, Wedge sampling for computing clustering coefficients and triangle counts on large graphs, Stat. Anal. Data Mining ASA Data Sci. J. 7 (4) (2014) 294–307.
[8] M.M. Wolf, J.W. Berry, D.T. Stark, A task-based linear algebra building blocks approach for scalable graph analytics, in: 2015 IEEE High Performance Extreme Computing Conference (HPEC), IEEE, 2015, pp. 1–6.
[9] T.M. Low, V.N. Rao, M. Lee, D. Popovici, F. Franchetti, S. McMillan, First look: linear algebra-based triangle counting without matrix multiplication, in: 2017 IEEE High Performance Extreme Computing Conference (HPEC), IEEE, 2017, pp. 1–6.
[10] S. Samsi, V. Gadepally, M. Hurley, M. Jones, E. Kao, S. Mohindra, P. Monticciolo, A. Reuther, S. Smith, W. Song, Graphchallenge.org: raising the bar on graph analytic performance, in: 2018 IEEE High Performance extreme Computing Conference (HPEC), IEEE, 2018, pp. 1–7.
[11] S. Samsi, J. Kepner, V. Gadepally, M. Hurley, M. Jones, E. Kao, S. Mohindra, A. Reuther, S. Smith, W. Song, Graphchallenge. org triangle counting performance, in: 2020 IEEE High Performance Extreme Computing Conference (HPEC), IEEE, 2020, pp. 1–9.
[12] J. Shun, G.E. Blelloch, Ligra: a lightweight graph processing framework for shared memory, in: Proceedings of the 18th ACM SIGPLAN Symposium on Principles and Practice of Parallel Programming, 2013, pp. 135–146.
[13] B. Schlegel, T. Willhalm, W. Lehner, Fast sorted-set intersection using SIMD instructions, in: ADMS@ VLDB, 2011, pp. 1–8.
[14] AMD, AMD64 Technology: 128-Bit SSE5 Instruction Set, 2007. http://developer.amd.com/wordpress/media/2012/10/AMD64_128_Bit_SSE5_Instrs.pdf.
[15] P.K. Tiwari, V.V. Menon, J. Murugan, J. Chandrasekaran, G.S. Akisetty, P. Ramachandran, S.K. Venkata, C.A. Bird, K. Cone, Accelerating ×265 with Intel® Advanced Vector Extensions 512, in: White Paper on the Intel Developers Page, 2018.
[16] J. Leskovec, A. Krevl, SNAP Datasets: Stanford Large Network Dataset Collection, 2014.
[17] J. Wang, J. Cheng, Truss decomposition in massive networks, Proceedings of the VLDB Endowment 5 (9) (2012), https://doi.org/10.14778/2311906.2311909.
[18] D. Nguyen, A. Lenharth, K. Pingali, A lightweight infrastructure for graph analytics, in: Proceedings of the Twenty-Fourth ACM Symposium on Operating Systems Principles, 2013, pp. 456–471.

[19] N. Sundaram, N.R. Satish, M.M.A. Patwary, S.R. Dulloor, S.G. Vadlamudi, D. Das, P. Dubey, Graphmat: high performance graph analytics made productive, arXiv preprint arXiv:1503.07241 (2015).

[20] S. Beamer, K. Asanović, D. Patterson, The GAP benchmark suite, arXiv preprint arXiv:1508.03619 (2015).

About the authors

Kaushik Ravichandran is a Computer Science undergraduate from PES University who is from Bangalore, India. His rich past research experience includes internships at Carnegie Mellon University and Microsoft Research Labs India. He is interested in the fields of Distributed Query processing for big data, Optimization and Graph Algorithms. He is keen on advancing scalable systems by tackling performance bottlenecks to empower researchers and data scientists with the compute power that they require.

Akshara Subramaniasivam is a Computer Science undergraduate from PES University who is from Bangalore, India. She is interested in the fields of Data Analytics, Machine Learning and related areas. Her past research in these fields have been lauded with multiple best paper awards - including one from the Government of India. She is keen on using her in-depth knowledge in advanced analytics to improve the quality of information retrieved from the data source and drive the research community forward.

Aishwarya Poomuttam Sreedas is a Computer Science undergraduate from PES University who is from Bangalore, India. She has accumulated vast knowledge and professes a diverse skill set which she sharpened in her past internships at Adobe and Intuit. Her interests are in the fields of Product Development, Algorithms and Optimizations. She is keen on using her knowledge of algorithms to build the next generation of customer centric products that are robust, efficient and guarantee an exemplarily smooth customer experience.

Kumar NS completed his Bachelors degree from UVCE, Bangalore in 1983 and Masters in Computer Science from IIT Madras in 1986. He worked as a Scientist in ISRO Satellite Center and Gas Turbine Research Establishment (DRDO). He is currently a visiting professor in the department of Computer Science at PESU Bangalore and is also on the board of studies of Computer Science at PESU. He offers consultancy in Object Oriented Analysis and Design, in porting legacy applications and interfacing software components and in system programming in the UNIX domain.

CHAPTER NINE

A comprehensive investigation on attack graphs

M. Franckie Singha and Ripon Patgiri
Department of Computer Science and Engineering, National Institute of Technology Silchar, Silchar, Cachar, Assam, India

Contents

Abstract

In the last few decades, the influence of the internet on human civilization is worth remembering. Today, starting from an individual to a big organization, everyone is dependent on the internet. The rise in numbers of computers, sensor devices, IoT devices, and the expansion of the network has made the internet and internet users prone to cyber-attacks. This chapter discusses attack Graph, a graph-based analysis of vulnerabilities and security in a network to discover the possible attack path that an intruder could take, and suggest preventive measures. Various attack graph models have been studied throughout the chapter; also the complexities and scalability are being discussed.

Advances in Computers, Volume 128
ISSN 0065-2458
https://doi.org/10.1016/bs.adcom.2021.10.004

1. Introduction

In this era of the internet, securing a network has been a concern. In simple term, network security is the practices and innovation a business sets up to ensure its IT foundation. This framework, thus, is comprised of all the information, programs, applications, web organizations, programming, and equipment. Network security breaks are typical, and a few happen throughout the planet consistently. Some are minor, with slight loss of information or money-related assets, yet a large number of them are major or even disastrous. The larger the network size, the more loopholes might generate in the network.

1.1 Necessity of securing a network

Attackers are ceaselessly searching for the new weaknesses to abuse. At the point when organizations are not gotten, data about associations and people, and surprisingly, our administration, are in danger of being uncovered or utilized against us. Organizational security is significant for home organizations, just as in the business world. Most homes with fast web associations have at least one remote switch, which could be exploited if not appropriately secured. A strong organizational security framework decreases the danger of information misfortune, robbery, and damage. Besides, network security is so significant just on the grounds as we live in a computerized first world. This computerized first world is simply ready to progress, as well, as an ever-increasing number of individuals anticipate regular administrations, exchanges, and data to be promptly accessible readily available, any place they are, out of nowhere.

1.2 Graph base cyber attack

With the increase of technology and researches in graph theory, an attacker now uses various tools that construct a graph showing the path from the least secured node to the primary nodes. The attack generally starts with collecting all the required information to generate the graph. The information includes the data like router configuration, machines connected, access control list, the network topologies. Once the path is generated, the attacker goes from one node to another, collecting all the credentials and finally reaching the main node, which controls the network system.

This same method can be applied to protect the network attacks from intruders. Since the network administrator will have more privilege and has more specific information about the whole network, a reverse path can be generated. Moreover, since the attacker will construct a graph with a small amount of specification that he has gathered, but the administrator will construct a graph with all required information giving him a full graph rather than a partial one that the intruder has constructed. Even if the attacker has a small amount of data and permission, they could generate a graph that could lead them to the admin node. From the defender's point of view, having all the data and resources, network configuration, and access privileges, the network defender can construct a graph that is more specific than the intruder's graph. This leads to the concept of attack graph and its uses in identifying the vulnerabilities in the network.

1.3 Attack graph

An Attack graph is a graph-based representation of a network attack. It is a representation of the network graphically, which helps in the analysis, prevention, and detection of the vulnerabilities in the network. To build an attack graph, the first component that is needed is the primary network configuration, credentials, access-level, network element information, and other data sources as demanded by the model to be generated. The Attack graph shows all the nodes with vulnerabilities associated with it. The node having highest the probability of attack may be called a vulnerable node. Also, it shows the machines that have admin privileges that would lead to complete control of the network when compromised. This information can be used to prevent the network from being attacked by simply removing those vulnerable nodes from the attack graph.

The model of attack graph differs from model to model. Some of the popular attack graph models have been discussed in the next subsection.

1.4 Attack graph models

Researchers have applied various methods to generate the attack graph to analyze the security issue in the network. Phillips and Swiler [1] uses an approach for generating the attack graph that is not an automatic generation of the graph but a manual approach. The nodes represent the attack state, and the edges represent the state change. Sheyner et al. [2] uses a model checking technique to build the attack graph. He uses NuSMV model checker for his experiment and uses XML to store all the specification of the network and passed onto a compiler that can translate it into NuSMV language. NuSMV is a tool that builds the attack graph. Xinming Ou et al. [3] also uses model checking to generate the logical attack graph. Boolean variables are used to modeling the graph and result of the tool is a graph generating all possible attack scenarios. The MulVAL tool is used to generate the logical attack graph. It is a network analyzer, which can perform multihost, multistage vulnerability analysis on a network. Ammann et al. [4] uses an algorithm based on a search graph to generate the attack graph with an assumption that the attack relies on monotonicity. Monotonicity states that once the precondition is satisfied, it cannot be unsatisfied. Kyle Ingols et al. [5] uses NetSPA tool builds multiple-prerequisite (MP) graph. NetSPA is used to import data from various sources. The data like root, configuration, credentials, network topologies are collected and thereby compute the reachability. Noel and Jajodia [6] use security conditions and aggregation rules to build the graph. An exploit dependency graph is used while generating the graph. The use of such a graph removes the duplicate paths while building the graph. Ghazo et al. [7], in their chapter, uses a model-based Automated Attack-Graph Generator and Visualizer (A2G2V) to generate the attack graph. Since it is complex, difficult, and error-prone to generate a large attack graph manually proposed a model that automatically generates the graph given the complete network configuration.

1.5 Methods for building and analyzing the attack graph

Sheyner et al. [2] model of Attack graph represents the network in the form of a finite state machine. Every transition in the graph shows the attack generated by the attacker. The architecture of the model is given in Fig. 1 below. As shown in Fig. 1, given the specification, the model checker NuSMV will produce the attack graph. The specifications are given in XML format to the compiler, specially designed which takes the input as XML and outputs it into the language understand by NuSMV. The output is then given to the model checker to generate the attack graph. Once the attack graph is generated, the vulnerability analysis can be done. The analysis can be done in two ways. First, determining the minimal set of atomic

Fig. 1 Sheyner's model of generating attack graph.

attacks that must be prevented so that the attacker does not go beyond that point. Second, finding the probability of being detected by the intruder.

Unlike the model suggested by Sheyner, Ammann [4] suggested a model of attack graph is generated in polynomial time rather than exponential. The graph is modeled such that the nodes represent the attributes (A) and the edges (E) represents exploits. These exploits can be a combination of pre-condition and post-condition of the attack. The construction of the graph is done layer-wise. Given the network specification, the attributes are divided into two sets as shown in Fig. 2: (1) Set of attributes that satisfy the initial state (2) Set of attributes which require further exploit, i.e., unsatisfied attributes. The circle in Fig. 2 represents the attributes. The straight line represents the initial separation layers. The unsatisfied attributes are also arranged into layers. Those attribute which is satisfied in single exploit, forms the layer1 and those which are satisfied in two exploits form the layer2 nodes, and so on. The goal node is a node having only one attribute. The layers are shown in Fig. 2, which is taken from the paper [4]. Once the graph is built, the analysis for the following can begin:

a)   Finding the minimal path: An attack is minimal if there is no exploit that can be deleted from the attack besides causing some different exploit's preconditions to become false.

b)   Finding the shortest attack path: To discover an attack that can be generated with the least number of steps from the initial state.

Generation of attack graph leads to a lot of complexity and scalability issue when the network size increases. The attack graph represented using a state-based graphs where the nodes define the attributes and edges define the exploits have severe scalability problems. Noel and Jajodia [6], in their method of generating the attack graph, uses exploit dependency graph

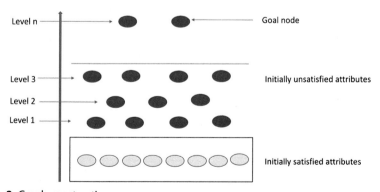

Fig. 2 Graph construction.

applies aggregation techniques to handle the complexity of attack graph representation. The exploits appear only once in the whole graph, which makes it simple. It gives a compact high-level view of the attack scenario. Hierarchical aggregation is used to join the subgraphs to a single graph. The aggregation is based on some common properties of attack graph elements. The rules for aggregation are as shown in Fig. 3, taken from the paper [6]. Fig. 3 shows the hierarchical aggregation of the nodes. As shown in Fig. 3, the leaf nodes exploit, and the condition are aggregated to form the exploit set and condition set, respectively. The condition set is aggregated to the machine for the machine satisfying the same condition, and some are aggregated to exploit set. The set of machines and exploit set together get aggregated to either machine exploit set or protection domain. This hierarchical aggregation of nodes helps in the interactive visual analysis of an attack graph. Since nonaggregated graphs and aggregated graphs with a single level view are very complex to view from a user's perspective, they have used a mixed level view. It represents the whole graph showing only the number of machines and exploits in a particular subnet. It also shows the

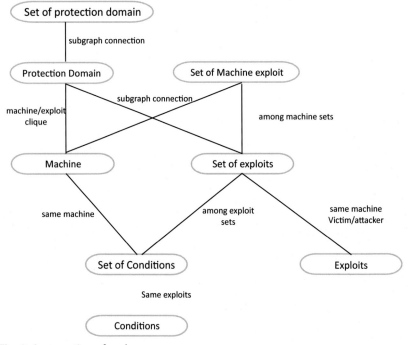

Fig. 3 Aggregation of nodes.

flow of attack from one subnet to another. It hides the individual machines and the associated attack on the machine explicitly, giving a high-level abstraction. The framework uses the Nessus vulnerability scanner to automatically build a network vulnerability model, which includes various modeled possible exploits. The framework creates an attack graph of exploits and their dependencies based on a starting machine and network attack target, enabling users to explore the attack graph through interactive visualization. Their research is the first one to represent the attack graph scenario in an interactive way.

Phillips and Swiler [1], while modeling the attack graph, where nodes represents the attack. The node is generated by a combination of various attributes or network specifications like machines, access levels. The edge represents the change in the state caused by a single attack. Each edge has weight which represents the success probability of going from one state to other. Moreover, applying shortest path algorithm on the graph gives a short path representing the low-cost attack. The weight can be viewed as a function of the configuration file, and attack profile. The attack graph here is a function of the configuration file, attack templates and attacker profile. Attack templates contain data like Operating system, version, configuration files contain the data of network, printers, routers while the attacker profile has the attackers capabilities to breach the system. The generation of a graph can be done from the start node or the goal node. All the generated nodes are kept in a queue. Initially, queue contains only the goal node, and nodes are added as they are created gradually. Analysis of the attack graph can be done to find the low-cost path, cost-effective paths to defend the attack and simulate dynamic attacks. Xinming Ou [3] uses a Logical attack graph, which is a directed graph that can also be represented in the form of a tree. It creates a logical correlation between the attack goals and the configuration of the network. A Logical attack graph has two nodes derivation nodes and fact nodes. The fact nodes are further divided into primitive nodes and derived fact nodes. Fact nodes represent the logical statement and derivation nodes represent the interaction rule for the derivation step. Edges in the graph represent the dependency relation between the nodes.

Fig. 4 represents a logical graph where nodes F1, F2, F3 ... are the fact nodes and D1, D2, D3 ... are the derivation nodes, and the point represents the primitive nodes. The fact nodes generate rules, and derivation nodes generate facts that satisfies the rules produced by fact nodes. Since facts can be derived using different rules, the node F3 in the above figure generates two derived nodes. For security assessment, MulVAL has been used.

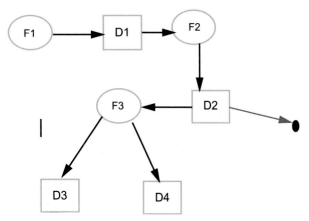

Fig. 4 Logical attack graph.

MulVAL is a reasoning tool for automatically figuring out protection vulnerabilities in networks. The key concept in the back of MulVAL is that most configuration records can be represented as Datalog tuples, and most assault strategies and OS protection semantics can be specific to the use of Datalog rules. The reasoning engine of MulVAL uses XSB, which is a Prolog system. It evaluates the Datalog interaction rules on input facts. The MulVAL engine records the trace of the assessment and forwards it to the graph builder, where it generates the logical attack graph as output. As shown in Fig. 5, all the necessary condition like network configuration, Machine configuration, and security measures are given to the MulVAl in the form of datalog. The MulVAL engine iteratively generates all the derivation steps or attack traces, based on the rules set for interaction. This attack traces are used for generating the attack graph.

kyle Ingols [5] model of attack graph uses NetSPA tool to extract data from various sources, which are required to construct the graph. The model builds a Multiple Prerequisite graph as it is faster and more expressible compared to a predictive graph. Nessus is used to show the vulnerabilities at the host, port, and interfaces. The reachability is calculated by finding the reachability matrix. The simplest way of computing it is to reach every known port and target IP address from each host in the network. Reachability matrix is built such that the row of which defines the source interface and the column defines the target post. A part of the matrix is collapsed into reachability groups which save large amounts of time and memory. This matrix generally contains a duplicate path, which is removed by collapsing the matrix such that the matrix contains only the non-redundant paths. The set of rules used for filtering are collapsed into

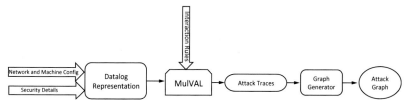

Fig. 5 The architecture of MulVAL.

Binary Decision Diagrams (BDDs), which reduce the traversal of filtering rules to a constant time. While constructing the MP attack graph, it considers three types of nodes:

a)   State nodes: It represents the access level of the attacker on a particular host.

b)   Prerequisite nodes: It represents a group of reachable nodes.

c)   Vulnerability instance: It represents the specific vulnerabilities present at a particular port.

The Breadth-first technique is used to construct the graph. Each node in graph is explored only once. Moreover, a node can appear in the graph if the attacker can obtain that node. The generated an MP graph contains no duplicate path, which reduces the complexity of the graph. Also, if we travel the MP graph in a depth-first manner, we get the full graph, and breadth-first traversal will give predictive graph. Unlike a predictive graph and full graph, MP graph can model the credentials. Ghazo [7], on the other hand, uses a model-based Automated Attack-Graph Generator and Visualizer (A2G2V) to generate the attack graph. A model checker is used, which is a tool that finds the security breach in a network. The model checker used in this paper is Jkind. While formulating the attack graph, the model requires the system's security configuration and the system-level properties of the system. The configuration is the data like the network elements and their connections, services used by the system. All these data are collected using Architecture Analysis and Design Language (AADL). It is an architecture description language made to construct for modeling both software and hardware components. The network is represented using Architecture Analysis and Design Language whereas, the vulnerabilities are represented using AGREE. A2G2V and AGREE (Assume Guarantee Reasoning Environment) will together generate the vulnerabilities, and the JKind model checker iteratively builds the new paths. The paths will then be combined and shown visually using the Graphviz tool. The working of Fig. 6 is described here. The security model is defined as $M = (S, E, s_0)$ where S is the locations and s_0 is the initial location which has the security information of

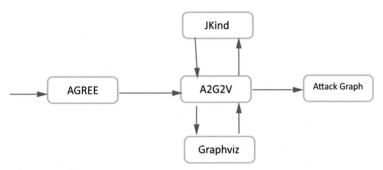

Fig. 6 A2G2V architecture.

the network, E represents the transition from one state to another after an attack. The security property is defined by φ. The input to the AGREE is the network description generated by AADL and Annex. The output of AGREE is in the form of Model M and security property φ, which is given to the A2G2V. Jkind on the satisfaction of the model M and the property φ will report that φ is satisfied, else Jkind will generate a path that the system will take to violate the property φ. This will continue, and gradually attack graph will be formed and visualization is done using the Graphviz tool.

The overall methods used in various papers stated above are listed in a tabular form in Table 1. Sheyner uses the model checking method to generate the graph, while Philips and Swiler model also uses model checking but uses search-based algorithm to find the path having the highest risk. Though Ammann has used a search-based graph model, the generation of the graph is done layer-wise. Noel and Jajodia use exploit dependency graph and Aggregation techniques to build the attack graph, which turns out to be more scalable and efficient. Kyle Ingols uses NetSPA to gather all the specifications and generates the attack graph, whereas Xinming Ou uses MulVAL, which generates attack traces for building the attack graph. Ghazo uses A2G2V along with Jkind model checker.

1.6 Scalability of models

Scalability in networking is a property to handle the growth of a network as the number of nodes in the network system increase. It has always been an issue to maintain the scalability and performance of a network together. The same has been the case in generating the attack graph. As the number of attributes in building the graph increases, the size of the attack graph increases. The drawback of the model suggested by Sheyner [2] is scalability. Even for small amount of host and vulnerabilities, the output states are

Table 1 Method of building Attack Graph.

Author	Method of attack graph generation
Sheyner et al.	Model checking
Philips and Swiler	Search graph-based model
Ammann et al.	Algorithm based on search
Noel and Jajodia	Exploit dependency graph and Aggregation techniques
Kyle Ingols	NetSPA tool on multiple-prerequisite
Xinming Ou	MulVAL and logical programming
Ghazo	A2G2V and Jkind model checker

extremely large, out of which only a few states are reachable. For 91 bits, the graph generates 2^{91} states out of which 101 states are reachable. To compute 229 bits, 5948 nodes and 68,364 edges are generated that took 2 h in the generation, which is an overhead. Phillips and Swiler [1], in their work, managed to handle the scalability for a small network, but with a large dataset, it leads to an exponential growth of the graph. The reason behind this exponential growth is the generation of duplicate paths while building the attack graph. Since most of the research published on attack graph uses a state-based graph to build the graph, which has a scalability problem, Noel and Jajodia [6], uses an exploit dependency graph instead. This has reduced the complexity to quadratic from exponential growth, but the model fails when the size of the network increases. Kyle Ingols [5] also considers the scalability problem while formulating the attack graph. Prior to their work, most of the researchers have used fully graph or predictive graph in order to build attack graph but have large time complexity. The full graph shows all the possible paths but takes a time complexity of $O(n!)$ as it contains duplicate paths. Some have used predictive graph, where the redundancy in the path is reduced to some extent. Also, building such a graph is much easier than the full graph, yet the path redundancy has existed considerably. The author here uses an attack graph generated based on an MP graph that increases linearly as the size of the network increases.

The scalability comparison of various models has been shown in Table 2.

1.7 Visualization of the attack graph

Graphviz is open source plan visualization software. Graph visualization is a way of representing structural data as diagrams of summary graphs and networks. It has necessary purposes in networking, software engineering,

Table 2 Scalability comparison of various models.

Author	Scalability
Sheyner et al.	Scalability problem, due to duplicate paths
Philips and Swiler	Scalability problem, exponential growth in the graph
Noel and Jajodig	Not feasible for large network
Kyle Ingols	Increases linearly as the size of the network increases

database and, web design, and visual interfaces for other technical domains. In Sheyner [2] model of attack graph, he uses GraphViz package to visualize the graph. Ghazo [7] has also used the same tool to visualize the graph generated by its model. Noel and Jajodia [6], on the other hand, uses Open source Nessus vulnerability scanner to visualize the graph. Nessus is one of the many vulnerability scanners used for the duration of vulnerability assessments and penetration checking out engagements, such as malicious attacks. From Nessus vulnerabilities, we discovered to be applicable to progressive network penetration. Not all these modeled exploits may appear in an assault graph, only those represented through genuine vulnerabilities on the scanned machines in the network.

Machine readability of attack graphs has always been an issue. Many researchers have tried to improve the visibility of the graph, but feasibility is an issue. Jooyong Lee et al. [8] suggested a method to increase the machine readability of a large attack graph. He uses ontology technology to describe the network configuration and represents the relationship among them based on the big data. The ontology is generated using (Resource Description Framework) schema and OWL (Web Ontology Language). Ontology helps machine in reasoning and learning the various relationship among the network components. The semantics are used on the edge of MP Graph to provide more precise information to the machine for increasing the readability and its inference. Mariam Ibrahim et al. [9] uses a windows visualizer application that builds a GUI for user to view the attack scenario properly. The GUI represents the graph with the number of rows of the produced attack scenarios and represents them in a visually pleasing, user-friendly manner.

1.8 Complexity analysis of various attack graph models

Complexity characterizes the behavior of a system or model whose components interact in multiple ways. Generation of attack graph of a system with low complexity is fairly simple, but if the system's components increase, then the complexity of generating the attack graph increases both in time and space. It is a general concept that as the configuration of the system, the attributes, the input to the graph generator model increase the output attack graph will grow exponentially if proper methods are not used. The complexity of attack graph generation in Sheyner's model is exponential. The reason behind this exponential increase is the generation of a large number of duplicate paths. This overhead of redundant paths has not been considered while modeling the architecture. As the size of the network is polynomial, the size of attack graph increases exponentially. Phillips and Swiler also experienced the exponential growth problem when the size of the network configuration increased. Unlike the model suggested in Refs. [1,2], Ammann, the graph is generated in polynomial time rather than exponential. The graph is modeled such that the nodes represent the attributes(A), and the edges(E) represent exploits. The time complexity for computation is given as $O(|A|^2.|E|)$, which can be simplified to $O(N^6)$. In the Xinming Ou model, the time complexity for generating the logical graph is $O(N^2)$. The experiment is run on a CPU of Pentium 4, 4.3 GHz, 1GB RAM. Microsoft Windows XP Professional is used to run the test. The experiment is run for various network topologies like star, ring, fully connected and the CPU time is found to be O(N2). Kyle Ingols model does not allow duplicate paths in the graph as it will considerably increase the complexity in graph generation, which is unnecessary. Thus, the model removes all the duplicate paths, thus reducing the time complexity to $O(E + N \log N)$. The test is done on 252 host with 3777 port and 8585 vulnerabilities on pentium-M 1.6 Hz,1 GB RAM machine having Linux kernel. The output MP graph has 8901 nodes and 23,315 nodes to it. The model not only reduces the time complexity but also reduces the space complexity.

The complexity analysis of the authors have been compiled in a tabular form and presented in Table 3.

Table 3 Complexity comparison of various models.

Author	Complexity
Sheyner's attack graph	Network size is polynomial and state is exponential
Philips and Swiler	State is exponential
Kyle Ingols	O(E + NlgN)
Ammann and Swiler	$O(N^6)$
Xinming Ou	$O(N^2)$

2. Attack graph in security and vulnerability assessment

Phillips and Swiler model of attack graph finds the vulnerabilities by building the attack graph, but due to the presence of configuration file and attacker profile in the template used, it may itself become a vulnerability to the system. Noel and Jajodia, in the paper [10], suggest analyzing the low-level vulnerabilities to achieve high level threat. In other words, the independent threat when combined may lead to a greater threat. Topological Vulnerability Analysis (TVA) tool is used to analyze the low-level threats in the network. TVA is composed of three components: a modeled exploits based on knowledge, description of the network, and an attack scenario. TVA uses these three components to generate the attack path, based on the combined model. Once the TVA model is formed, the analysis of critical path can be done. Based on the attack paths generated the vulnerabilities can be found out. It is a tool that also automates penetration testing. Though TVA seems useful, but it is not feasible for large network as the attack graph will not be scalable. To improve the network security Yimin Cui et al. [11] has suggested an attack graph based on a probabilistic model which enhances the security by adding various other factors in analyzing the network. A Classical probabilistic approach is used to analyze the graph, given the information about the network components like devices, vulnerabilities. The probabilistic graph is generated from attribute graph by removing the loops, assuming a monotonic attack. Nodes are added to the graph after scanning the security properly. The analysis of security is done by quantifying the network vulnerabilities and then determining the security measures based on each scenario of vulnerabilities. Their experiment shows considerable improvement in security analysis. Attack graph has been used in IOT based industries in recent years to find the vulnerability. Huan Wang et al. [12] suggest finding the vulnerabilities by using attack graph and maximum flow. Common Vulnerability Scoring System (CVSS) is used to find the security measures which will increase the quantification degree in the attack path. The paper focuses on finding the optimal attack path using maximum loss flow, as more the maximum loss, more is the risk in the network. Analysis of the network is done by building a global attack graph with no loop. Maximum loss is calculated using CVSS based on which the best attack path is determined. Then, the vulnerabilities are determined, which are causing that path insecure. The generation of the attack graph takes $O(n_{ep}.n_p)$, and the complexity of finding the optimal path is $O(n_n^2)$, where n_{ep} is attack instances, n_p is number of reachable elements and n_v is number of vulnerable nodes. Rashmi Sahay et al. [13], on their paper uses an attack graph to find the vulnerabilities caused due to misappropriation in the rank property of RPL (IPv6 Routing Protocol over Low power and Lossy network). The attack graph is generated by finding out all the threats in Rank Properties. RPL is a routing protocol for wireless networks with low power consumption. It generates a DAG (directed cyclic graph). Each node in the graph is associated with a property called rank property. Rank keeps on increasing its value as it moves far away from the root node. RPL can further be subdivided into DODAG (Destination Oriented Directed Acyclic Graphs). Rank is one of the property which is used to identify the nodes and DODAG. Rank is used to find the loop in a network, generating routes, construction of DAG, path validation and discovery of neighboring nodes. If there is any misappropriation in rank value then it could compromise with security issues. To prevent this, an attack graph is used to generate a graph that could show all the vulnerabilities and possible attack paths caused due to rank property. Fig. 7 is taken from the paper [13], which represents an attack graph, showing all the vulnerabilities. Fig. 7 represents an attack graph where nodes are the state when a

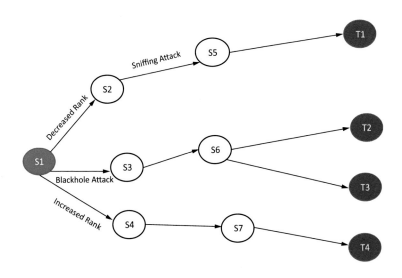

S1: Malicious Node
S2,S3 … S7 : Various Vulnerabilities
T1,T2 … T4 : Terminal compromised nodes

Fig. 7 Attack graph representing the exploits.

certain vulnerability is exploited while the edges show the postcondition of the previous state when that vulnerability is being exploited. As shown in the figure, S2 node has been generated when S1 node has been exploited by Decrease rank attack. When S2 is again compromised with the sniffing packet attack it goes to the state S5 and so on.

3. Case study of attack graph

Attack graph has been used extensively in various industries and firm to protect their network from intruders. Attack graph has provided a game-changing way of analyzing the network vulnerabilities and to protect and defend the network. It has made one of the best use of graph to save and protect the data from getting manipulated, corrupted, or from handling in illegal ways from cyber attackers.

3.1 Case-1: Detection of vulnerabilities in cyber physical system using Attack graph

Mariam Ibrahim et al. [9] uses architecture analysis and design method is applied on model-based attack graph to determine the threat in cyber physical systems. The paper has presented how the basic unit of Cyber Physical System (CPS) is compromised when loopholes exist in the unit and visualizing the system using an attack graph. Studies have been made on Pressurized water Nuclear Power Plant (NPP), Industrial Control System (ICS), Vehicular Network System (VNS). The specification required to build the model have generated using AADL and the security verification is done by Jkind model checker. The experiment is executed roughly based on the paper [7].

3.1.1 Pressurized water nuclear power plant(NPP)

Fig. 8 is an architecture of a water reactor taken the reference from paper [9]. The NPP has three parts, Primary which does heating and pressurizing of water, Secondary is a steam generator, and lastly Cooler condenses the steam back to liquid. Besides these, it has other layers, which include the field layer consisting of sensors, protection system, and heat emitter. The Control layer contains PLC and RTU, and Supervisor layer contains workstation, control room and other elements as shown in Fig. 8, Enterprise layer contains the management computer, access points. The network backbone contains control network which links control layer and supervisor layer while enterprise layer links supervisor layer and enterprise layer.

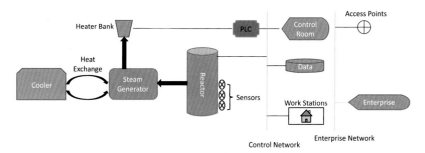

Fig. 8 Water reactor based on nuclear power plant.

The NPP has some vulnerabilities present that can be exploited. Those are the unsecured firewall which does not monitor data flow between supervisor and control layer, the main control room, workstations has OS vulnerabilities present and the PLC, sensor and the actuators used have firmware vulnerabilities. The attack model here uses AADL and AGREE to generate the system elements and security properties. This generated information are given to the Jkind model checker, which checks the system with a security property [7] and submits counterexample (CE). When all the CE are generated, it is exported in the form of CSV file and given to the windows visualizer. Once it is submitted the visualizer, the attack scenario can be seen by selecting "View Attack Scenario(s)," "Insert Attack Spread-Sheet," and "Generate Attack Graph" options from the GUI option. During the experiment, 16 CSV files were generated for 28 vulnerabilities. The CE-16 generated by Jkind has 220 rows which is very large. The windows visualizer condensed it by exporting only the attack scenarios and showing the simplified view in the GUI.

3.1.2 Industrial control system (ICS)

While conducting an experiment on the ICS the vulnerabilities which are taken into account is vulnerability on PLC, i.e., COTS and firmware. Some of the notable attacks that could be possible on ICS due to these vulnerabilities are: Malware injection, social engineering, sniffing, Dos attacks. Given the security property, the goal of the attacker is to generate system breakdown by either inducing a delay from actuators, control logic, and operating systems or changing the data entering them. The counterexamples will be generated by Jkind in the form of a spreadsheet. The experiment here generates six CEs, and it is observed that the most vulnerable component of the system is PLC which generally controls the actuators, control logic, and operating systems of the network.

3.1.3 Vehicular network system (VNS)

For vehicular network system, the vulnerabilities that were considered are Exhaustion of resources, Lack of Authentication in the VNS, Multicast Messaging which can be accessed by all access point in the network. The possible network attacks in the system are Trojan-Horse which can manipulate user data and exploit bluetooth connections, DoS attack, Sniffing, Attack to user's credentials.

Given the formal definition of the model like network description, pre- and postcondition of attacks, the vulnerabilities and the security properties, the attack scenarios can be generated. The JKind model checker creates the counter-example as a spread-sheet. The generate spreadsheet has six instances which are when encoded by gives 20 CEs, generating the complete attack graph. One of the attack path shows the attacker gaining the root access of user's phone and eventually gaining access to the main unit of the network and thereby knowing all the functions of the network send message to the VNS as a privileged user.

3.2 Case-2: Optimizing IoT devices using attack graph

It is an increasing concern that the rise in IoT devices has increased the risk of attack in the network system and devices. Moreover, IoT enabled devices are prone to attack due to their low resources, cost and firmware. Also, the position of the devices plays a major role in securing the devices from any attack.

Noga Agmon et al. in the paper [14] uses attack graph to deploy the IoT devices in a network such that it locates all the locations where IoT devices can be installed to analysis the risk of attack when a new device is added in the network. Also, the attack graph is used to optimize the following issues:

a)   Minimal security problem when a new device is to be installed in the system.

b)   The maximum number of device deployment without compromising the risk.

The vulnerabilities that exist while working with IoT devices are:

a)   Most of the IoT devices are connected to a single app, due to which it becomes very difficult to secure all the devices. Besides, all the devices are from a different vendor, and the network policies also differ.

b)   The number of IoT devices is very large, so implementing a security protocol that could secure each and every device is a hard task.

c) Due to the requirement of low resources and low functionality, it is very easy to breach an IoT device.

Moreover, the wireless communications are also not secured due to the focus on minimizing the Capital expenditure which has compromised for low security protocol.

According to the author, the network model is such that all the IoT devices are group according to their types. Also, it is made sure that the not more than one device of a type should reside in a particular location. Before the generation of attack graph and after the generation, while analyzing the graph all the communication protocols and the location constraints are considered. For each device, when the perfect location, communication range and a connectivity map is generated. To generate the attack graph MulVAL [3] is used. The attack graph model is inspired from the paper [3]. The experiment is run with a network having three types (detector, refrigerator, camera) of IoT devices, nine IoT devices (four detectors, two cameras, and three refrigerators) and eight locations. Into this environment, three detectors having four different possible locations, one camera for having two possible locations, and two refrigerators having two possible locations are to be deployed. When these information along with the constraints of communication, location and, vulnerabilities are given to the model, it generates an attack graph. From the graph generated, it is concluded that for optimizing minimal risk problem for deploying six IoT devices, the risk increases by 19% and the average number of IoT devices that can be deployed without compromising the risk is 4.40. But when the devices are plotted randomly without using attack graph the minimal risk problem has increased to 44% and the average number of IoT devices that can be deployed without compromising risk is 4.40.

3.3 Case-3: Risk analysis and security management using Attack graph

Denis Ivanov et al. in the paper [15] uses attack graph to analyze the risk and, security of a system. To generate the attack graph, the data required are the network configuration, the domain name of the host, the OS used. Besides this information, the attack graph is supplied with other information that will help in analyzing the risk and security of the graph. To know the importance of the nodes, it needs to know the type of the nodes and the services running. To identify the type of nodes Nmap security scanner is used. Each of the types has its own value, which represents the importance of the node.

The services running has also been assigned some value ranging between 0 and 100. The value of the node type and the value of the services will together calculate the criticality level of a certain node. After that, the downward risk value of all the nodes is calculated, whose value represents how much a node is vulnerable to affect other nodes. Finally, the risk of the whole system is calculated from the downward risk values. The prevention from attack is done by minimizing the risk of the whole by finding the node which is maximizing the risk value of the system.

The implemented system model has four functionalities as: processing of the data, assessment of the risk, module selection and visualization. The attack graph is generator is given the input in graph description language (DOT) from the module selector which is processed by the processing function. The risk assessment function calculates the overall risk of the system by iteratively comparing the risk value with the previous risk value by eliminating the risk causing nodes from the graph. The visualization function will generate the attack graph with minimum risk value.

4. Conclusion

This chapter focuses on the various attack graph models, their working and analyzes the various use cases of attack graph in practical scenario. Also, a comparative analysis of the various models in terms of time complexity, scalability issues has been done while generating the graph. From the survey of these papers, it is seen that the even though the attack graph is generated, but if the efficiency of the graph is not satisfactory, the graph cannot be implemented practically. To build an attack graph for efficient detection, protection and prevention of the cyber-attacks, the attack graph has to be less complex in terms of time and space. Moreover, the visualization of the graph plays an important role in analyzing the vulnerabilities in the network.

References

[1] C. Phillips, L.P. Swiler, A graph-based system for network-vulnerability analysis, in: Proceedings of the 1998 workshop on New security paradigms, 1998, pp. 71–79.
[2] O. Sheyner, J. Haines, S. Jha, R. Lippmann, J.M. Wing, Automated generation and analysis of attack graphs, in: Proceedings 2002 IEEE Symposium on Security and Privacy, IEEE, 2002, pp. 273–284.
[3] X. Ou, W.F. Boyer, M.A. McQueen, A scalable approach to attack graph generation, in: Proceedings of the 13th ACM conference on Computer and communications security, 2006, pp. 336–345.

[4] P. Ammann, D. Wijesekera, S. Kaushik, Scalable, graph-based network vulnerability analysis, in: Proceedings of the 9th ACM Conference on Computer and Communications Security, 2002, pp. 217–224.
[5] K. Ingols, R. Lippmann, K. Piwowarski, Practical attack graph generation for network defense, in: 2006 22nd Annual Computer Security Applications Conference (ACSAC'06), IEEE, 2006, pp. 121–130.
[6] S. Noel, S. Jajodia, Managing attack graph complexity through visual hierarchical aggregation, in: Proceedings of the 2004 ACM workshop on Visualization and data mining for computer security, 2004, pp. 109–118.
[7] A.T. Al Ghazo, M. Ibrahim, H. Ren, R. Kumar, A2g2v: automated attack graph generator and visualizer, in: Proceedings of the 1st ACM MobiHoc Workshop on Mobile IoT Sensing, Security, and Privacy, 2018, pp. 1–6.
[8] J. Lee, D. Moon, I. Kim, Y. Lee, A semantic approach to improving machine readability of a large-scale attack graph, The Journal of Supercomputing 75 (6) (2019) 3028–3045.
[9] M. Ibrahim, Q. Al-Hindawi, R. Elhafiz, A. Alsheikh, O. Alquq, Attack graph implementation and visualization for cyber physical systems, Processes 8 (1) (2020) 12.
[10] S. Jajodia, S. Noel, B. O'berry, Topological analysis of network attack vulnerability, in: Managing Cyber Threats, Springer, 2005, pp. 247–266.
[11] Y. Cui, J. Li, W. Zhao, C. Luan, Research on network security quantitative model based on probabilistic attack graph, in: ITM Web of Conferences, Vol. 24, EDP Sciences, 2019, p. 02003.
[12] H. Wang, Z. Chen, J. Zhao, X. Di, D. Liu, A vulnerability assessment method in industrial internet of things based on attack graph and maximum ow, Ieee, Access 6 (2018) 8599–8609.
[13] R. Sahay, G. Geethakumari, K. Modugu, Attack graph—based vulnerability assessment of rank property in rpl-6lowpan in iot, in: 2018 IEEE 4th World Forum on Internet of Things (WF-IoT), IEEE, 2018, pp. 308–313.
[14] N. Agmon, A. Shabtai, R. Puzis, Deployment optimization of iot devices through attack graph analysis, in: Proceedings of the 12th Conference on Security and Privacy in Wireless and Mobile Networks, 2019, pp. 192–202.
[15] D. Ivanov, M. Kalinin, V. Krundyshev, E. Orel, Automatic security management of smart infrastructures using attack graph and risk analysis, in: 2020 Fourth World Conference on Smart Trends in Systems, Security and Sustainability (WorldS4), IEEE, 2020, pp. 295–300.

About the authors

Mr. M. Franckie Singha is pursuing his Ph.D. degree at the Department of Computer Science & Engineering, National Institute of Technology Silchar, Assam, India. His research interest spans from Machine Learning, Big Graph, Security and Computer Networking.

Dr. Ripon Patgiri has received his Bachelor Degree from Institution of Electronics and Telecommunication Engineers, New Delhi in 2009. He has received his M.Tech. degree from Indian Institute of Technology Guwahati in 2012. He has received his Doctor of Philosophy from National Institute of Technology Silchar in 2019. After M.Tech. degree, he has joined as Assistant Professor at the Department of Computer Science & Engineering, National Institute of Technology Silchar in 2013. He has published numerous papers in reputed journals, conferences, and books. His research interests include distributed systems, file systems, Hadoop and MapReduce, big data, bloom filter, storage systems, and data-intensive computing. He is a senior member of IEEE. He is a member of ACM and EAI. He is a lifetime member of ACCS, India. Also, he is an associate member of IETE. He was General Chair of 6th International Conference on Advanced Computing, Networking, and Informatics (ICACNI 2018, http://www.icacni.com) and International Conference on Big Data, Machine Learning and Applications (BigDML 2019, http://bigdml.nits.ac.in). He is Organizing Chair of 25th International Symposium on Frontiers of Research in Speech and Music (*FRSM 2020*, http://frsm2020.nits.ac.in) and International Conference on Modelling, Simulations and Applications (CoMSO 2020, http://comso.nits. ac.in). He is convenor, Organizing Chair and Program Chair of 26th annual International Conference on Advanced Computing and Communications (ADCOM 2020). He is guest editor in special issue "Big Data: Exascale computation and beyond" of EAI Endorsed Transactions on Scalable Information Systems. He is also an editor in a multi-authored book, title "*Health Informatics: A Computational Perspective in Healthcare*", in the book series of *Studies in Computational Intelligence*, Springer. Also, he is writing a monograph book, titled "*Bloom Filter: A Data Structure for Computer Networking, Big Data, Cloud Computing, Internet of Things, Bioinformatics and Beyond*", Elsevier.

CHAPTER TEN

Qubit representation of a binary tree and its operations in quantum computation

Arnab Roy, Joseph L. Pachuau, and Anish Kumar Saha
Department of Computer Science and Engineering, National Institute of Technology Silchar, Silchar, India

Contents

Abstract

Based on quantum mechanics, Quantum computing allows states of superposition and entanglement on qubits. Using parallelism, quantum computing performed better in various well-known algorithms. Quantum computing deals with a different data representation called qubit, a challenging format to transfer classical computing into quantum computing. This chapter proposed an architecture for representing Binary trees in quantum computing. Two main tree operations, insertion and deletion are explained in this architecture. Various merits and computation complexity of these quantum circuits are measured and analysed.

1. Introduction

Quantum computing is the use of quantum mechanics to perform computation. Due to its ability of superposition and parallelism, quantum computing offers a higher computational power. Several well-known algorithms have been developed in quantum computing such as Deutch,

Deutsch–Jozsa, quantum Fourier transform (QFT), Factoring, Shor's algorithm, Grover Search [1, 2]. Deutch, Deutsch–Jozsa algorithm are examples that perform the same task with fewer queries than a classical computer. Shor's algorithm shows that it is possible to solve factorization in polynomial time that is impossible in classical computers due to longer computational time. Grover search, a quantum search algorithm has a complexity of $O(\sqrt{N})$ while similar classic search algorithm has $O(N)$ complexity [3]. QFT is exponentially faster than its classical counterpart [4]. Although all Fourier transform tasks cannot be implemented as QFT works on quantum states and classical Fourier transform works with vectors. Quantum computation is reversible while classical computation is irreversible in nature.

Translation of classical algorithm to quantum is challenging mostly due to data representation and reversibility. Quantum computer uses qubit or quantum bits for data representation. Encoding data into qubits is a preliminary step in quantum computing. Representation of real numbers was implemented with a qubit rotation [5]. To construct a database of classical information, quantum random access memory was proposed in [6]. In quantum image processing, different methods for quantum image representation like flexible representation for quantum images (FRQI) [7], multichannel representation for quantum images (MCQI) [8], novel enhanced quantum representation (NEQR) [9], $(n + 1)$-qubit normal arbitrary quantum superposition state (NAQSS) [10], etc. are proposed. In [11], classical real-world data are represented by searching for a compatible quantum state.

For efficient translation of classical algorithms to quantum, data must be represented for efficient retrieval and modification. Quantum computing approaches for data structure can be broadly classified into classical data structure with quantum access and Fully quantum data structure [12]. The first approach uses classical data structure which is accessed by quantum computers. The second is a fully quantum data structure where data is encoded into qubits. Quantum walks, which evolved from classical random walks, have been implemented with both quantum access and fully quantum data structure [13]. A hardware-oriented model for data structure implementation take advantage of parallel computing capabilities. This was achieved with the use of qubit processor [14]. One important aspect of data structure is the analysis of algorithms in terms of complexity. Quantum circuits can be measured in terms of quantum cost. The quantum cost of a circuit is the number of primitive gates needed to realize the circuit [15]. In the case of classical computing, there are several existing data structures for different purposes. These data structures are not all applicable or even necessary quantum computing. But the concept proves of these data structures are useful for designing quantum algorithms. This chapter proposed a quantum implementation of binary tree. The quantum circuit implementation for deletion and insertion operation is presented and discussed.

2. Related works

In a quantum data structure, the quantum walk method was implemented to solve graph problems. Quantum walks are of two types: continuous and discrete-time quantum walk. Both of these types are used to solve the binary welded tree problem and the triangle finding problem [16]. Coin operator, phase shift, oracle and timestep are the operators used in the circuit to solve the triangle finding and binary welded tree problem. For a discrete-time quantum walk on any underlying graph, there is a relation between the Staggered walk and Grover walk. A Grover walk is said to be periodic, if and only if a Staggered walk is periodic [17]. Staggered walk performs tessellation cover on a line graph by inducing Hoffman graphs. It uses a time evolution operator, consisting of two unitary operators. For a continuous time-quantum walk, a graph exhibiting fractional revival, a phenomenon for entanglement, is constructed using the adjacency matrix model [18]. Based on a quantum walk three models, arc-reversal model, shunt-decomposition model and two-reflection model are proposed [19]. The performance of these models gets affected, by the change in graph characteristics. These models are related to the concept of rotation and factorization applied on a graph. The impact is mainly due to two parameters, i.e., trace and entropy. Matrices and operators used in these models are permutation matrix, unitary matrix, coin operator, shift operator, and normalized matrices. The quantum complexity is studied in [20], where the set membership problem is considered. The quantum bit probe model is formulated, in which answers to the queries are generated using limited bit probes. Oracle operators are used in this model. A deterministic finite automaton is a machine that accepts or rejects a given string following a particular language. This concept of a finite automaton is used in quantum computing, named as quantum finite automaton [21].

Quantum computing approach to solve graph coloring problems was presented in [22]. All possible vertex coloring of a graph is represented by creating superposition. Using the depth-first traversal (DFT) method, a tree is generated from a graph, consisting of two edges: tree edge and back edge. These edges help in determining the possibility of coloring a graph. Unitary operators and projection operators are used to determine the correct color combination of the graph. Another approach to the graph coloring problem uses Grover's algorithm to synthesize the coloring problem on a disjoint graph in a binary quantum system [23]. This algorithm is implemented using two operators: Oracle and Diffusion operator. Oracle represents the states describing different possibilities, to check whether graph coloring is possible or not. Next, the diffusion operator amplifies the probability of resulting states to maximum, while keeping the other states near to zero. In [24], a D-wave quantum computer solves a random spanning tree problem, to test its performance. All possible random spanning trees are generated, and the outputs are resolved and produced. A problem of graph connectivity on a planar graph is solved using a quantum algorithm [25]. For the quantum query, a new upper bound on the complexity is provided, showing exponentially better than other classical query complexity. A quantum-based decision tree is investigated in [26]. Here a model is designed that grow a quantum decision tree to classify quantum objects. Neumann entropy is used as a splitting criterion to construct this decision tree. Quantum techniques are also used to perform sampling of a quantum binary tree [27]. It is inspired by an efficient classical algorithm named Markov Chain Monte Carlo (MCMC), with a speedup in process.

Apart from data structure and Graph theory, there are available other research papers focused on importance, design, applications, and performance in quantum computing. In a circuit, the more the amount of loss in information, more is the heat dissipation. According to the author, if the circuit is designed in a reversible manner the loss of heat can be minimized [28]. In quantum computing, works related to reversible circuits are performed in [29–31]. In these works, various models such as logically reversible Turing machines, $N \times N$ reversible quantum multiplexer and n:2^n reversible decoder are proposed. These models focus on efficiency, by reducing the overhead of the circuits. Pauli X, Toffoli, reversible quantum ripple quantum adders are some of the gates used in the circuit.

3. Representation of a binary tree

In a binary tree, there are four possibilities of zero, left, right, and both child for a parent node. Here a qubit represents a parent node and the state of the qubit denotes the number of child nodes. State $|0\rangle$ or $|1\rangle$ denotes the presence of a left or right child for the parent node. Super position state $\frac{|0\rangle - |1\rangle}{\sqrt{2}}$ or $\frac{|0\rangle + |1\rangle}{\sqrt{2}}$ denotes zero or both children for the parent node.

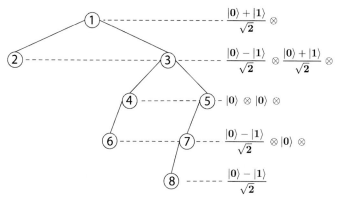

Fig. 1 Representation of quantum binary tree.

Superposition state is obtained by Hadamard gate (H) by $H|0\rangle = \frac{|0\rangle+|1\rangle}{\sqrt{2}}$ and $H|1\rangle = \frac{|0\rangle-|1\rangle}{\sqrt{2}}$. The tree parses from top to down and traverses from left to right direction. An example is given in Fig. 1 for equivalent binary tree representation. Nodes 1 and 3 have two child, thereby state of the qubit is $\frac{|0\rangle+|1\rangle}{\sqrt{2}}$. Nodes 2, 6, and 8 have zero child, so the state is $\frac{|0\rangle-|1\rangle}{\sqrt{2}}$. Nodes 4 and 7 have only left child with state of the qubit $|0\rangle$ each. The equivalent quantum tree is $\left(\frac{1}{\sqrt{2}}\right)^{5} [(|0\rangle+|1\rangle) \otimes (|0\rangle-|1\rangle) \otimes (|0\rangle+|1\rangle) \otimes |0\rangle \otimes |0\rangle \otimes (|0\rangle-|1\rangle) \otimes |0\rangle \otimes (|0\rangle-|1\rangle)]$ in the sequence of 1, 2, 3, 4, 5, 6, 7, and 8, respectively.

4. Operations on quantum binary tree, algorithm, and quantum circuits

4.1 Insertion of a node

In this section, we explain an insertion procedure for the proposed quantum binary tree. For inserting a new node, the location of parent and new node are taken from input qubits $|P_1...P_n\rangle$ and $|L_1...L_n\rangle$. As the new node has zero child, the state of it will be $\frac{|0\rangle-|1\rangle}{\sqrt{2}}$. The new node $\frac{|0\rangle-|1\rangle}{\sqrt{2}}$ is now placed using swap operator in the child location. All swap operations are controlled by child $|L_1...L_n\rangle$ qubits. The parent node is now another child node. The state of the parent node needs to changed from $\frac{|0\rangle-|1\rangle}{\sqrt{2}}$ to $|0\rangle$ or $|0\rangle$ to $\frac{|0\rangle+|1\rangle}{\sqrt{2}}$, respectively. The change is possible to obtained by Hadamard(H) followed by Pauli X.

$$X.H|0\rangle = \frac{|0\rangle + |1\rangle}{\sqrt{2}} \qquad X.H\frac{|0\rangle - |1\rangle}{\sqrt{2}} = |0\rangle$$

The approach is applicable for inserting as a left child. For right child insertion, the small change is required in definition of node levels, in which super position state $\frac{|0\rangle+|1\rangle}{\sqrt{2}}$ or $\frac{|0\rangle-|1\rangle}{\sqrt{2}}$ denotes presence of zero or both children in the parent node. After inserted as right child, parent node needs to change from $\frac{|0\rangle+|1\rangle}{\sqrt{2}}$ to $|1\rangle$ or $|1\rangle$ to $\frac{|0\rangle-|1\rangle}{\sqrt{2}}$ respectfully. Same quantum circuit is capable of handling both changes. Note that, there is a global factor($e^{-i.\pi}$) in the desired state, which is possible to omit as explained below,

$$X.H\frac{|0\rangle+|1\rangle}{\sqrt{2}} = |1\rangle$$

$$X.H|1\rangle = \frac{|1\rangle - |0\rangle}{\sqrt{2}} = e^{i.\pi}.\frac{|0\rangle - |1\rangle}{\sqrt{2}} \approx \frac{|0\rangle - |1\rangle}{\sqrt{2}}$$

An example is given in Fig. 2 for inserted left node(n) for Fig. 1. Fig. 3 shows quantum circuit for insertion. The quantum circuit takes three inputs as quantum tree $|N_1...N_8\rangle$, new node location $|L_1...L_8\rangle$ and parent node location $|P_1...P_8\rangle$. New node to be inserted is shown in the first qubit state with input $|1\rangle$. To understand the input format, assume parent location=4 and child location=6, therefore,

$$|P_1P_2P_3P_4P_5P_6P_7P_8\rangle = |11110000\rangle$$
$$|L_1L_2L_3L_4L_5L_16_7L_8\rangle = |11111100\rangle.$$

The cascading set of swap gates will now place the new node $\frac{|0\rangle-|1\rangle}{\sqrt{2}}$ in 6th location, where as the parent node at 4th location is changed to $\frac{|0\rangle+|1\rangle}{\sqrt{2}}$.

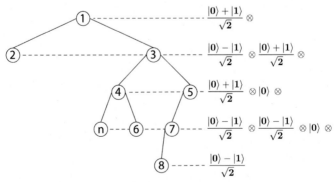

Fig. 2 Quantum binary tree after insertion operation.

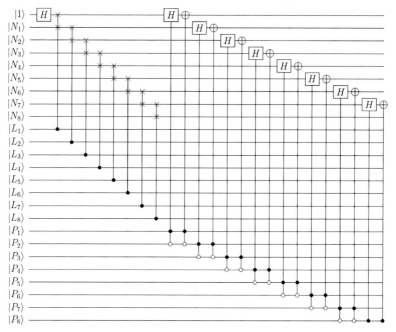

Fig. 3 Quantum circuit for insert operation.

Algorithm 1 Insertion(NewNode $|1\rangle\otimes$ Tree $|N_1N_2...N_n\rangle\otimes$ LeafLocation $|L_1L_2...L_n\rangle\otimes$ ParentLocation $|P_1P_2...P_n\rangle$)

$|Q_{initial}\rangle \equiv |1\rangle \otimes |N_1...N_n\rangle \otimes |L_1...L_n\rangle \otimes |P_1...P_n\rangle$

$|Q_0\rangle \equiv \mathrm{H}|1\rangle \otimes |N_1...N_n\rangle \otimes |L_1...L_n\rangle \otimes |P_1...P_n\rangle$

$\equiv \frac{|0\rangle - |1\rangle}{\sqrt{2}} \otimes |N_1...N_n\rangle \otimes |L_1...L_n\rangle |P_1...P_n\rangle$

for $i=1$ To n **do**

$\quad |Q_i\rangle \equiv |N_1...N_{i-1}\rangle \otimes SWAP(\frac{|0\rangle - |1\rangle}{\sqrt{2}}), |N_i\rangle, |L_i\rangle) \otimes |N_{i+1}...N_n\rangle |L_1...L_n\rangle |P_1...P_n\rangle$

$\quad \equiv |N_1...N_i\rangle \otimes \frac{|0\rangle - |1\rangle}{\sqrt{2}} \otimes |N_{i+1}...N_n\rangle |L_1...L_n\rangle |P_1...P_n\rangle$

for $i=1$ To n **do**

$\quad |Q_{n+i}\rangle' \equiv |N_1...N_{i-1}\rangle \otimes HCC(|N_i\rangle, |P_i\rangle, |\overline{P_{i+1}}\rangle) \otimes |N_{i+1}\rangle ... \otimes \frac{|0\rangle - |1\rangle}{\sqrt{2}} \otimes$

$\quad ...|N_n\rangle |L_1...L_n\rangle |P_1...P_n\rangle$

$\quad |Q_{n+i}\rangle \equiv |N_1...N_{i-1}\rangle \otimes XCC(|N_i\rangle, |P_i\rangle, |\overline{P_{i+1}}\rangle) \otimes |N_{i+1}\rangle ... \otimes \frac{|0\rangle - |1\rangle}{\sqrt{2}} \otimes$

$\quad ...|N_n\rangle |L_1...L_n\rangle |P_1...P_n\rangle$

Return $|N_1\rangle ... \otimes \frac{|0\rangle - |1\rangle}{\sqrt{2}} \otimes ...|N_n\rangle |L_1...L_n\rangle |P_1...P_n\rangle$;

Notation:

- SWAP(T_1, T_2, C_1): Qubits T_1 and T_1 is to be swapped under control line C_1. Complement above a qubit represents active zero control line.
- XCC(T_1, C_1, C_2): Pauli X is applied on first qubit T_1 under control lines C_1 and C_2.
- HCC(T_1, C_1, C_2): Hadamard H is applied on first qubit T_1 under control lines C_1 and C_2.

4.2 Deletion of node

This section explains the process of deletion in the proposed quantum binary tree. Quantum circuits follow reversible characteristics, for this if the circuit of insertion is mirrored-flipped, then the circuit will perform the deletion operation. Same as earlier, location of the node to be deleted and the parent are provided as input qubits $|L_1 \ldots L_{n-1}\rangle$ and $|P_1 \ldots P_{n-1}\rangle$. The deleted node is placed in the first location using swap operation. The swap operation is controlled by $|L_1 \ldots L_n\rangle$. The state of parent node needs to change from $\frac{|0\rangle+|1\rangle}{\sqrt{2}}$ to $|0\rangle$ or $|0\rangle$ to $\frac{|0\rangle-|1\rangle}{\sqrt{2}}$. This is a reverse of the insert operation. Therefore the change is made by using Pauli X followed by Hadamard.

$$H.X\frac{|0\rangle+|1\rangle}{\sqrt{2}} = |0\rangle \tag{1}$$

$$X.H|0\rangle = \frac{|0\rangle - |1\rangle}{\sqrt{2}} \tag{2}$$

An example is given in Fig. 4 for deleting node 6 from Fig. 1. Fig. 5 shows quantum circuit for deletion. The quantum circuit takes three inputs: quantum tree $|N_1 \ldots N_8\rangle$, new node location $|L_1 \ldots L_8\rangle$ and parent node location $|P_1 \ldots P_8\rangle$. Node to be deleted is moved to the first qubit with state $\frac{|0\rangle-|1\rangle}{\sqrt{2}}$. Same as earlier input format, assume Parent Location=4 and Child Location=6, therefore,

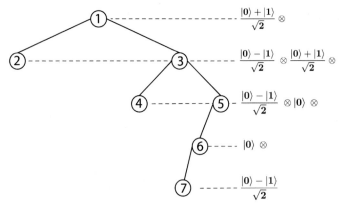

Fig. 4 Quantum binary tree after deletion operation.

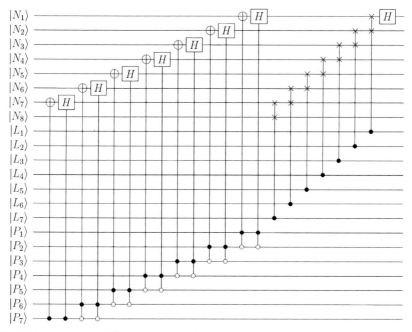

Fig. 5 Quantum circuit for delete operation.

$$|P_1 P_2 P_3 P_4 P_5 P_6 P_7\rangle = |1111000\rangle$$
$$|L_1 L_2 L_3 L_4 L_5 L_1 6_7\rangle = |1111110\rangle$$

The cascading set of swap gates will now place the node 6 in 1st location, where as the parent node at 4th location is changed to $\frac{|0\rangle - |1\rangle}{\sqrt{2}}$.

Algorithm 2 Deletion(Tree $|N_1...N_n\rangle \otimes$ *LeafLocation* $|L_1...L_{n-1}\rangle \otimes$ *ParentLocation* $|P_1...P_{n-1}\rangle$)

$|Q_{initial}\rangle \equiv |N_1...N_n\rangle \otimes |L_1...L_{n-1}\rangle \otimes |P_1...P_{n-1}\rangle$

for $i=n\text{-}1$ *To* 1 **do**

$\quad |Q_i\rangle' \equiv |N_1...N_{i-1}\rangle \otimes XCC(|N_i\rangle, |P_i \overline{P_{i+1}}\rangle) \otimes |N_{i+1}...N_n\rangle |L_i...L_n\rangle |P_1...P_n\rangle$

$\quad |Q_i\rangle \equiv |N_1...N_{i-1}\rangle \otimes HCC(|N_i\rangle, |P_i \overline{P_{i+1}}\rangle) \otimes |N_{i+1}...N_n\rangle |L_i...L_n\rangle |P_1...P_n\rangle$

for $i=n\text{-}1$ *To* 1 **do**

$\quad |Q_{n+i}\rangle \equiv$

$\quad |N_1...N_{i-1}\rangle \otimes SWAP(|N_i N_{i+1}\rangle, |L_i\rangle) \otimes |N_{i+2}...N_n\rangle \otimes |L_1...L_n\rangle \otimes |P_1...P_n\rangle$

$|Q_f\rangle \equiv H^{\frac{|0\rangle - |1\rangle}{\sqrt{2}}} \otimes |N_1...N_{k-1}\rangle |N_{k+1}...N_n\rangle |L_1...L_{n-1}\rangle |P_1...P_{n-1}\rangle$

Return $|1\rangle \otimes |N_1...N_{k-1}\rangle |N_{k+1}...N_n\rangle |L_1...L_{n-1}\rangle |P_1...P_{n-1}\rangle$

5. Output measurement, analysis of quantum circuit merits, and complexity

The state of each node is either $|0\rangle$, $|1\rangle$, $\frac{|0\rangle+|1\rangle}{\sqrt{2}}$ or $\frac{|0\rangle-|1\rangle}{\sqrt{2}}$. For determining the states, the quantum circuit is run multiple time and observes the change in the output. Qubits with same state is either $|0\rangle$ or $|1\rangle$. Remaining are superposition state, i.e., $\frac{|0\rangle+|1\rangle}{\sqrt{2}}$ or $\frac{|0\rangle-|1\rangle}{\sqrt{2}}$, as because of showing both output 0/1 for various execution. To distinguish the phase in the superposition, the circuit once more run followed by Hadamard gates placed at end to obtain the superposition into pure state $|0\rangle$ and $|1\rangle$. $\frac{|0\rangle+|1\rangle}{\sqrt{2}}$ or $\frac{|0\rangle-|1\rangle}{\sqrt{2}}$ is identified for classical output 0 or 1.

Various analysis of merits for the quantum circuits is presented in Table 1. In Table 1, quantum cost (QC) is the number of primitive gates in it. Primitive gates (1×1 and 2×2) cost 1, regardless of the internal structure. For calculating the QC of XCC/Toffoli gate and controlled swap gate are broken into primitive gates, and obtaining QC of XCC, HCC and Swap gates are 5, 5, and 7, respectively, as in Fig. 6 [32].

For insertion, if we have a tree of n nodes, we need n number of controlled swap gates with QC$=7 \times n$. We also need n number of HCC and XCC gates with a total of QC$=5 \times 2 \times n = 10 \times n$, making the total QC for insertion is $17n$. In deletion with n nodes, the number of swap, XCC and HCC gates are $(n - 1)$ each. Therefore the total QC for deletion $= 17 \times (n - 1) = 17n - 17$. For a single insertion operation with a worse case QC of $17n$, the operational cost for this is in the order of n and thus the complexity of the insert operation is $O(n)$. Similarly, the delete operation takes

Table 1 Merits of proposed circuits.

Analysis	Insertion	Deletion
QC	$17n$	$17n - 17$
Garbage output	$2n$	$2n - 2$
No. of control lines	$2n$	$2n - 1$
Gate count	$3n$	$3n - 3$
Critical path delay (worse case)	$(4n + 5)\Delta$	$(4n + 5)\Delta$
Complexity	$O(m \cdot n)$	$O(m \cdot n)$

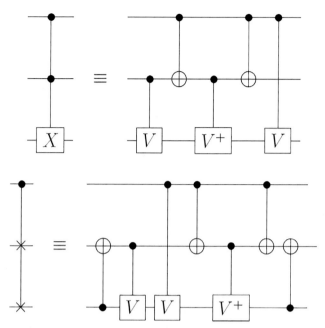

Fig. 6 Decomposition of Toffoli and controlled swap gate.

$(17n + 7)$ QC, and the complexity is $O(n)$. If the measurement is performed m times for the same operation, then the complexity for the entire operation is $O(m \cdot n)$.

Another merit, critical path delay calculates the path with maximum delay in Δ unit time. This unit is introduced as the delay of primitive gates are different depending on the technology. The delay of all 1×1 and 2×2 gates are taken as 1. For larger gates, the delay is calculated from its implementation in primitive gates. The delay of XCC/Toffoli gate is 4Δ. HCC gate can be broken down in the same way as XCC and thus has the same delay 4Δ. The controlled swap gate is a Fredkin with delay 4Δ. If we see the circuit, all qubits in cascading SWAP operation are dependent on each other, while cascading XCC and HCC gates are not dependent on each other qubits. For this, the critical delay path consists of $(n + 2)$ gates and a Hadamard gate in a worst-case scenario. Therefore the Critical path delay is calculated as $[4(n + 2) + 1]\Delta = (4n + 5)\Delta$.

The garbage output is the number of unwanted qubits in the output. Qubits $|P_1...P_n\rangle$ and $|L_1...L_n\rangle$ are garbage output as the measurement of these values does not contribute to output tree. The gate count measures performance as the number of quantum gates in the circuit.

6. Conclusion

In this chapter, a representation of binary is proposed in quantum computation. Each node is considered as a qubit and its various state $|0\rangle$, $|1\rangle$, $\frac{|0\rangle+|1\rangle}{\sqrt{2}}$ or $\frac{|0\rangle-|1\rangle}{\sqrt{2}}$ says the number of children of it. Two main operations like insertion and deletion are designed in quantum circuit. The same circuit is possible to use for both insertion and deletion. The pattern of the circuit structure is helpful for designing large size trees. Algorithms and various merits are analysed with an explanation. Future works include insertion and deletion without providing both child and parent location.

References

[1] M.A. Nielsen, I. Chuang, Quantum Computation and Quantum Information, American Association of Physics Teachers, 2002.
[2] C.P. Williams, Explorations in Quantum Computing, Springer London, 2010, ISBN: 9781846288876. https://books.google.co.in/books?id=QE8S-WjIFwC.
[3] L. Tarrataca, A. Wichert, Tree search and quantum computation, Quantum Inf. Process. 10 (4) (2011) 475–500.
[4] L. Hales, S. Hallgren, An improved quantum Fourier transform algorithm and applications, in: Proceedings 41st Annual Symposium on Foundations of Computer Science, IEEE, 2000, pp. 515–525.
[5] N. Wiebe, V. Kliuchnikov, Floating point representations in quantum circuit synthesis, N. J. Phys. 15 (9) (2013) 093041.
[6] D.K. Park, F. Petruccione, J.-K.K. Rhee, Circuit-based quantum random access memory for classical data, Sci. Rep. 9 (1) (2019) 1–8.
[7] P.Q. Le, F. Dong, K. Hirota, A flexible representation of quantum images for polynomial preparation, image compression, and processing operations, Quantum Inf. Process. 10 (1) (2011) 63–84.
[8] B. Sun, A.M. Iliyasu, F. Yan, F. Dong, K. Hirota, An RGB multi-channel representation for images on quantum computers, J. Adv. Comput. Intell. Intell. Inform. 17 (3) (2013) 404–417.
[9] Y. Zhang, K. Lu, Y. Gao, M. Wang, NEQR: a novel enhanced quantum representation of digital images, Quantum Inf. Process. 12 (8) (2013) 2833–2860.
[10] H.-S. Li, Q. Zhu, R.-G. Zhou, L. Song, X.-J. Yang, Multi-dimensional color image storage and retrieval for a normal arbitrary quantum superposition state, Quantum Inf. Process. 13 (4) (2014) 991–1011.
[11] R.B. de Sousa, E.J.S. Pereira, M.P. Cipolletti, T.A.E. Ferreira, A proposal of quantum data representation to improve the discrimination power, Nat. Comput. 19 (3) (2020) 577–591.
[12] M. Fillinger, Data structures in classical and quantum computing, arXiv e-prints (2013) arXiv–1308.
[13] S. Jeffery, R. Kothari, F. Magniez, Nested quantum walks with quantum data structures, in: Proceedings of the Twenty-Fourth Annual ACM-SIAM Symposium on Discrete Algorithms, SIAM, 2013, pp. 1474–1485.

[14] V. Hahanov, I. Hahanova, O. Guz, M.A. Abbas, Quantum models for data structures and computing, in: Proceedings of International Conference on Modern Problem of Radio Engineering, Telecommunications and Computer Science, IEEE, 2012, p. 291. 291.

[15] M. Mohammadi, M. Eshghi, On figures of merit in reversible and quantum logic designs, Quantum Inf. Process. 8 (4) (2009) 297–318.

[16] A. Chakrabarti, C. Lin, N.K. Jha, Design of quantum circuits for random walk algorithms, in: 2012 IEEE Computer Society Annual Symposium on VLSI, IEEE, 2012, pp. 135–140.

[17] S. Kubota, E. Segawa, T. Taniguchi, Y. Yoshie, A quantum walk induced by Hoffman graphs and its periodicity, Linear Algebra Appl. 579 (2019) 217–236.

[18] A. Chan, G. Coutinho, C. Tamon, L. Vinet, H. Zhan, Quantum fractional revival on graphs, Discrete Appl. Math. 269 (2019) 86–98.

[19] C. Godsil, H. Zhan, Discrete-time quantum walks and graph structures, J. Comb. Theory A. 167 (2019) 181–212.

[20] J. Radhakrishnan, P. Sen, S. Venkatesh, The quantum complexity of set membership, in: Proceedings 41st Annual Symposium on Foundations of Computer Science, IEEE, 2000, pp. 554–562.

[21] A. Nayak, Optimal lower bounds for quantum automata and random access codes, in: 40th Annual Symposium on Foundations of Computer Science (Cat. No. 99CB37039), IEEE, 1999, pp. 369–376.

[22] E. D'Hondt, Quantum approaches to graph colouring, Theor. Comput. Sci. 410 (4–5) (2009) 302–309.

[23] A. Saha, A. Chongder, S.B. Mandal, A. Chakrabarti, Synthesis of vertex coloring problem using grover's algorithm, in: 2015 IEEE International Symposium on Nanoelectronic and Information Systems, IEEE, 2015, pp. 101–106.

[24] J.S. Hall, M.A. Novotny, T. Neuhaus, K. Michielsen, A study of spanning trees on a D-Wave quantum computer, Phys. Procedia 68 (2015) 56–60.

[25] S. Jeffery, S. Kimmel, Quantum algorithms for graph connectivity and formula evaluation, Quantum 1 (2017) 26.

[26] S. Lu, S. Braunstein, Quantum decision tree classifier, Quantum Inf. Process. 13 (3) (2014) 757–770, https://doi.org/10.1007/s11128-013-0687-5.

[27] D. Provasoli, B. Nachman, C. Bauer, W.A. de Jong, A quantum algorithm to efficiently sample from interfering binary trees, Quantum Sci. Technol. 5 (3) (2020) 035004.

[28] C.H. Bennett, Logical reversibility of computation, IBM J. Res. Dev. 17 (6) (1973) 525–532.

[29] H. Thapliyal, N. Ranganathan, S. Kotiyal, Design of testable reversible sequential circuits, IEEE Trans. Very Large Scale Integr. VLSI Syst. 21 (7) (2012) 1201–1209.

[30] S. Kotiyal, H. Thapliyal, N. Ranganathan, Circuit for reversible quantum multiplier based on binary tree optimizing ancilla and garbage bits, in: 2014 27th International Conference on VLSI Design and 2014 13th International Conference on Embedded Systems, IEEE, 2014, pp. 545–550.

[31] A.K. Saha, K. Sambyo, C. Bhunia, Design and analysis of n: 2 n reversible decoder, IETE J. Educ. 57 (2) (2016) 65–72.

[32] R. Wille, R. Drechsler, Towards a Design Flow for Reversible Logic, Springer Science & Business Media, 2010.

About the authors

Arnab Roy received B.Tech Degree from Central Institute of Technology, Kokrajhar, Assam, India, M. Tech Degree from North Eastern Hill University, Shillong, Meghalaya, India. He is currently pursuing Ph.D in Computer Science and Engineering at National Institute of Technology, Silchar, India.

Joseph L. Pachuau received B.Tech Degree from North Eastern Regional Institute of Science and Technology, M. Tech Degree from National Institute of Technology, Silchar, India. He is currently pursuing Ph.D in Computer Science and Engineering at National Institute of Technology, Silchar, India.

Anish Kumar Saha received B.E from National Institute of Technology Agartala, India, M.Tech from Tripura University, India and Ph.D. from National Institute of Technology Arunachal Pradesh, India in Computer Science and Engineering. He is currently working as Assistant Professor in the department of Computer Science & Engineering at National Institute of Technology Silchar, India. His research interests in Optimization, Software Defined Networking, and Quantum Computing.

CHAPTER ELEVEN

Modified ML-KNN: Role of similarity measures and nearest neighbor configuration in multi-label text classification on big social network graph data

Saurabh Kumar Srivastava[a], Ankit Vidyarthi[b], and Sandeep Kumar Singh[b]
[a]College of Computing Sciences & IT, Teerthanker Mahaveer University, Moradabad, U.P., India
[b]Department of CSE & IT, Jaypee Institute of Information Technology, Noida, India

Contents

Advances in Computers, Volume 128
ISSN 0065-2458
https://doi.org/10.1016/bs.adcom.2021.10.006

Abstract

Determining labels of textual content is a kind of challenging task. Classification systems are helping to identify different aspects of life especially in prediction of personality, elections, and healthcare related insights, etc. Our work presented a unique modified multi-label k nearest algorithm (Modified-MLKNN) in the direction to provide multiple labels of textual content that can help in early alarming systems and especially in case of monitoring infectious disease propagation. The study in this direction showed that text related pre-processing configurations play an important role and largely impact the classification task. The overlapping of terms during the learning phase leads to incorrect classification and requires more powerful ways to correctly classify text documents into multiple labels. The proposed Modified MLKNN algorithm which gave a minimum of 5% improved result over traditional ML-KNN. The algorithm is experimented and validated over two different multi-label text corpus Disease and Seattle.

1. Introduction

Social networks like Twitter and Facebook provide open platform where people share their text responses. Generally, this activity is termed as text postings. These postings carry information and can help in certain specific domains for its betterment. Text-based classification system can utilize such postings in view to produce summaries. Textual content over social media belongs to either unstructured or semi-structured category. The emergence growth of Web 3.0 online information sources has grownup enormously. The researcher's tried to utilize these sources for identifying meaningful pattern to improve the convenience of human being. Therefore, some automatic tools can help to analyze such large collection of text data. The increasing volume of document demands the relation with some concrete category in a mutually exclusive manner. To address this problem it is found that text categories may belong to many labels. For example the posting related to health might be related to other topics like–disease, diagnosis process, or education. In real world scenario text content may have multiple labels. So, multi-label classification algorithm can suggest possible labels for a given text document.

Now days, researchers are attracted in dealing with text analysis to get some valuable insights as a result. Generally, multi-label classification is critical, as it belong to exponential number of possible labels. Text-mining is also a domain of research where only textual data is considered for analysis. Short text messages (like: Twitter data) attracted most of the researchers because of its simplicity, verbosity and character length.

The use of social media (Twitter) postings related to healthcare has been modeled by researchers in many ways. Authors in Ref. [1,8] proposed a dynamic real-time architecture to track early disease outbreak prediction. Support Vector Machine (SVM) based system achieved 88% accuracy in terms of performance. Authors did not raise multi-label context of texts.

Authors in Ref. [2] modeled the problem to overcome the 1–3 week delay of manually aggregating and processing the clinical information. The study proposed a real-time influenza tracking model based on auto regression and Google's search data. The results showed superior approach in terms of accuracy. The study presented in Ref. [3] is very similar to Ref. [2], here author hypothesize the relation between seasonal influenza epidemics as a major public health concern for respiratory illness. The approach uses health queries to detect influenza spread areas.

Authors in Ref. [4,5] presented a real-time flu and cancer related surveillance system based on Twitter data. The model tracks of flu and cancer related activities on twitter. The model has aided visual support which showed the results in US disease surveillance map. The system was helpful for early prediction of seasonal disease outbreaks. The study showed disease (flu and cancer) related monitoring distribution in terms of its types, symptoms and treatments.

In the context of health surveillance using social media, paper [6] proposed a website as a platform for health surveillance using twitter (healthtweets.org). The surveillance mines twitter data and showed health trends in three main domains which are temporal, location and maps. Authors in Ref. [7] devised an auto regression model to predict influenza like illness (ILI) level in population and proposed a framework to predict spread of influenza. The authorized United States of America body, center of disease control (CDC) has reported that Flu is highly correlated with ILI cases. The Framework Social Network Enabled Flu Trends (SNEFT) monitors messages with a mention of Flu indicators. The study claimed substantially improved model in predicting ILI cases using social media. Author in Ref. [9] examine twitter streams to track public sentiments in context of H1N1. Authors have used twitter to measure the actual disease activity in the context of health-related events.

The study presented in Ref. [10] used a regression model with ridge regularization and showed improved prediction accuracy. The experimentation combines twitter messages and CDC's ILI data.

In machine learning multi-class-based classification is common problem. However, text content may have multiple labels. Multi-label classification task is not a new concept; it is generalized form of multiclass problem. In multi-label context, study proposed in paper [11,29] demonstrated its application domains like images and videos annotation, music categorization into emotions, functional genomics, etc. Authors exhaustively showed the places where multi-label classification is used widely.

While in paper [12,13] authors have made comprehensive reviews of multi-label learning algorithms which are frequently used by the researchers. Mainly two categories of algorithms are used for classification, transformation and adaptation. Both the papers have discussed on evaluation metrics of multi-label learning. Paper [13] reported the problem of exploiting label correlation by various multi-label learning techniques. Authors in paper [14] presented the most frequently used techniques used to deal with ML problems in pedagogical manner which basically highlighted the similarity measures and differences between techniques.

Work in Ref. [15] signifies the concept of learning objects and validated the findings on four different MLL algorithms. The results were obtained using transformation and adaptation algorithms. Ensemble of Classifier Chains (ECC), Random K-Label sets, Ensemble of Pruned Sets and Multi-label K Nearest Neighbors (MLKNN) were used for experimentation. Authors have used 16 evaluation measures and concluded that ECC perform best among all selected algorithm category. In Ref. [16] authors have used a new ensemble based on set cover problem and demonstrated its effectiveness with constraints, and represented each individual labels with coverage of inter-label correlations. The method performs highly efficient over RAKEL and other ensemble methods.

Authors in Ref. [17] compared 12 MLL methods using 16 evaluation measures over 11 benchmarking dataset and concluded that random forest of predictive clustering trees (RF-PCT) and hierarchy of multi-label classifiers (HOMER) outperforms among all. In Ref. [18] the work addressed learning problem using method multi-label Naive Bayesian (MLNB) which adapts traditional Bayes classifiers and dealt multi-label instances. Two things are incorporated in the proposed algorithm: (1) Feature selection techniques based on PCA and (2) Subset selection technique based on GA. The algorithm helps in finding of appropriate subset of features for prediction.

The MLNB suggested that incorporating feature selection is helpful in multi-label learning.

Authors in Ref. [19] proposed Multi-instance multi-label extreme learning machine (MIML-ELM) framework to conceptualize complex objects. The study showed the concept of minimizing high computational cost of SVM when used as a classifier. In Ref. [20] authors described the framework for generating multi-label dataset named as MLDataGen. The result illustrates that hypercube strategy provide better classification than based on hyper-sphere.

In literature there also exists work on hierarchical multi-label classification of social texts [21]. The author proposed the idea to resolve the problem of short text message and its complicated relationships among the classes. The chunk based structural learning framework is used for hierarchical multi-label classification. Study presented in Ref. [22] performed multi-task multi-label (MTML) classification for sentiments and topic classification and tried to overcome the sparsity of microblog data. In the process authors have utilized latent association between tweets sentiment and topics for classification. The proposed MTML algorithm has shown 5% accuracy improvement on sentiments and 12% improvement on topic classification. Multi-label classification for short text messages (Twitter) is a critical task. Twitter data is sparse in nature, it lacks with enough information. Dealing sparsity in multi-label learning is an essential aspect for correct classification.

Health related aspects are very common on social media. People usually share experiences related to their health issues, symptoms and diagnosis related processes frequently. These discussions can be used for identifying health related knowledge from the social networks. In this regard, paper [23] explores the methodology to overcome the limitations of traditional methods, and with the help of n-gram analysis, the study, tried to identify health related insights from social media. The author reported the importance of feature selection algorithm in order to identify best possible features in five heterogeneous feature set. In Ref. [24] the author performs emotion analysis in which belongs to many categories and for this he utilizes graphic emotions, punctuation expressions together using MEC Multi-label emotion classification algorithm, the evaluated results are used for single and multi-label classification using Weibo (a twitter in china) data set.

Author in Ref. [25] have used transformation and adaptation algorithms for ML classification. They have used semi-structured data for ML classification and reported results showed that BR performs better over LP and MLKNN. The author in Ref. [26] proposed a tool to annotate twitter messages related to healthcare, it can also collect tweets. The tool prepares feature set and automatically uses these feature set for relevant text filtering. The authors in Ref. [27] proposed a mechanism to identify incident related information using social media (Twitter). The results are based on situational information and can be used in decision support systems.

Transformation algorithm binary relevance (BR), label power set (LP) and classifiers chain (CC) has used for performance evaluation and SVM is used as a base classifier. Authors have used precision and recall values to compare the results.

With the above reported work, we can say that twitter sparsity play vital role in short text classification. The researchers working in this domain tried to utilize this aspect to model real-time data for text-based expert system. The text-based system can perform surveillance over social media, and can help many ways to raise the quality of real-time decisions. Twitter or any other social media services can be a source for such real-time surveillance. In current multi-label research, researchers did not link text relate preprocessing configurations with ML classifiers. The effectiveness of the configuration can be identified for efficient use of ML Classifiers.

The paper is divided into sections. Section 2 presents the methodology used for multi-label algorithm evaluation. Section 3 explains Disease dataset and Seattle dataset used for experimentation. The measures used for the results evaluation are discussed in Section 4. Section 5 presents architecture for experimenting multi-label algorithms. The data related configuration setups are discussed in Section 6. Section 7 demonstrates the results, its graphical implications and modified ML-KNN algorithm. In last, the study presents conclusion and future work.

2. Methodology

In this work, we targeted multi-label perspective of text classification system. We considered two well-known categories of ML algorithms transformation and adaptation. Three algorithms in the category of transformation (BR, LP and CC) and two algorithms from adaptation (MLKNN and BPNN) have used for result evaluation over Disease and Seattle datasets. Both the datasets are prepared using tweets.

2.1 Problem transformation methods

The problem transformation methods transform the problem into one or more single label classification or regression problem. The well-known categories of transformation methods are: Binary Relevance (BR), Label Power-set (LP), and Classifier Chain (CC) methods.

2.1.1 Binary relevance

The algorithm breaks the entire multi-label learning task into q independent binary classification problems, and each binary classification problem corresponds to a possible label in the label sets [5]. Binary Relevance (BR) [11] is a popular approach of transformation method that actually creates k datasets $(k = |L|$, total number of classes), each for one class label, and classifier train themselves on each of these datasets. The datasets contains the same number of instances as per original data, but each dataset $d\lambda_j$, $1 \leq j \leq k$ positively labels instances that belong to class λ_j and negative otherwise.

The simplest strategy one-against-all strategy converts multi-label problem into several binary classification problems. This approach is known as the binary relevance method (BR) of classification.

2.1.2 Label power set

This algorithm works with the concept of label combination method, which has been targeted by several studies [2,15,22]. The method combines entire label sets into single labels to form a single-label problem. For this single-label problem, the set of possible single labels represents all distinct label subsets from the original multi-label representation. In this way, LP based methods work with the label correlations. However, the possible label subsets can be very large.

2.1.3 Classifier chain

Classifier Chain method is closely related to the BR method proposed by Ref. [16]. This method uses Q binary classifiers linked along a chain.

2.2 Algorithm adaptation methods

These multi-label methods have ability to adapt, extend and customize an existing machine learning algorithm [28] for multi-label learning. The literature presented multi-label methods based on adaptation of following machine learning algorithms: boosting, k-nearest neighbors, decision tree and neural networks.

2.2.1 ML-KNN

The ML-KNN is one of the popular K-nearest neighbor (KNN) lazy learning algorithms [3–5]. The retrieval of KNN is same as in the traditional

KNN algorithm. The main difference is the determination of the label set of an unlabeled instance. The algorithm uses prior and posterior probabilities of each label within the k-nearest neighbors.

2.2.2 BPNN

BPNN is an example of neural network adaptation for multi-label learning [6,7]. BPNN uses back-propagation criteria for its learning. The main modification is the introduction of a new error function that works on multiple labels.

3. Datasets preparation and description

We have evaluated result on our gold standard healthcare related Disease dataset and to compare the configuration parameter we have used a standard Seattle dataset [27]. Both the datasets are multi-label dataset. We have used Twitter's tweepy API [30] to prepare Disease dataset. We collected tweets and prepared our corpus using disease-based keywords mentioned in Ref. [31]. Several preprocessing steps are considered to reduce noise in textual content. Following steps are involved for the data preprocessing steps:.

Step 1: Special characters of programming languages (e.g. ":", "/") are removed.

Step 2: Duplicate tweets which were repeated in the entire dataset are removed.

Step 3: Common English-language stopwords (e.g., "the", "it", "on") are removed. To perform this step a standard stopwords list has used mentioned at Stanford University website. We have not prepared any customized stopwords list.

Step 5: Word stemmer (porter's stemmer) is applied to find the root of each word.

Step 6: Results are evaluated on lower case data set to minimize the discrepancy.

These preprocessing steps are used to prepare noise free dataset as special symbols/syntaxes, duplicates and stopwords are viewed as noise and will degrade the performance of classifiers. Two preprocessing configurations we considered for the experimentation. With the literature we understood that concrete terms play an important role in classification. We have used two text related configuration (C_0 & C_3) for MLL task. Here in the experimentation we manually annotated all the tweets into its multi-label category with guidance of healthcare expert. For the analysis we have used C_0 configuration to clearly visualize the impact of our study.

We considered a standard dataset mentioned in the paper [27] named as Seattle dataset which is a multi-label dataset used for analysis for impact measurement of configuration parameters. The dataset comprises incident-related tweets and labeled as: Fire, Shooting, Crash and Injury. We have experimented C_0 and C_3 configurations for both the datasets. The authors have used Twitter Search API and collected English language public tweets within 15 km radius around Seattle, Washington and Memphis Tennessee.

4. Measures used

Following measures are used in context of performance evaluation of ML Classifiers. All the above-mentioned algorithms are experimented using Disease and Seattle dataset for measuring impact of configuration parameters.

4.1 Subset accuracy

The subset accuracy shows fraction of correctly classified examples [17]. It a measure to evaluate correct predicted label set which is identical to the ground-truth label set. Subset accuracy can be regarded as a multi-label counterpart of traditional accuracy metric, and this metric especially used when the size of label space (q) is large.

4.2 Hamming loss

The hamming loss is a measure to evaluate the fraction of misclassified instance-label pair, calculated when a relevant label is missed or an irrelevant is predicted [17]. When each example in S is associated with only one label, hloss $S(h)$ will be $2/q$ times of the traditional misclassification rate; it is defined as Eq. (1).

$$hloss(h) = 1/p \sum_{i=1}^{p} 1/q |h(x_i) \Delta Y_i| \qquad (1)$$

here, Δ refers for the symmetric difference between two sets.

4.3 Example based precision

Example based precision is defined as Eq. (2).

$$\text{Precision } (h) = 1/N \sum_{i=1}^{N} |h(x_i) \cap y_i|/|y_i| \tag{2}$$

4.4 Example based recall

Example based recall is defined as Eq. (3).

$$\text{Recall } (h) = 1/N \sum_{i=1}^{N} |h(x_i) \cap y_i|/|h(x_i)| \tag{3}$$

4.5 Example based F-measure

F1 score is the harmonic mean between precision and recall and is defined as Eq. (4).

$$F1 = 1/N \sum_{i=1}^{N} 2*|h(x_i) \cap y_i|/|h(x_i)|+|y_i| \tag{4}$$

F1 refers example based metric and its value is an average overall examples in the dataset. The score 1 shows its best and worst score at 0.

4.6 Micro averaged precision

Micro averaged precision (precision averaged based on all example/label pairs) is defined as Eq. (5).

$$\text{MicroPrecision} = \sum_{j=1}^{Q} tp_j/ \sum_{j=1}^{Q} tp_j+ \sum_{j=1}^{Q} fp_j \tag{5}$$

where, tp_j, fp_j are defined as for macro-precision.

4.7 Micro averaged recall

Micro average recall (recall averaged based on all example/label pairs) is defined as Eq. (6).

$$\text{MicroRecall} = \sum_{j=1}^{Q} tp_j/ \sum_{j=1}^{Q} tp_j+ \sum_{j=1}^{Q} fn_j \tag{6}$$

where, tp_j, fn_j are defined as for macro-recall.

4.8 Micro averaged F-measure

Micro averaged F-Measure is based on harmonic mean between micro-precision and micro-recall. Micro averaged F-Measure is defined as Eq. (7).

$$MicroAveragedFMeasure = \frac{2*MicroPrecision*MicroRecall}{MicroPrecision+ MicroRecall} \quad (7)$$

5. Architecture used for analysis

The presented generic model shown in Fig. 1 is used for empirical analysis, we have used multi-label dataset for analyzing impact of pre-processing configuration, we have considered two main preprocessing configurations raw category (C_0) and the best configuration based on IR Work evaluation (C_3) for the analysis multi-label dataset. Five well known algorithms belonging to problem transformation (BR, CC and LP) and

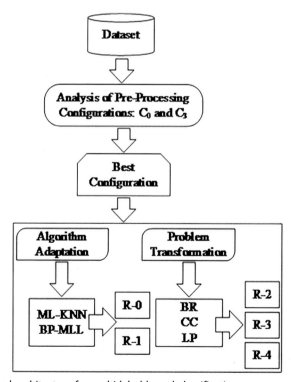

Fig. 1 General architecture for multi-label based classification.

algorithm adaptation (MLKNN and BPNN) is used for comparative analysis. Results R1, R2, R3, R4 and R5 are compared based on eight evaluation measures.

6. Configuration setup and its requirement

As per the empirical analysis of our previous work we have found that text preprocessing configuration plays an important role in classification. We have used IR based classifiers that have significant impact on classification with C_3 (Stemmed and removed stopwords) configuration. To test and validate the impact in reference to multi-label classification, we have used C_0 (Raw dataset) configuration as a reference for impact evaluation with C_3 configuration. We have used our own five Disease data set for result evaluation and to validate our finding we have used standard Seattle dataset mentioned in Ref. [27]. We have used 2010 tweets in disease category and 1823 tweets in Seattle data category.

6.1 C_0 & C_3 configuration

In this configuration we have used raw dataset of 2010 tweets and 1823 tweet (Disease and Seattle dataset) for the analysis. With the analysis of C_0 (raw corpus) configuration, we found data C_3 configuration performs best in both the cases problem transformation and algorithm adaptation. This C_3 configuration performs best in both the dataset category (Disease and Seattle). MLKNN outperforms in Seattle category with C_3 configuration. While in Disease dataset category C_3 configuration have significant impact on MLKNN and BPMLL. Precision and recall values are significantly improve in both the cases. The details of results are mentioned with the graphical measures which are presented in the result section.

With the results we can conclude that preprocessing configuration plays an important role in multi-label text classification and the classifiers mentioned above have significant impact over their evaluation (Tables 1 and 2).

Table 1 Disease data set results on configuration C_0 & C_3.

Algorithms	Subset ACC	Subset HL	ExBased PRE	ExBased REC	ExBased FMEA	MicAv PRE	MicAv REC	MicAv FMEA	Cfg.
BR	0.88±0.011	0.03±0.004	0.89±0.013	0.90±0.013	0.89±0.013	0.97±0.011	0.90±0.013	0.93±0.009	C_0
LP	0.94±0.014	0.03±0.006	0.94±0.014	0.94±0.014	0.94±0.014	0.94±0.014	0.94±0.014	0.94±0.014	
CC	0.91±0.024	0.03±0.009	0.92±0.024	0.92±0.024	0.92±0.024	0.91±0.024	0.92±0.024	0.92±0.024	
MLKNN	0.83±0.022	0.04±0.006	0.84±0.020	0.85±0.018	0.84±0.019	0.93±0.013	0.85±0.018	0.89±0.015	
BPMLL	0.00±0.000	0.56±0.262	0.15±0.076	0.58±0.440	0.22±0.135	0.16±0.078	0.58±0.441	0.22±0.135	
BR	0.89±0.016	0.02±0.003	0.90±0.015	0.91±0.014	0.90±0.015	0.97±0.005	0.91±0.014	0.94±0.008	C_3
LP	0.94±0.012	0.02±0.005	0.94±0.012	0.94±0.012	0.94±0.012	0.94±0.012	0.94±0.012	0.94±0.012	
CC	0.92±0.020	0.03±0.008	0.92±0.019	0.93±0.018	0.92±0.019	0.92±0.020	0.93±0.018	0.92±0.019	
MLKNN	0.91±0.014	0.02±0.003	0.91±0.014	0.91±0.014	0.91±0.014	0.96±0.013	0.91±0.014	0.94±0.008	
BPMLL	0.00±0.00	0.65±0.088	0.19±0.012	0.70±0.20	0.29±0.033	0.19±0.014	0.70±0.198	0.30±0.035	

Table 2 Seattle dataset results on configuration C_0 & C_3.

Algorithms	Subset ACC	Subset HL	ExBased PRE	ExBased REC	ExBased FMEA	MicAv PRE	MicAv REC	MicAv FMEA	Cfg.
BR	0.50±0.032	0.25±0.014	0.75±0.015	0.75±0.020	0.74±0.017	0.75±0.013	0.75±0.019	0.75±0.015	C_0
LP	0.62±0.036	0.23±0.024	0.77±0.023	0.77±0.025	0.77±0.024	0.77±0.023	0.77±0.024	0.77±0.024	
CC	0.59±0.029	0.25±0.020	0.75±0.020	0.75±0.022	0.75±0.020	0.75±0.019	0.75±0.020	0.75±0.020	
MLKNN	0.46±0.017	0.27±0.009	0.75±0.012	0.72±0.032	0.72±0.017	0.74±0.012	0.72±0.032	0.73±0.014	
BPMLL	0.01±0.022	0.52±0.032	0.47±0.029	0.57±0.353	0.47±0.160	0.47±0.033	0.57±0.352	0.47±0.159	
BR	0.58±0.011	0.23±0.005	0.78±0.008	0.78±0.010	0.77±0.008	0.77±0.007	0.78±0.010	0.77±0.006	C_3
LP	0.63±0.011	0.23±0.010	0.77±0.009	0.78±0.011	0.77±0.010	0.77±0.009	0.77±0.011	0.77±0.010	
CC	0.62±0.020	0.23±0.014	0.77±0.012	0.77±0.015	0.77±0.014	0.77±0.012	0.77±0.015	0.77±0.014	
MLKNN	0.54±0.021	0.25±0.015	0.76±0.016	0.74±0.023	0.74±0.018	0.76±0.017	0.74±0.022	0.75±0.016	
BPMLL	0.049±0.045	0.57±0.050	0.43±0.046	0.43±0.181	0.41±0.107	0.43±0.046	0.43±0.181	0.41±0.106	

7. Results, graphical representation C_0 & C_3 and modified ML-KNN algorithm

7.1 C_0 configuration

We applied C_0 (raw corpus, i.e., tweets were taken without any pre-processing except removal of duplicate tweets and URLs) configuration in which all the documents in both the datasets are belonging to raw category. We did this intentionally because we want to utilize maximum features for multi-label classification. Problem transformation and algorithm adaptation methods are used for performance evaluation based on eight different measures. With the results we can conclude that LP algorithm performs best with 93% accuracy in Disease dataset and 62% accuracy in Seattle dataset and the corresponding hamming losses are 2.4% and 2.3%. The corresponding bar plots based on Disease and Seattle datasets are mentioned in Figs. 2–5.

7.2 C_3 configuration

C_3 (Tweets where both stop word removal and Stemming is done) shows that LP in this category outperforms among datasets. It improves performance over C_0 category with a value of 94% and 63% respectively and corresponding hamming losses are 2.3% and 2.2%. But in case of algorithm adaptation method it improves performance by minimum 3% and maximum 8% accuracy for both the datasets. This could possibly be due to good features in both the categories. Finally we can conclude that algorithm adaptation improves the performance in a higher pace if the datasets are having good terms in the corpus. The corresponding bar plots based on Disease and Seattle datasets are mentioned in Figs. 2–5.

7.3 Conventional ML-KNN

Multi-label k nearest neighbor (ML-KNN) is a derivation of popular KNN algorithm [10]. The KNN algorithm follows a non-parametric and lazy learning approach. The ML-KNN adapts this approach and works in two phases. The first phase identifies K nearest neighbors of each test instance in training. Further, in second phase, maximum a posteriori (MAP) principle is utilized as per number of neighboring instances belonging to each possible class to determine the label set for the test instance. The original ML-KNN uses Euclidean similarity measure and its default nearest neighbor size 8. In our work the effectiveness of ML-KNN is evaluated based on four similarity measures (Euclidean, Manhattan, Minkowski, and Chebyshev) mentioned in Ref. [11] and their variations with number of nearest neighbors.

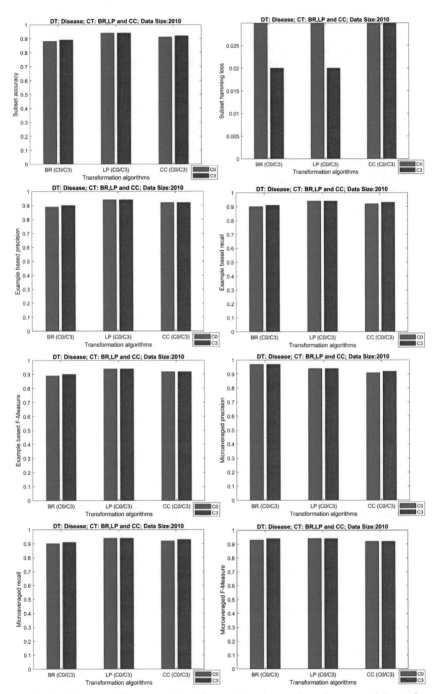

Fig. 2 Bar plot of transformation algorithms with measures based on C_0 and C_3 configurations on disease dataset.

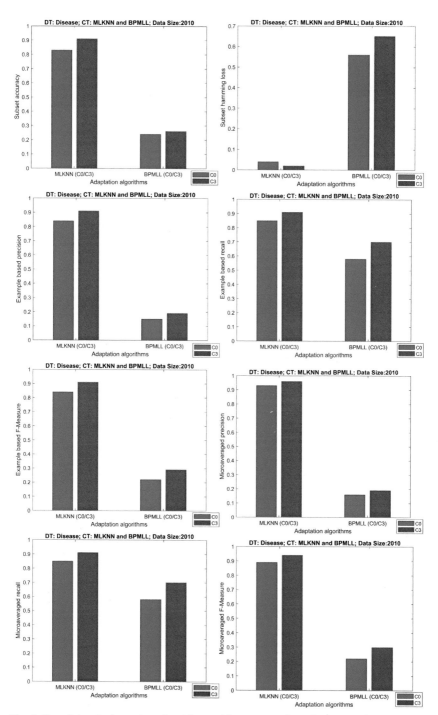

Fig. 3 Bar plots of adaptation algorithms with measures based on C_0 and C_3 configurations on disease dataset.

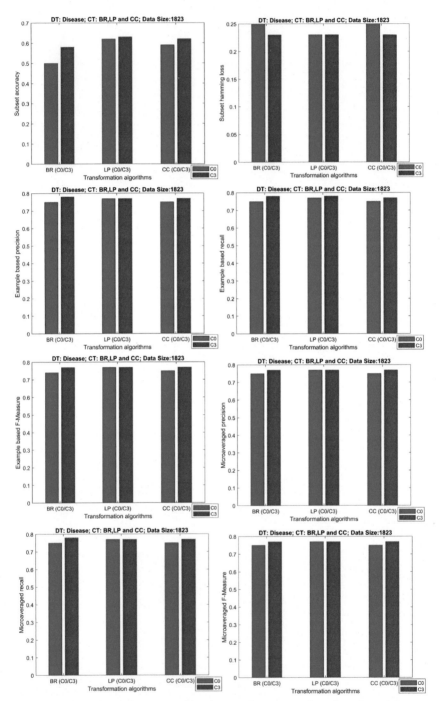

Fig. 4 Bar plots of transformation algorithms with measures based on C_0 and C_3 configurations on Seattle dataset.

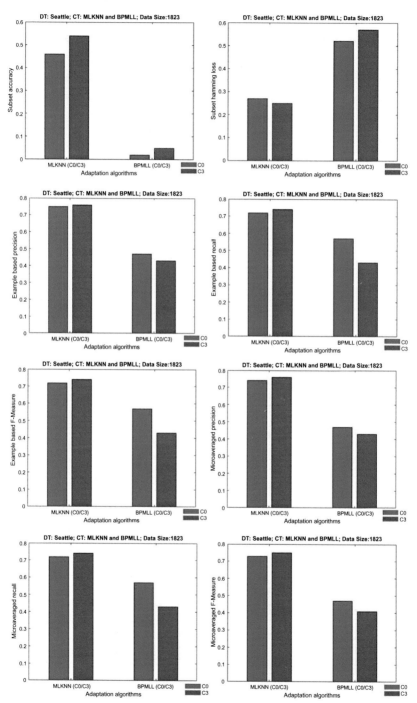

Fig. 5 Bar plots of adaptation algorithms with measures based on C_0 and C_3 configurations on Seattle dataset.

7.4 Modified MLKNN

Modified ML-KNN is advancement in exiting ML-KNN. Where four types of similarity measures (Euclidean, Manhattan, Minkowski, and Chebyshev) is used with different nearest neighbor parameter (5, 8, 11, and 14) which is used for evaluation robustness in MLKNN. The evaluation framework is mentioned in Fig. 6. Experiment showed that the performance of MLKNN can be improved by selecting some well-experimented similarity measures and appropriate number of nearest instances belonging to each possible class.

7.5 Algorithm

Require: Datasets D_k, Pre-processing configuration P_l and multi-label text classifiers C_m.

Mathematical representation for a set of datasets, pre-processing configurations, and classifiers is given below.

$D = \{d_1, d_2, \ldots, d_k\}$ d_q: $1 \leq q \leq k$, $|D| = k$, i.e., number of datasets.

$P = \{p_1, p_2, \ldots, p_l\}$ p_r: $1 \leq r \leq l$, $|P| = l$, i.e., number of preprocessing configurations.

$C = \{c_1, c_2, \ldots, c_m\}$ c_s: $1 \leq s \leq n$, $|C| = m$, i.e., number of classifiers.

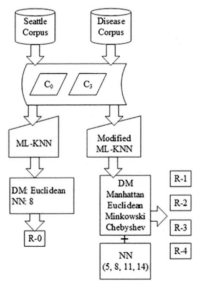

Fig. 6 Modified MLKNN evaluation framework.

for (q = 1; q ≤ k; q++) {
for (r = 1; r ≤ l; r++) {
for (s = 1; s ≤ m; s++) {
    On each dataset d_q apply p_r preprocessing configuration followed by classifier c_t. Compute Subset ACC, Subset Hamming Loss, ExBased PRE, ExBased REC, ExBased FMEA, MicAvPRE, MicAvREC, MicAvFMEA of classifier c_s. Where, c_s belongs to transformation and adaptation category algorithms.
}}}
for (q = 1; q ≤ k; q++) {
for (r = 1; r ≤ l; r++) {
    On each dataset d_q apply p_r preprocessing configuration, based on similarity measures (Euclidean, Manhattan, Minkowski, Chebyshev) and Nearest Neighbor (5, 8, 11, 14). Compute Subset ACC and Subset Hamming Loss of classifier ML-KNN.
}}

7.6 Result evaluation based on disease and Seattle dataset

See Tables 3 and 4.

Table 3 Disease dataset results based on NN and distance measures with C_3/C_0 configuration.

Algorithm	Subset accuracy	Subset Hamming loss	Configuration (C_3/C_0)	Distance measure	NN value
M-MLKNN	91.44	2.34	C_3	Manhattan	8
	83.73	4.16	C_0		11
M-MLKNN	91.44	2.34	C_3	Euclidean	8
	83.73	4.16	C_0		11
M-MLKNN	91.44	2.34	C_3	Minkowski	8
	83.73	4.16	C_0		11
M-MLKNN	11.45	17.93	C_3	Chebyshev	5
	05.13	19.03	C_0		11

Table 4 Seattle dataset results based on NN and distance measures with C_3/C_0 configuration.

Algorithm	Subset accuracy	Subset Hamming loss	Configuration (C_3/C_0)	Distance measure	NN value
M-MLKNN	52.83	25.33	C_3	Manhattan	8
	02.25	35.19	C_0		8
M-MLKNN	52.83	25.33	C_3	Euclidean	8
	48.49	26.85	C_0		5
M-MLKNN	52.83	25.33	C_3	Minkowski	8
	48.49	26.85	C_0		5
M-MLKNN	11.45	17.93	C_3	Chebyshev	5
	02.25	35.19	C_0		8

8. Conclusion

With the result discussed above we can say that feature configuration plays an important role in multi-label text classification. We proposed a general framework for analyzing preprocessing configurations over multi-label classifier. We have used this framework to conduct our empirical studies on tweet classification when multiple labels can be used. We found that the preprocessing configuration of a classifier has a significant impact on its performance and label power set algorithm performs best among all in both the dataset category. The concept is validated over own prepared disease dataset as well as standard Seattle dataset. We found that prior work on tweets multi-label classification research does not use optimal preprocessing configurations in multi-label work. Here we also concluded that algorithm adaptation methods are having greater impact with the good feature ratio which gives a minimum of 3% and maximum of 8% accuracy improvement in both the datasets. We can conclude on the basis or results that qualitative text features plays an important role in classification whether it is related to single class, multi-class or multi-label classification problem. Our results substantially advance the state of the art in multi-label tweet classification. Practitioners can use our results to accelerate the task of multi-label classification problem.

9. Future work

Since multi-label tweet classification has gained the attention of researchers now days. The most obvious future work in this direction is the addition of other ML classifier families to extend the presented framework for the classification where hierarchy of label is maintained or a text instance is having larger number of labels. Ensemble based ML classifiers can be used to extend our work. Query expansion techniques may be a useful in preprocessing step for ML classifiers. We can use the similar concepts for validating more benchmark dataset.

Acknowledgments

We are thankful toward all the individuals who ever helped us in completion of this work in any form. We are also thankful toward academic organizations, i.e., Jaypee Institute of Information Technology Noida and Galgotias University Greater Noida for providing permission for doing collaborative work and all valuable feedbacks.

Conflicting of interests

The author(s) declared no potential conflicts of interest with respect to the research, authorship, and/or publication of this article.

Funding

The author(s) received no financial support for the research, authorship, and/or publication of this article.

References

[1] M. Sofean, M. Smith, A real-time disease surveillance architecture using social networks, Stud. Health Technol. Inform. 180 (2012) 823–827.
[2] S. Yang, M. Santillana, S.C. Kou, Accurate estimation of influenza epidemics using Google search data via ARGO, Proc. Natl. Acad. Sci. U. S. A. 112 (47) (2015) 14473–14478.
[3] J. Ginsberg, M.H. Mohebbi, R.S. Patel, L. Brammer, M.S. Smolinski, L. Brilliant, Detecting influenza epidemics using search engine query data, Nature 457 (7232) (2009) 1012.
[4] K. Lee, A. Agrawal, A. Choudhary, Real-time disease surveillance using twitter data: demonstration on flu and cancer, in: Proceedings of the 19th ACM SIGKDD International Conference on Knowledge Discovery and Data Mining, 2013, pp. 1474–1477.
[5] K. Lee, A. Agrawal, A. Choudhary, Real-time digital flu surveillance using twitter data, in: The 2nd Workshop on Data Mining for Medicine and Healthcare, 2013.
[6] M. Dredze, R. Cheng, M.J. Paul, D. Broniatowski, HealthTweets.org: a platform for public health surveillance using Twitter, in: Workshops at the Twenty-Eighth AAAI Conference on Artificial Intelligence, 2014.

[7] H. Achrekar, A. Gandhe, R. Lazarus, S.H. Yu, B. Liu, Predicting flu trends using twitter data, in: 2011 IEEE Conference on Computer Communications Workshops (INFOCOM WKSHPS), 2011, pp. 702–707.

[8] M. Sofean, M. Smith, A real-time architecture for detection of diseases using social networks: design, implementation and evaluation, in: Proceedings of the 23rd ACM Conference on Hypertext and Social Media, 2012, pp. 309–310.

[9] A. Signorini, A.M. Segre, P.M. Polgreen, The use of Twitter to track levels of disease activity and public concern in the US during the influenza A H1N1 pandemic, PLoS One 6 (5) (2011) e19467.

[10] H. Hirose, L. Wang, Prediction of infectious disease spread using Twitter: a case of influenza, in: 2012 Fifth International Symposium on Parallel Architectures, Algorithms and Programming, IEEE, 2012, pp. 100–105.

[11] G. Tsoumakas, I. Katakis, I. Vlahavas, Mining multi-label data, in: Data Mining and Knowledge Discovery Handbook, Springer, Boston, MA, 2009, pp. 667–685.

[12] A.C. DeCarvalho, A.A. Freitas, A tutorial on multi-label classification techniques, in: Foundations of Computational Intelligence, vol. 5, Springer, Berlin, Heidelberg, 2009, pp. 177–195.

[13] M.S. Sorower, A Literature Survey on Algorithms for Multi-label Learning, vol. 18, Oregon State University, Corvallis, 2010, pp. 1–25.

[14] M.L. Zhang, Z.H. Zhou, A review on multi-label learning algorithms, IEEE Trans. Knowl. Data Eng. 26 (8) (2013) 1819–1837.

[15] A. Aldrees, A. Chikh, Comparative evaluation of four multi label classification algorithms in classifying learning objects, Comput. Appl. Eng. Educ. 24 (4) (2016) 651–660.

[16] L. Rokach, A. Schclar, E. Itach, Ensemble methods for multi-label classification, Expert Syst. Appl. 41 (16) (2014) 7507–7523.

[17] G. Madjarov, D. Kocev, D. Gjorgjevikj, S. Dzeroski, An extensive experimental comparison of methods for multi-label learning, Pattern Recognit. 45 (9) (2012) 3084–3104.

[18] M.L. Zhang, J.M. Peña, V. Robles, Feature selection for multi-label naive Bayes classification, Inform. Sci. 179 (19) (2009) 3218–3229.

[19] Y. Yin, Y. Zhao, C. Li, B. Zhang, Improving multi-instance multi-label learning by extreme learning machine, Appl. Sci. 6 (6) (2016) 160.

[20] J.T. Tomas, N. Spolaor, E.A. Cherman, M.C. Monard, A framework to generate synthetic multi-label datasets, Electron. Notes Theor. Comput. Sci. 302 (2014) 155–176.

[21] Z. Ren, M.H. Peetz, S. Liang, W. Van Dolen, M. De Rijke, Hierarchical multi-label classification of social text streams, in: Proceedings of the 37th International ACM SIGIR Conference on Research & Development in Information Retrieval, 2014, pp. 213–222.

[22] S. Huang, W. Peng, J. Li, D. Lee, Sentiment and topic analysis on social media: a multi-task multi-label classification approach, in: Proceedings of the 5th Annual ACM Web Science Conference, 2013, pp. 172–181.

[23] S. Tuarob, C.S. Tucker, M. Salathe, N. Ram, An ensemble heterogeneous classification methodology for discovering health-related knowledge in social media messages, J. Biomed. Inform. 49 (2014) 255–268.

[24] J. Yang, L. Jiang, C. Wang, J. Xie, Multi-label emotion classification for tweets in Weibo: method and application, in: 2014 IEEE 26th International Conference on Tools with Artificial Intelligence, 2014, pp. 424–428.

[25] H. Sajnani, S. Javanmardi, D.W. McDonald, C.V. Lopes, Multi-label classification of short text: a study on Wikipedia barn stars, in: Workshops at the Twenty-Fifth AAAI Conference on Artificial Intelligence, 2011.

[26] M. Sofean, K. Denecke, A. Stewart, M. Smith, Medical case-driven classification of microblogs: characteristics and annotation, in: Proceedings of the 2nd ACM SIGHIT International Health Informatics Symposium, 2012, pp. 513–522.

[27] A. Schulz, E.L. Mencia, T.T. Dang, B. Schmidt, Evaluating multi-label classification of incident-related tweets, in: Proceedings of the Making Sense of Microposts (# Microposts 2014), Seoul, Korea, 2014, pp. 7–11.

[28] J. Alzubi, A. Nayyar, A. Kumar, Machine learning from theory to algorithms: an overview, J. Phys. Conf. Ser. 1142 (1) (2018) 012012. IOP Publishing.

[29] G. Tsoumakas, I. Katakis, Multi-label classification: an overview, Int. J. Data Warehous. Min. 3 (3) (2007) 1–13.

[30] J. Roesslein, Tweepy Documentation, 2009. Online http://tweepy.readthedocs.io/en/v3, 5.

[31] P. Velardi, G. Stilo, A.E. Tozzi, F. Gesualdo, Twitter mining for fine-grained syndromic surveillance, Artif. Intell. Med. 61 (3) (2014) 153–163.

About the authors

Prof. Saurabh Kumar Srivastava currently working as Professor in the College of Computing Sciences & IT, Teerthanker Mahaveer University, Moradabad, U.P. India. He has around 14+ years of experience in academic and research. His current areas of interest include Text Mining, computer vision, AI and Healthcare, Recommender Systems in Software Engineering, and Social Media Analytics. He has successfully guided and completed several B.Tech Projects, M.Tech thesis and guiding PhD in different areas of Computer Science Engineering. He has to his credit 20 publications in reputed peer reviewed International Journals and International Conferences.

Ankit Vidyarthi presently holds a post of Assistant Professor (Sr. Grade) in the Department of Computer Science Engineering & Information Technology, Jaypee Institute of Information Technology Noida. He joined the Institute in June 2018 and from then onwards he is associated with academics and research activities at the university level. He had completed his Postdoc (Machine Learning) from Bennett University in 2018. He obtained his Ph.D. from the Department of CSE,

Malaviya National Institute of Technology Jaipur in 2017. He had several research papers in reputed SCI/(E) indexed journals with good impact factors. He had also published 20 research articles in various peer-reviewed conferences of IEEE/Springer/ACM/Elsevier which were indexed in Scopus. He is a member of ACM with professional members and SIGACT membership. He is also associated with various journals as a reviewer which are of high standards and indexed in SCI/(E). He is also a member of the Technical Program Committee in various conferences and workshops. He is associated with one journal (JIEEE) as an associate Editor-in-Chief, Senior Editor with AI Foundation Trust, India, and Associate Editor (GE) with IEEE Transactions on Industrial Informatics, and Interdisciplinary Sciences: Computational Life Science Springer Journal.

Prof. Sandeep Kumar Singh is currently working as Professor in the department of CSE & IT. He has around 20+ years of experience in academic and corporate training. His current areas of interest include Mining Software Repositories, Data Mining and Healthcare, Software Code Quality, Recommender Systems in Software Engineering, Software Fault Prediction, and Search based Software Engineering, Social Media Analytics. He has successfully guided and completed several B.Tech Projects, M.Tech thesis and five Ph.D Thesis in different areas of Computer Science Engineering. He has to his credit 70 publications in reputed peer reviewed International Journals and International Conferences.

CHAPTER TWELVE

Big graph based online learning through social networks

Rahul Chandra Kushwaha
Department of Computer Science and Engineering, Rajiv Gandhi University, Doimukh, Itanagar, India

Contents

Abstract

This research work presents a Big Graph based novel computational technique for the enrichment of online learning through Social Learning Networks based recommendations from Twitter, Instagram, YouTube, LinkedIn, etc. The research work is based on NCERT secondary class mathematics e-Textbook. The data mining is used to the enrichment of the e-textbook using the relevant e-resources available from online social media. An algorithmic framework has been developed to extract the mathematical concepts from the e-textbook accessing by the user and match these extracted concepts using phrase graph technique from the twits from the social media and recommend the relevant users to the enrichment of the e-textbook using big graph-based recommendations.

Abbreviations

ICT	information and communication technology
LMS	learning management system
MOOCs	massive open online courses
NCERT	National Council of Educational Research and Training

1. Introduction

Big data are the huge data set to store, process and analyze. These data are generated through social media and large set of networking devices. These devices are connected through the network which can be represented by the big graph. Big graph is a set of nodes and edges of huge number of nodes and edges. In Big data, each device is represented as a node and their connection as the edge of the Big Graph. Social media can be represented by interconnection of large number of texts, audio or video data. Each file which are shared by the users in social media can be represented by the nodes and their interconnection as an edge.

Our society has advanced a lot over last decade due to emergence of ICT. Now a day, tradition learning has been shifted through online learning. Students' dependency on single textbook has already declined; There are vast variety of educational resources available on Internet in the form of open educational resources, video lectures and multimedia contents, etc. Students prefer to read reference books for enhancing in-depth understanding of any topic through various online platforms. Students often, want to explore and find other resources of learning such as online materials available on the Internet. The social media is now becoming popular among the students to exchange information and to connect each-other. Social media platforms like YouTube, Facebook, Quora, etc. are becoming popular platform for online learning. The emergence of online learning like Open Educational Resources, Open Courseware, Learning Management System and MOOCs has shifted the traditional learning into online learning. Thus, the traditional textbooks are now converting into e-textbooks.

Now a days online learning has been become new normal for the education. Major coverage of higher education has been shifted from offline to online. The online learning technologies has been enabled a democratization of education, allowing everyone to receive the same high-quality education whether they live in any part of the world. The notion is that various MOOCs platforms such as edX, Coursera, Udacity and SWAYAM has been able to overcome many of the serious constraints in the developing world and grant smart and motivated students to access teaching resources that they would never have been able to receive before [1–3]. Hence due to popularity of online learning, there is a big challenge to provide quality learning materials to the learners for online learning. Many textbooks specially from the developing region of the world are not well written and have a lot of deficiency in the clarity of concepts. Thus, there feels to provide some technological methods to identify the deficiency of the textbooks and improve its deficiency using algorithmic way. Further improve these learning contents using available resources from the online media.

In this article, an algorithmic technique to identify the learning concepts from the section of the textbooks has been proposed. Further an algorithmic technique to enrich these learning concepts using the augmentation of the contents available from the social media like YouTube, Twitter, *Instagram* and *Quora*, etc. Thus, it will be better if we can have an enrichment of e-textbook. Here, we have proposed a novel approach for enrichment of e-textbook incorporating various e-resources. The proposed model extracts the important learning concepts from the textbooks and enriches these learning concepts through the recommendation of the e-resources available on the Internet. The earlier models are based on page ranked based content recommendation while this model is the enhancement of the model. This model uses graph-based content recommendation. The ranking of the recommended links is based on the cosine similarity with the extracted concepts and based on the cosine similarity the content is recommended. Thus, this model produces better recommendations in comparison to the existing models (Fig. 1).

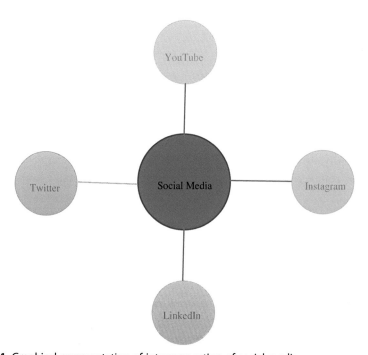

Fig. 1 Graphical representation of interconnection of social media.

2. Literature review

Many research has been done to improve learning pedagogy through online media. Due to digitalization of education, huge amount of data are generating during online pedagogy. Hence, Educational Data Mining and Learning Analytics is using for improving online learning [2,4–8]. Stephen Downes and George Siemens has proposed connectivism theory to improve online learning for the digital age. Connectivism theory is based on learning through social networks. According to Stephen Downes the knowledge is distributed across the network of connections. Knowledge is shared through the nodes and edges in the networks. These nodes are connected through the edges and therefore that learning happens using traverse the networks [1].

Connectivism is representation of society into the social networks where peoples are connected through the networks and share their thoughts through the social networks. Society is much complex, global, and mediated by increasing advancements in technology. Due to advancement of social media, a huge amount of data generating through various social media. People share their ideas through social media and hence now social media has become a medium of social learning. Connection of ideas through social media in large scale may forms big graph, where each source idea acts as a node of the big graph and connection of the ideas acts as an edge of the big graph [5].

It is the arrangement of a complex structure of ideas which are networked to form specific information sets. Knowledge can be derived from a diversity of opinions. The connection of their opinions is a collaboration of current ideas. The core skill of big graph based social learning is the ability to see connections between various information sources and establish connection to facilitate collaborative and continual learning [6].

The first point of connectivism theory is the individual person. The personal knowledge delivered by the person to their connective organizations, is transferred to the system from one person to another person which enrich the system through the knowledge. Individual person is source of knowledge, who share their knowledge to the social networks. In any system, there are some groups of peoples with common goal. They carry forward the knowledge and extend the knowledge system [7,8].

Decisions are based on continuously growing new climate of thinking. Knowledge can be come from outside learners. The constant update or shift of knowledge can be external database or some specialized source of information. The learner is more important for his existing knowledge in place of some outside knowledge. George Siemens stated that the exponentially developing knowledge and knowledge based complex society can be represented by nonlinear model of learning (process) and knowing (state). Any person cannot be independent by learning and knowing. Due to emergence of information technology and Internet, the concept of digital cities has evolved that collaborated vide range of collective works through networks which are linked locally as well as globally. It constructs learning and knowing based communities [1].

Recommender system is an emerging artificial intelligence-based technique using now a days to recommend contents based on user's preference or choice. The system tracks the user's activities and provide recommendations of items similar to the other items which users visited earlier. The same technology can be used in learning technologies to recommend similar contents which are relevant to the users. The course recommender system is gaining attention of research community [9]. There are two approaches of course recommender system; Content based and Collaborative filtering. In a content-based approach, only those courses are recommended to a user that are similar to the courses liked by him earlier. The courses in content-based recommendation approach are usually textual in nature. Collaborative filtering, also recommends courses to a user based on his past search history, however, it does not consider the content of the items for recommendation decision [10,11].

The student might face problems as the textbooks have deficiency like clarity of concept, lack of textual explanations, deficient graphical elaborations, language deficiency, etc. [12]. Many researchers are working since last few years to improve such types of problems in digital textbooks [2]. There are many formulae has been developed to calculate the readability of the paragraph. Concept graph can be drawn to check the conceptual enrichment of the paragraph. Rakesh Agarwal et al. has implemented data mining technique to enrich textbooks [13–17]. The techniques for algorithmically augmenting different sections of a book with links to selective content mined from the world wide web [18]. For finding relevant articles, first identify the set of key concept phrases contained in a section. Using these phrases, those web articles are identified that represent the central concepts presented in the section and augments the section with links to them [16]. There are some algorithmic techniques has been developed to the enrichment of the textual concepts using the recommendation of relevant images and videos contents available on the world wide web [15].

The online learning contents may also enrich through the recommendations of social networks for collaborative learning or social learning networks. Massive peoples are connected through different places through the networks which can be considered as the nodes of the big graph which are connected through the various edges of the big graph. The Big graph can be analyzed, and data mining and analytics is used to provide the relevant contents through the social network [3,9,19–21].

3. Proposed model

The proposed model is based on the implementation of the text mining approach to extract the learning concepts of the e-textbook and enrichment of these learning concepts using the augmentation of the contents available on the World Wide Web. The textbooks are called readable when each sentence in paragraph well written and easy to understand. There are different readability formulae has been developed to calculate readability of the page/paragraph. Concept graph is used to illustrate the concepts in the paragraph in graphical representation. A sentence is considered informative if it has dense concept graph. That is, in paragraph it contains many concepts and these concepts are related to each other [12].

3.1 Concept extraction

The concept extraction has two phases. In first phase, the page/paragraph of the textbook is parsed, and the set of concepts is generated. The parsing can be done by any available parser. Here, Stanford POS (Parts of Speech) tagger [22,23] is considered for this purpose. Finding important concepts in a textual content is a challenging problem. Each sentence is tagged using Stanford POS Tagger. The tagger assigns a unique part-of-speech to each word in a sentence, by processing the entire sentence.

The next part is to form a candidate set of concepts by determining the terminological noun phrases present in the text. The concepts of interest here is consist of noun phrases containing adjectives, nouns, and sometimes prepositions. Sometimes, it contains other parts of speech such as verbs, adverbs, or conjunctions but it is rare. The following three grammatical patterns (P1, P2, and P3) are considered for determining terminological noun phrases [13,24]. It can be expressed by these three patterns using regular expressions as:

$$P1 = C*N$$

$$P2 = (C*NP)?(CN)$$
$$P3 = A*N+$$

where A an adjective, P a preposition, N refers to a noun, and $C = A \mid N$. P1 refers to a sequence of zero or more adjectives or nouns, ending with noun. P2 is extension of P1, which is relaxation of P1 which contains two such patterns connected with a preposition. Examples of P1 are "equivalent real number," "positive number," and "Pythagoras theorem." Examples of P2 are "laws of exponents" and "degree of accuracy." P3 represents a sequence of zero or more adjectives, ending with more than one nouns. This pattern is a restrictive version of P1, where an adjective occurring between two nouns is not allowed. The example of P3 is "nonterminating non-recurring" and "decimal expansion."

3.2 Determining key concept

In second phase, refined set of concepts is extracted. To extract concepts, a set of important keywords set [25] of the textbook is constructed in prior. Then the similarity among the extracted concepts and the keyword sets is measured using following formulae: [26].

$$Sim(X, Y) = \frac{|X \cap Y|}{|X \cup Y|} \tag{1}$$

where, X is the concept phrase extracted from the e-textbook, Y is the concept phrase available in mathematics keywords data set constructed for the school education system. $|X \cap Y|$ is the terms common in both phrases whereas, $|X \cup Y|$ is the union of the terms available in the both phrases. $|X|$ is the number of terms in a phrase. $Sim(X, Y)$ represents the similarity score of phrase X and Y. The Algorithm 1 outlines the concept extraction process.

Algorithm 1. (concept extraction).
  Input (e-textbook page/paragraph)
1) Identify the page/paragraph of the textbook to be enriched.
2) Parse the textbook page/paragraph.
3) Use Stanford POS tagging for concept extraction
4) List Concept extracted from the textbook
5) Input(concept keyword data set)
6) Match the extracted concepts with the available concept keyword set using the formulae $Sim(X, Y) = \frac{|X \cap Y|}{|X \cup Y|}$

If (Sim(X,Y) > threshold), then X is a valid concept phrase.
Output:(concept extracted to be recommendation).

3.3 Content Recommendation

The next part of the work is to provide e-content recommendation. We have used the web crawler to crawl the e-contents from the social media like Twitter, Instagram, YouTube and GeoGebra. It results a list of available contents from the social media on the world wide web. A phrase graph is generated based on the keywords [27]. The following example shows how the phrase graph is generated.

Take an Example to explain as followed:

Mathematics Concept Keyword: probability distribution function
    probability → distribution → function

Phrase graph G_k contains $n(G_k) = \{n_1, n_2, n_3\}$; where $n_1 =$ probability, $n_2 =$ distribution and $n_3 =$ function. $e(G_k) = \{e_1, e_2\}$ where $e_1 =$ probability → distribution, $e_2 =$ distribution → function where, n and e are nodes and edges of the keyword graph G_k, respectively.

Recommendations:

1) probability distribution function

2) cumulative distribution function

3) probability distribution

4) distribution function

    ...

Recommendation Phrase Graphs G_{r1}: probability → distribution → function

$$n(G_{r1}) = \{n_1, n_2, n_3\}; e(G_k) = \{e_1, e_2\};$$

G_{r2} : cumulative → distribution → function $n(G_{r2}) = \{n_4, n_2, n_3\}; e(G_{r1}) = \{e_3 e_2\}$ where $n_4 =$ cumulative and $e_3 =$ cumulative → distribution;

G_{r3} : probability → distribution $n(G_{r3}) = \{n_1, n_2\}; e(G_{r3}) = \{e_1\}$

G_{r4} : distribution → function $n(G_{r4}) = \{n_2, n_3\}; e(G_{r4}) = \{e_2\};$

Here, $n_4 =$ binary and $e_3 =$ binary → number;

Now the graph similarity can be used to match the keyword with the recommendation phrase. According to the cosine similarity the ranking of the links provided. The higher ranked link is provided for the recommendation.

Calculate cosine similarity [24,26] between the phrase query, with the recommendation phrase.

$$Sim(G_k, G_r) = Cos\theta = \frac{G_k, G_r}{|G_k|.|G_r|} \qquad (2)$$

Thus, when $Sim(G_k, G_r) = Cos0$, $Sim(G_k, G_r) = 0$; as both phrase vectors are same. Thus, G_r will be chosen for recommendation.

3.4 Determine relevant content

To determine relevant contents the content summarization technique can been implemented. The content summarization method identifies the relevant contents from the recommended contents and provide the ranking to the recommended contents. The Algorithm 2 outlines the content recommendation process.

Algorithm 2. (e-content recommendation).

Input (concept keyword extracted from the textbook)
1) K[ ] = {k$_1$,k$_2$,. . . k$_m$} *// Concept extracted with Algorithm 1*
2) SN[ ] = {Twitter, Instagram, YouTube, GeogebraTube}
   For i = SN[0]: SN[3]
   a) Go to the root directory of SN[i]
   b) Crawl the all pages of SN[i]
   c) Use google API to recommendation of the econtents
   d) Generate the phase graph Gk based on keywords
   e) Generate a phase graph Gr based on extracted concepts for the recommending the links
   f) Find out the similarity between the phrase query graph with the list of recommendation query graph

$$Sim(G_k, G_r) = Cos\theta = \frac{G_k, G_r}{|G_k|.|G_r|}$$

   g) Rank the recommended links based on graph similarity
   h) Define threshold to be recommendation If (G_r > Threshold) Then
      G_r is the valid recommendation link *// Choose a valid Threshold value*
End For loop
3) Generate the list of recommended contents Result[i]
Output: final e-content recommendation link from social media.

The graphical representation of the conceptual representation of the proposed method is depicted in Fig. 2.

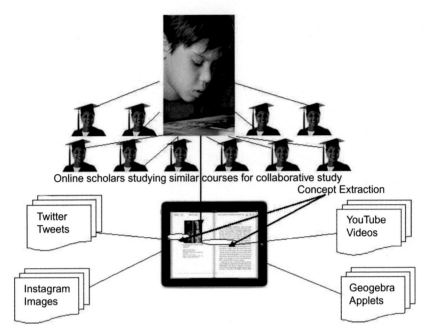

Online scholars studying similar courses for collaborative study

Concept Extraction

Twitter
Tweets

YouTube
Videos

Instagram
Images

Geogebra
Applets

Fig. 2 Enrichment of online learning through social media.

4. Experimental results

The NCERT secondary class (class IX) mathematics textbook has been taken in pdf format. The textbook contains 12 chapters thus 12 pdf file has been taken for the experiment. The Complete textbook is available at NCERT website [10,11,25,28]

In first phase, the textbook file has parsed using text mining tools and concepts extracted using Stanford POS tagging. Now, the mathematics domain concepts have filtered out by matching with the secondary grade mathematics keyword data set. Normally, the threshold value chosen between 0.5 and 0.6 to get best result.

Table 1 shows, the sample of mathematics keyword concepts relevant to the secondary class mathematics. The full keyword concepts set is available at GitHub [8]. The keyword sets are constructed through the NCERT Textbook.

The keyword concepts are filtered out using the above-mentioned formulae. This filtering technique provides good results. Now, the next phase is to recommendation of the e-resources, the google publicly available google API has been used. Let us take an example to show the experimental result.

Table 1 Mathematics keywords data set.

Number line	Radius	Perpendicular
Natural number	Equivalent rational numbers	Sulabhsutras
Whole number	Operations on Real Numbers	Right angled triangle
Negative integer	Laws of exponents	Circle
Pythagoras theorem	Degree of accuracy	Triangle

Textbook Paragraph: *"You are already familiar with the Pythagoras Theorem from your earlier classes. You had verified this theorem through some activities and made use of it in solving certain problems."*

Concept Extracted: Pythagoras theorem we take this concept as a keyword to demonstrate recommendations from the social media.

Recommendation Result for the concept:

a) *Twitter*

1) Bee @Bee mufc Sep 16

Spurs beat United, Watford beat Spurs, United beat Watford. Football is not mathematics, To those that already did pythagoras Theorem b4 d game.

2) Sophie Wall @ sophiewall Aug 30 More

School is a load of shit I remember stressing over algebra and pythagoras' theorem n the only math I've used since gcses is figuring out someone's change from a tenner in work

3) Shikimik Acid zw @DuncanKush Nov 12 More Another day, I haven't used the: 1. The Pythagoras Theorem, 2. Simultaneous Equation, 3. Mupanda High school

...

b) *Quora*

1) Mathematics: What is the new Pythagoras Theorem? Answer 1 of 9 View All

Anonymous—The phrase "New Pythagoras Theorem" has cropped up in a series of anonymous answers here on Quora recently. However, there is no theorem that goes by that name. . . . The answer...

(more)

2) What is the proof of Pythagoras' Theorem? Answer 1 of 16 ·View All

Theja Srinivas, Favourite subject—There are many proofs for Pythagoras theorem. . . . I think the easiest among them is proof using similar triangles. . . . This proof is based on the proportionality of the two.

similar... (more)

3) What is Pythagoras theorem?

Answer 1 of 15 View All.

Denis N. Matei, B.A Computer-Generated Imagery & 3D Art, Grafisch Lyceum Rotterdam (2016)—Q: What is Pythagoras theorem? It is a theorem that lets you calculate the length of c of a right triangle— recognized by the two legs that meet at 90 degrees,

which is where... (more)

...

c) *Instagram*

1) notetakingg Maths notes. . . . . planalogthese are cloudyiswritingSo neat ¡3 oumstudiesthis is so neat and pretty omg meghchan OMG SO SO SO GORGEOUS ...

2) marquezfnm

#ThanksMath #pythagorastheorem kimerakoffee...

3) learningmaderadical

- learningmaderadicalLove the maze activities! Place these pages in a dry erase sleeve and you save on paper while kids can save their papers from possible erasing multiple times. This maze has students practice finding the hypotenuse, only! Link in profile; http://bit.ly/pthypotenuse.

...

d) *YouTube*

1) Algebra - Pythagorean Theorem yaymath 874 K views 5 years ago

Sure, The Pythagorean Theorem is technically it's a Geometry topic, but why not learn about it in Algebra? Some people prefer...

2) How do we Derive the Pythagoras Theorem? Part 1

Don'tMemorise

79 K views 4 years ago

A massive topic, and by far, the most important in Geometry. To view all videos, please visit https://DontMemorise.com .

3) Pythagorean Theorem — #aumsum #kids #education #science #learn It'sAumSumTime • 448 K views 3 years ago

Pythagorean Theorem. In Pythagorean theorem triangle should be right angled. Longest side of triangle is opposite to the 90 ...

...

e) *Geogebra*

1) Pythagoras's Theorem

Author:JudithHohenwarter,MarkusHohenwarter,

StephenJull.

*Topic:*Geometry, Pythagoras or Pythagorean Theorem Move the green and blue sliders. What happens?

2) Proofs Without Words

Author:StevePhelps *Topic:*Geometry, Pythagoras or Pythagorean Theorem.

Here is a GeoGebraBook of Proofs Without Words for the Pythagorean Theorem.

3) Pythagoras Theorem Author: matteLena,S.S.

Kaushik Topic: PythagorasorPythagoreanTheorem.

Pythagoras Theorem

...

The concept extraction technique provides the significant results. The keyword data set filters the relevant concepts in mathematics domain. In the recommendation phase, the recommendations presented from social media. The result presents the relevant recommendations of twitter twits, Instagram images, Youtube videos and Geogebra applets. This technique provides social networks of learners from the social media and recommend the precise and relevant recommendations of e-contents shared by the learners.

5. Conclusion

This research work has presented the novel big graph-based approach to enrich the e-textbook by the augmentation of the e-resources from social media. NCERT textbooks are widely used across India. Thus, this technique will be benefited to the NCERT e-textbooks users. Now social media are becoming very popular among the students for online and collaborative learning. The social media is widely used to share the e-resources among the people. The popular social media like Twitter is used to share micro twits, this technique will use to make social networks of the twitter users to share micro e-contents on twitter. Similarly, Instagram is popular for image and micro video sharing. Therefore, a social network of Instagram users can be made and follows the contents share by the users. YouTube videos are widely used for sharing lecture videos. A social network of similar interest learners may be recommended to share and follows the relevant contents. Geogebra applets are very popular for dynamic mathematics pedagogy. Thus, a social network of Geogebra users can be constructed to share and follows the relevant e-contents of Geogebra applets. Thus, this technique will be benefited to the scholars for making social networks of similar interests of learners and learning relevant contents from social media. The enrichment of e-textbooks will reduce the learning deficiencies and enhance the understanding capability of the learners.

References
[1] G. Siemens, Connectivism: A Learning Theory for the Digital Age, 2013. http://www. itdl.org/journal/jan 05/article01.htm.
[2] E. Cutrell, C.B. SrinathBala, A. Cross, etc., Massively Empowered Classroom: Enhancing Technical Education in India, Microsoft Research India, TechReport, 2013.
[3] TEA, Learning Analytics and Technology Enhanced Assessment (TEA). http://bristol. ac.uk/education.
[4] Office of Educational Technology, Enhancing Teaching and Learning through Educational Data Mining and Learning Analytics: An Issue Brief, U.S. Department of Education, Office of Educational Technology, Washington, (October, 2012. http://www.ed.gov/technology.
[5] Z. Zdrahal, D. Gasevic, A. Wolff, C. Rose, G. Siemens, Learning analytics and machine learning, in: Proceedings of the Fourth International Conference on Learning Analytics And Knowledge, 2014, pp. 287–288, https://doi.org/10.1145/2567574.2567633.
[6] S. Slade, P. Prinsloo, Student perspectives on the use of their data: between intrusion, surveillance and care, in: Challenges for Research into Open & Distance Learning: Doing Things Better – Doing Better Things, 2014.
[7] G. Siemens, P. Tittenberger, Handbook of emerging Technologies for Learning - emerging Technologies for Learning, Soc. Theory 2 (2009) (Online). Available: http://ltc.umanitoba.ca/wikis/etl/index.php/Handbook_of_Emerging_Technologies_ for_Learning).
[8] R. Baker and G. Siemens, "Educational Data Mining and Learning Analytics Ryan S.J. d. Baker, Teachers College, Columbia University George Siemens, Athabasca University 1.," Cambridge Handb. Learn. Sci, 2013.
[9] C. Romero, S. Ventura, E. Garcia, Data mining in course management systems: Moodle case study and tutorial, Comput. Educ. 51 (2008) 368–384.
[10] F. Ricci, L. Rokach, B. Shapira, P.B. Kantor, Recommender Systems Handbook, 2010 (ISBN 978-0-387-85819-7).
[11] A. Klasnja-Milicevic, M. Ivanovic, A. Nanopoulos, Recommender systems in e-learning environments: a survey of the state-of-the-art and possible extensions, Artif. Intell. Rev. 44 (4) (2015) 571–604. https://doi.org/10.1007/s10462-015-9440-z.
[12] R. Agrawal, S. Gollapudi, A. Kannan, K. Kenthapadi, Studying from electronic textbooks, in: Conference on Information and Knowledge Management (CIKM), 2013.
[13] Rakesh Agrawal, Sreenivas Gollapudi, Anitha Kannan and Krishnaram Kenthapadi, Data Mining for Improving Textbooks, SIGKDD Explor., 13(2), 7–19.
[14] R. Agrawal, S. Gollapudi, K. Kenthapadi, N. Srivastava, R. Velu, Enriching Textbooks through Data Mining, ACM, DEV, 2010.
[15] Rakesh Agrawal, Sreenivas Gollapudi, Anitha Kannan and Krishnaram Kenthapadi (2011). "Enriching textbooks with images". Conference on Information and Knowledge Management (CIKM) 2011.
[16] R. Agrawal, S. Gollapudi, A. Kannan, K. Kenthapadi, Enriching education through data mining, in: S.O. Kuznetsov, D.P. Mandal, M.K. Kundu, S.K. Pal (Eds.), LNCS, vol. 6744, 2011, pp. 1–2.

[17] R. Agrawal, M. Christoforaki, S. Gollapudi, A. Kannan, K. Kenthapadi, A. Swaminathan, Mining videos from the web for electronic textbooks, in: ICFCA, 2014, pp. 219–234.

[18] R. Agrawal, S. Gollapudi, A. Kannan, K. Kenthapadi, Identifying enrichment candidates in textbooks, in: WWW: Proceedings of the 20th international conference companion on World wide web, 2011.

[19] D.N. Mithili, S.R. Kasireddy, Graph analysis and visualization of social network big data, in: Social Network Forensics, Cyber Security and Machine Learning, Applied Sciences and Technology, Springer, Singapore, 2019. https://doi.org/10.1007/978-981-13-1456-8_8.

[20] W.M. Campbell, C.K. Dagli, C.J. Weinstein, Social network analysis with content and graphs, Lincoln Lab. J. 20 (1) (2013) 62–81. https://www.ll.mit.edu/sites/default/files/page/doc/2018-05/20_1_5_Campbell.pdf.

[21] J. Maldonado-Mahauad, M. Pérez-Sanagustín, R.F. Kizilcec, N. Morales, J. Munoz-Gama, Mining theory-based patterns from big data: identifying self-regulated learning strategies in massive open online courses, Comput. Hum. Behav. 80 (2018) 179–196. ISSN 0747-5632, https://doi.org/https://doi.org/10.1016/j.chb.2017.11.011.

[22] K. Toutanova, C.D. Manning, Enriching the knowledge sources used in a maximum entropy part-of-speech tagger, in: Proceedings of the Joint SIGDAT Conference on Empirical Methods in Natural Language Processing and Very Large Corpo-ra (EMNLP/VLC-2000), 2000, pp. 63–70.

[23] K. Toutanova, D. Klein, C. Manning, Y. Singer, Feature-rich part-of-speech tagging with a cyclic dependency network, in: Proceedings of HLT-NAACL, vol. 2003, 2003, pp. 252–259.

[24] V.K. Singh, R. Piryani, A. Uddin, D. Pinto, A content based e-resourse recom-mender system to augment e-book-based learning, in: Multidisciplinary Trends in Artificial Intelligence, LNCS, 2013, pp. 257–268. ISBN 9783642449482.

[25] Mathematics Concept Keyword Data Set https://github.com/RAHULBHU/NCERT-Mathematics-Concept-Keyword-Data-Set/wiki.

[26] R. Agrawal, S. Gollapudi, A. Kannan, K. Kenthapadi, Similarity search using concept graphs, in: CIKM, 2014, pp. 719–728.

[27] R.C. Kushwaha, A. Singhal, A. Biswas, Textbook enrichment using graph based E-content recommendation, J. Comput. Theor. Nanosci. 17 (1) (2020) 492–498. American Scientific Publishers. DOI: https://doi.org/10.1166/jctn.2020.8696.

[28] NCERT Secondary Class Mathematics Textbook. http://ncert.nic.in/textbook/textbook.htm?iemh1=0-15.

About the author

Rahul Chandra Kushwaha is currently working as Assistant Professor with the Department of Computer Science and Engineering, Rajiv Gandhi University (Central University), Doimukh (Itanagar), India. Earlier he has worked as Assistant Professor with Banaras Hindu University, India and as senior research associate (SA/SA-SD) at National Council of Educational Research and Training (NCERT), New Delhi, India. He has received Ph.D. degree in Computer Science from Banaras Hindu University, Varanasi, India, in 2019. He has published several research articles in reputed international journals, conferences, and book chapters. His research interests include machine learning, deep learning, data mining, learning science and technologies.

CHAPTER THIRTEEN

Community detection in large-scale real-world networks

Dhananjay Kumar Singh and Prasenjit Choudhury
Department of Computer Science and Engineering, National Institute of Technology, Durgapur, India

Contents

Advances in Computers, Volume 128
ISSN 0065-2458
https://doi.org/10.1016/bs.adcom.2021.10.007

329

Abstract

Large-scale real-world networks (i.e., web graphs) having hundreds of billions of vertices and trillions of edges usually exhibit inhomogeneity, resulting in densely inter-connected nodes (i.e., community structures). Identification of communities in real networks is important because it allows to classify the functions of nodes by their structural positions in their communities. This chapter recognizes the importance of communities and their detection in real networks. It discusses several methods for detecting network community structures. The issues and challenges associated with community detection are assessed thoroughly. Once the community structures from the network are identified using community identification methods, the next step is to assess the identified communities' quality. Therefore, several community evaluation metrics are briefly discussed. Further, some of the application areas of community detection are also specified.

1. Introduction

Large-scale real-world networks (i.e., big networks or big graphs) having hundreds of billions of vertices and trillions of edges are ubiquitous across many different scientific domains (such as technological network, information network, biological network, social network, language network, software network, etc.) [1–6]. Networks (or graphs) consist of vertices (or nodes) and links (or edges), where the edges connect the pair of nodes. Generally, vertices and links in networks represent entities and connection between vertices, respectively [7]. For example, in social networks, people can be represented as nodes and the social connections (i.e., friendships) between the people might be represented as edges; in citation networks, vertices represent scientific articles and edges represent citation relationships between scientific articles; in collaboration networks, the nodes are authors and two authors are considered associated if they co-authored a paper together or if they wrote paper together or if they collaborated together; in protein–protein interaction network, proteins are represented as vertices and interactions among proteins are represented as edges; in transport networks, vertices represent the cities and links represent the connections (e.g., road, airlines, etc.) between the cities; in World Wide Web (WWW), vertices represent webpage and edges represent hyperlinks between webpages. Networks could be either static or dynamic. If the networks are static networks, then its structures do not change over time. And when the networks are dynamic networks, then its structures evolve through time.

Community, that is, set of nodes that have strong connections internally but weak connections externally, is one of the crucial features of large-scale networks [8,9]. For example, communities represent groups of friends in social networks, modules of functionally associated proteins in protein–protein interaction networks, sets of websites dealing with related topics in web graph, groups of people who work together on same topic in collaboration networks, and so on. Communities in the large-scale real networks (or graphs) are of several types: non-overlapping or disjoint community structures (for example, students in an institute belonging to different disciplines), overlapping community structures (for example, persons having different social group membership in Facebook), hierarchical (for example, the human body cells form tissues, tissues form organs, and so on), etc.

Identification (or detection) of communities in networks are important, and there are a tremendous number of methods, such as topology based methods [10], random-walk based methods [11], modularity optimization [12], label propagation [13], clique percolation [14], greedy algorithms [15], etc., that have been introduced to find communities in the last two decades. These traditional methods (or algorithms) for detecting network community structures require the knowledge of the entire network structure. Sometimes we only need to detect the communities of a certain part of the network. Therefore, the detection of local community structures became a fascinating problem and has attracted the attention of networks' research communities. And, recently, several local community finding methods have been introduced to uncover local network communities, such as Clauset [16], LWP [17], LCDMC [18], RTLCD [19], etc. These local methods for detecting local communities are two-stage methods. In first stage, these methods take a source vertex or a set of core vertices as the initial community. In second stage, they expand the initial communities by greedily selecting the optimal nodes based on quality function.

In real networks, overlapping community structure is a natural phenomenon, i.e., at the same time, a vertex can associate with multiple network communities. For example, persons having different social group membership (e.g., family group, friend group, colleague group, etc.) on social networks, a node might have multiple functions in biological networks, an employee of a company may be involved in multiple projects in a company, a person may participate in the discussion of multiple forums or blogs, a researcher might work with variety of research groups in collaboration networks, etc. The problem of community detection becomes more challenging when we consider the overlap of nodes, that is, during the design of an overlapping community detection algorithm, we need to consider both the internal connections of the nodes and the overlap of the nodes. In the last two decades, there are several overlapping community discovering methods have been introduced, such as Clique Percolation Method (CPM) [14], Speaker-listener Label Propagation Algorithm (SLPA) [20], Order Statistics Local Optimization Method (OSLOM) [21], Community Overlap PRopagation Algorithm (COPRA) [22], etc. The primary goal of the overlapping community identification method is to uncover overlapping nodes and communities in the network.

Nowadays, most of the real networks (such as communication network, social network, information network, transportation network, etc.) are dynamic in nature [23]. In dynamic networks, the nodes or edges are added to or removed from the network along time. For example, changes in social networks are frequently made by social users leaving or joining one or more communities or groups. Dynamic networks' communities are community structures that can evolve or change with time. For example, community of a football team players, as time goes by, some football players may leave football team, while new football players may join the football team. After long time, may be no old football players of the team will be present in the football team, however the football team may exist. The identification of dynamic community in networks is the process of uncovering the relevant community structures that changes or evolves along time. Several dynamic methods have recently been introduced to discover the networks' dynamic communities, such as QCA—Quick Community Adaptive [24], BatchInc [25], GreMod [26], LBTR—Learning Based Targeted Revision [27], etc. The basic idea of these dynamic community detection techniques is to update communities continuously by taking into account results from most recent time slices.

Once the community structures from the network are identified using the community identification methods, the next step is to assess the identified communities' qualities. The evaluation of identified communities becomes easier, when the networks' "ground-truth" communities (i.e., original or actual communities) are known a priori. The validation metric (e.g., NMI, ARI, etc.) estimates similarity between network's identified communities and "ground-truth" communities. The scoring metric (for example, modularity) assigns the score to the network communities without having the knowledge of underlying "ground-truth" communities of the network. Another evaluation metrics that is F-measure are also widely used metrics to evaluate the detected community structures.

Community structures play a significant role in many application areas (e.g., epidemic spreading, link prediction, information diffusion, recommendation system, terrorist group detection, suspicious events detection, and many more). The strong community structures in the networks are useful in minimizing the danger, brought by the epidemic prevalence, if special attention paid to the vertices that are near to the infected seeds. The concept of community structure is also used to predict missing links. For example, detecting hidden associations between members of a criminal network, discovering potential collaborators in collaboration networks, etc. Communities may be potentially useful in resolving the problem of maximizing influence. The inter-community edges between the nodes in different community structures are important for information diffusion because these inter-community edges provide more crucial information. For example, we can use the concept of community homophily to swiftly assess the users' influence. We can also choose the seed nodes from various community structures to reduce the overlap and increase the influence. In recommendation systems (RSs), the concept of community is used to generate user neighborhoods that helps in improving the accuracy of RSs. The overlapping communities play a significant role in message forwarding strategies in communication networks by significant reduction of duplicate messages. At a higher level, complex networks are used to represent software structures. And, the identification of community in dependency network is used to refactor the class level object-oriented software. In telecommunication networks, community detection helps in the identification of suspicious events. Community evaluation might be helpful in highlighting the outlier's groups of customers (i.e., suspicious communities) in terms of usage. In criminal networks, Community identification techniques are used to detect terrorist or criminal groups. With the help of centrality measures, the identification of important key members of the criminal communities is possible and that members might be targeted to disrupt the criminal groups.

The rest of chapter is organized as follows: Section 2 introduces community structures and different kinds of communities in real networks. Section 3 mentions the issues and challenges in detecting the community structures.

Section 4 discusses several methods for detecting the communities from real-world networks. Section 5 introduces various evaluation metrics that is used after the detection of community structures from the network. Section 6 identifies some of important applications of community detection. Section 7 concludes the chapter.

2. Community structure

Communities in a network are sets of network vertices with strong internal connections but poor external connections. In large-scale real-world networks, a variety of community structures exist. Broadly, community structures can be classified as follows: (i) Non-overlapping (or disjoint) communities vs. Overlapping communities, (ii) Local communities vs. Global communities, and (iii) Static communities vs. Dynamic communities.

2.1 Disjoint community

In non-overlapping or disjoint community structures, each vertex belongs to only one community structure of the network. For example, students in an institute belonging to different disciplines, people in a particular region support different political parties, and so on.

2.2 Overlapping community

In overlapping communities, the vertices in a network can be part of many different communities. In most cases, a node in a real-world network belongs to several community structures. For example, persons having different social group membership (e.g., colleagues, friends, family, etc.) on social networks, a researcher may be working in different research domains, a research paper may include more than one research topic, a person may participate in the discussion of multiple forums or blogs, an employee of a company may be involved in multiple projects in a company, a node might have multiple functions in biological networks, etc. Furthermore, the number of communities an individual can join in an online social network is essentially unlimited as a person can belong to as many groups as they want at the same time.

2.3 Local community

The local community structures are tightly interconnected vertex sets that are uncovered on the basis of local information of a network. For large-scale real-world networks having millions or billions of vertices or edges, it very difficult to identify or visualize the community structures of the networks. In this scenario, local community structures can be used to extract a smaller, relevant sub-structure of the network for further analysis.

2.4 Global community

Global community detection is possible only if the whole network information is available in advance. However, real networks (such as web graph) are massive networks and it is very hard or sometimes not possible to get whole information of these types of networks. If the network size is small, then one can think of applying global community detection methods.

2.5 Static community

Static communities are communities in the networks that does not change or evolve over time. Static community exists only in the static networks. Once the networks are recorded for a particular period of time, it can be considered as the static networks.

2.6 Dynamic community

Dynamic community is a community structure in the networks that can evolve or change over time. For example, community of a football team players, with the passage of time, some of the football players may leave the football team, while new players may join the team. After long time, may be no old football players of the team will be present in the football team, however the football team may exist.

3. Issues and challenges in detecting communities

In real-world networks, community identification is used to understand the underlying network structure and gain insight into its structure. Defining what are good communities in networks is already a challenge in itself. Designing an effective and efficient community identification algorithm for massive scale networks is extremely challenging task. There are several factors that need to be considered, such as (i) size of the networks, (ii) dynamic nature of the networks, (iii) overlapping nature of the networks, (iv) validation of detected communities, etc.

3.1 Network size

With the continuous expansion of massive-scale real-world networks, the network size has become enormous. These days, real networks have billions of nodes and trillions of edges. Analyzing these huge real-world networks is a big challenge for network research community. The computation required to analyze whole network communities of such massive-scale networks is unimaginable. To solve the above problem, scientists or researchers can design distributed or parallel techniques for community identification in the massive-scale real networks.

3.2 Networks' dynamic nature

Massive-scale real networks (such as Twitter, Facebook, etc.) are inherently dynamic. The nodes or edges in dynamic network are added to or removed from network along time. Discovering the dynamics of communities in real-world networks help us to trace how communities frequently evolve with time. However, comprehending these communities is extremely difficult, especially in the networks that are always changing where the network community often changes with time. Tracking the community evolution along time in evolving complex network is much crucial to know how new community structures are originated and how interaction happens between them. The rapid and unpredictable changes that occur in dynamic networks make this a very difficult problem.

3.3 Overlapping nature of communities

Many people may join or be involved in multiple groups or communities in social networks, so network community structures may overlap. The problem of community identification becomes more challenging when we consider the overlap of nodes, that is, during the design of an overlapping community detection algorithm, we need to consider both the internal connections of the nodes and the overlap of the nodes.

3.4 Scalability

Real-world applications often generate large-scale networks that require efficient processing. The scalability of community identification methods is a serious issue as real networks are growing rapidly. The concept of parallelization of community identification algorithms can be used to solve the scalability issue. Algorithm parallelization leverages distributed computing resources by splitting the whole network into several local subnetworks and requires substantial cluster environment management costs and a large number of computing resources.

3.5 Estimation of number of communities

The community identification is one of the most effective tools to understand and interpret the network topology, and there exist many methods to find community structures in different networks (e.g., artificial networks, real-world networks, etc.). But most methods require a number of communities beforehand which reduces the usefulness of community identification as an analytical tool.

3.6 Validation of detected communities

The lack of a unique definition of community structure is one of important issues with the problem of community identification. Scientists or researchers generally define certain features that a good community must have in studies dealing with this issue. Over the past two decades, a variety of methods have been introduced to find communities in the network. But, how to assess these community identification methods is still a challenging problem. From the literature, the most commonly used evaluation metric for community identification methods is normalized mutual information (NMI) which requires ground-truth communities. Another most widely used evaluation metric is modularity. Modularity is used to evaluate community detection methods if the network does not contain ground-truth communities.

4. Methods for detecting communities

Communities are very crucial for understanding the network topology and how networks function. The identification of community structures in the real-world networks has many important applications, such as discovering people who share similar hobbies, finding criminal or terrorist organizations, etc. To address the issue of uncovering community structures in real networks, a variety of methods for detecting network communities have been introduced over the past two decades. Here, in the following subsection, we briefly discussed some of the important or popular (in literature) methods or techniques to detect large networks communities.

4.1 Local methods

The methods for detecting local community aim to detect a group of nodes with dense connections that includes the given query node. From the literatures, some of the important methods for detecting local community are discussed below.

Radicchi et al. [28] introduced a method to detect local communities that iteratively merges adjacent vertices into the community after meeting the clustering conditions. The main goal of this method is to maximize the local modularity. Clauset et al. [16] put forward the local modularity R to evaluate the local communities of graph. Local modularity R is defined as the total boundary links (or edges) divided by the total links that are connected with the boundary vertices (or nodes).

Luo, Wang and Promislow (LWP) proposed a locally optimized method to identify local community in massive-scale networks [17]. This method has several advantages such as (i) there is no need to pre-determine the number of members in the community, (ii) This locally optimized approach takes substantially less time to execute than global optimization methods, (iii) this method also allow overlapping vertices between local communities, etc. The computational time complexity of LWP locally optimized method is $O(K^2 d)$, where 'K' denotes number of vertices in sub-network to be discovered and 'd' denotes average degree of vertices in sub-network.

A maximum clique extension based local community detection method, known as LCD-MC, was proposed by Fanrong et al. [18]. In the LCD-MC method, it first discovers all maximum cliques that contain the source vertex and then initializes the discovered maximum cliques as the initial local communities. And, it then uses greedy optimization to extend each unclassified local community until a specific goal has been met. And lastly, anticipated local community structures will be acquired until all the maximum cliques are allocated to a community.

A local two-stage method was introduced by Ding et al. [19] to identify local communities which uses the concept of community core detection and community extension. The first stage (i.e., communities' core members identification) replaces seed node with target community core members and extension of identified community core members is based on the relationship with the neighboring nodes.

Nascimento et al. [29] proposed a special heuristic to discover communities based on local clustering coefficient. An approach for local community identification was proposed by Wu [30] which is based on the number of links that are shared across identified communities and community neighbor vertices. This method gives priority to the vertices having a high link similarity value.

4.2 Global methods

Methods for detecting global community require a thorough understanding of the entire network structure. Some of the most important global community identification techniques are listed as follows: (a) *Modularity-based:* Louvain [12], FastGreedy [31], Leading eigenvector [32]; (b) *Vertex similarity-based:* WalkTrap [33]; (c) *Diffusion-based:* Label propagation [13]; (d) *Compression-based:* InfoMap [11], etc.

4.2.1 Louvain

The Louvain method is a greedy optimization method for detecting community structures in massive-scale graphs [12]. It aims to optimize the modularity of network's communities. Modularity [34] is a metric that compare the relative density of edges within community structures to edges outside community structures. The Louvain method first assigns a vertex of the graph as a different community. And, then a node is moved to the one of its neighbor's community whose amalgam results in the greatest gain in modularity. Repeat the second step for all nodes until a maximum of modularity is achieved (i.e., no further improvement can be achieved). The computational complexity of Louvain algorithm is $O(n \times log n)$, where 'n' represents number of nodes in network.

4.2.2 FastGreedy

FastGreedy was proposed by Clauset et al. [31]. FastGreedy algorithm is a greedy-based algorithm that optimizes modularity value. In FastGreedy algorithm, each node is first assigned to a singleton community and then communities are merged iteratively, with each merge yielding the greatest gain in the modularity value. The above step is repeated until there is no further gain in modularity score. The computational time complexity of FastGreedy algorithm for sparse and hierarchical network is $O(n \times log^2 n)$, where 'n' represents number of nodes in network.

4.2.3 Leading eigenvector

Newman [32] proposed the leading eigenvector community detection method, which uses the modularity matrix eigenvalues and eigenvectors to optimize the modularity score. In this method, the eigenvector of modularity matrix is first determined, and then network is divided into two parts so that the modularity score is maximum. The above step is repeated until there is no further increment in value of modularity. The computational time complexity of leading eigenvector method for sparse network is $O(n^2)$, where 'n' represents number of vertices in network.

4.2.4 WalkTrap
WalkTrap was introduced by Pons et al. [33]. This algorithm is a hierarchical algorithm which is based on random walk. The WalkTrap algorithm first computes the distance between all adjacent vertices, and then merges two adjacent nodes into a single new one (i.e., community) and the distance between other nodes are updated. The computational time complexity of WalkTrap is $O(mn^2)$, where 'n' and 'm' denotes number of vertices and links in network respectively. And for sparse networks, the computational time complexity of WalkTrap is $O(n^{2 \times logn})$, where 'n' represents number of vertices in network.

4.2.5 Label propagation
Raghavan et al. [13] presented label propagation. Each node in this algorithm is initially assigned one of 'k' unique labels, and then the algorithm proceeds iteratively and reassigns the labels to the nodes in such a way that every vertex has the majority of its neighbors' labels. Label propagation algorithm will terminate once every vertex has same label as most of its neighbors currently have. Label propagation has a computational time complexity of $O(E^2)$, where 'E' represents total number of edges in network.

4.2.6 InfoMap
Rosvall et al. [11] proposed InfoMap algorithm. This algorithm is an information theoretic principle based algorithm that begins with encoding the network into groups in order to maximize the amount of information. And then the InfoMap sends signal to decoder through a limited capacity channel. Decoder makes an attempt to decode the message and generate a list of possible candidates. If number of candidates are less, then more information has been transferred about the network. The InfoMap has a computational time complexity of $O(E)$, where 'E' represents total number of edges in network.

4.3 Disjoint methods

The detection of disjoint (or non-overlapping) communities in real network is very crucial, and there are a variety of methods, such as hierarchical clustering, modularity maximization, statistical mechanics, random walks, etc., that have been introduced to uncover disjoint communities in the massive-scale real network. Here, we have briefly discussed some of the important or popular (in literature) methods (or techniques) to discover disjoint communities in networks.

4.3.1 Divisive hierarchical method

Girvan et al. [8] introduced a hierarchical approach for identifying disjoint community structures in real networks. It finds community structures by steadily eliminating edges (based on edge-betweenness) from the network. Edge betweenness of an edge e in network is number of shortest paths that cross through edge 'e'. Edges connecting community structures in the network will have high edge betweenness value and by successively removing these edges leads to the isolation of groups of vertices and the underlying communities of the network is disclosed. In sparse networks, the computational time complexity of Girvan-Newman approach is $O(n^3)$, where 'n' represent number of vertices in networks. This method has advantage that number of communities need not be required in advance.

Newman [35] introduced a modularity matrix based bisection spectral method that first divides network nodes into two modules based on signs of modularity matrix leading eigenvector elements and then subdivides those modules recursively based on generalized modularity matrix. Using the modularity matrix concept, Newman derived several algorithms to detects community structure in networks.

4.3.2 Agglomerative hierarchical methods

Newman [32] introduced a greedy-based agglomerative hierarchical community detection technique. In this approach, each vertex of graph is assigned as a singleton community, and then the communities are repeatedly joined together in pairs in such a way that each join produces a greatest gain in the modularity value. This technique has a computational time complexity of $O((m+n)n)$ and for sparse networks, the computational cost of this technique is $O(n^2)$, where 'n' and 'm' represent number of vertices and links in networks, respectively.

Clauset et al. [31] presented a hierarchical agglomeration method for identifying communities in networks. It works by greedily optimizing the modularity. The computational time complexity of this method is $O(md\log n)$, where 'n' and 'm' represent number of vertices and edges in network, respectively, and d represents the dendrogram's depth that describe the communities.

Pons and Latapy [33] proposed a random walks based hierarchical agglomerative method to find the communities of the network. In this method, after calculating the distance between all adjacent vertices, two adjacent communities are chosen to merge into a new community, and distance between them is updated. The worst case computational time complexity of this method is $O(mn^2)$ and the space complexity is $O(n^2)$. In real-world cases, this method has a computational time complexity of is $O(md\log n)$, and space complexity is $O(n^2)$. where 'n' and 'm' denote number of vertices and connections (or links) in network, respectively.

4.4 Overlapping methods

In large networks, identification of overlapping communities is a difficult task and has attracted the attention of networks' research communities over the past two decades. There are several methods for analyzing overlapping communities in the networks have been introduced. Below, we have briefly discussed some of well-known methods for finding overlapping communities in real networks.

4.4.1 CPM

Palla et al. [14] presented CPM (Clique Percolation Method), which is based on k-clique concept, that is, complete subgraphs (fully connected) of k-nodes. This method starts by finding all k-cliques in given network and then forms overlapping communities by merging cliques with common vertices. The CPM method is appropriate for networks that have densely connected regions and it produces decent results when the size of cliques, i.e., k, is small (say, between 3 and 6) [14,20,36]. The CPM was first implemented by CFinder [37] and in many real-world applications, its time complexity is polynomial [14]. However, the CPM method fails to perform as well in many massive-scale real-world network.

4.4.2 COPRA

Gregory [22] presented the Community Overlap PRopagation Algorithm (COPRA) to find overlapping communities in massive-scale real networks. Label propagation approach, introduced by Raghavan et al. [13], is the foundation of the COPRA algorithm and this COPRA algorithm capable of discovering community structures that overlaps. In the COPRA algorithm, each node is initially labeled by a unique identifier and a belonging coefficient that indicates strength of a node's community membership (i.e., summation of all belonging coefficient of a vertex is equal to 1). And then, each vertex updates its belonging coefficients in a synchronous manner at each time step by averaging belonging coefficients from all of its neighbors. The 'v' parameter determines the maximum number of communities a node can be associated to. COPRA has a computational time complexity of $O(vm\log(vm/n))$ per iteration (where, 'n' and 'm' are number of vertices and links in network, respectively). COPRA algorithm is suitable for large and dense networks.

4.4.3 OSLOM

Lancichinetti et al. [21] presented the OSLOM (Order Statistics Local Optimization Method) to identify overlapping modules in networks. OSLOM method finds the communities in a network by looking at their statistical significance. It improves the statistical significance of community structures that is described in terms of a global null model. The worst case computational cost of the OSLOM is $O(n^2)$, where, 'n' denotes number of network's nodes. Computational cost of OSLOM is depend on the communities of the underlying network being studied. OSLOM performs well in discovering highly overlapping communities in directed networks. OSLOM can be used as a refinement approach in massive-scale real networks to clean up results of the initial partition produced by a faster method.

4.4.4 SLPA

The SLPA (Speaker Listener Label Propagation Algorithm) was presented by Xie et al. [20] for uncovering overlapping communities in very massive networks. This algorithm is a Speaker-listener based label propagation algorithm that propagates labels based on dynamic interaction rules. SLPA able to find nested overlapping hierarchy in both unipartite and bipartite networks. Computational cost of SLPA is $O(tm)$, where, 'm' denotes number of links in network and 't' represents maximum number of iterations (for example, $t \geq 20$). By generalizing the interaction rules, SLPAw can be applied to weighted and directed networks.

4.5 Static methods

In the last one and a half decade, there are several static methods for detecting communities that have been introduced, such as Modularity Optimization

[12] method tries to identify community structures based on their modularity score; CPM (Clique Percolation Method) [14] builds up communities by using k-cliques concept (i.e., fully connected complete modules of k-nodes); Label Propagation [13] algorithm first assigns labels to every node and then iterates and reassigns node labels to the nodes in such a manner that every node of the network takes frequent node labels of its neighbors in a synchronous way; and Greedy algorithm [15], a hierarchical bottom-up algorithm that optimizes the quality function (such as modularity) in a greedy manner.

4.6 Dynamic methods

Detecting dynamic community structures means tracking how a group of densely connected nodes changes over time. Recently, a few techniques have been presented to identify dynamic modules in real networks. Below, some of the most important (or well-known in literature) dynamic community identification methods are briefly discussed.

4.6.1 QCA

Quick Community Adaptation (QCA) [24] is an adaptive and fast algorithm for efficiently discovering network communities in dynamic network where changes are often made over time. This method is a modularity-based method. QCA also traces the evolution of communities over time. In QCA method, only the previously detected communities and processes on network changes are taken into account. Therefore, it remarkably reduces processing time and computational cost. Because of its fast running time, this adaptive approach is well suited for rapidly changing massive real dynamic networks.

4.6.2 BatchInc

Chong et al. [25] proposed an incremental batch algorithm in which directly affected nodes (due to network changes) are first assigned to a singleton community, and then in the next step it uses Louvain [12] algorithm. When the graphs evolve significantly then this method is much more effective. Furthermore, this technique ensures a high level of precision. If the number of changes notably rises, the computing cost of BatchInc algorithm increases.

4.6.3 GreMod

GreMod was introduced by Shang et al. [26]. It is a two-stage incremental modularity based algorithm for identifying community structures and tracking the communities' evolution in networks. To obtain an initial community, this method first uses the Louvain [12] algorithm to discover static communities in the network, and then applies an incremental updating strategy to keep track of communities of evolving networks. Shang et al. [26] claim that the computational complexity of GreMod is as low as $O(1)$ that makes it possible to keep track of dynamic communities in a fine-grained way.

4.6.4 LBTR

LBTR (Learning Based Targeted Revision) was introduced by Shang et al. [27]. By filtering out the vertices that remain unchanged in the network, this technique aims to make the better community identification method efficiency. There are two versions of LBTR—(a) LBTR-SVM, and (b) LBTR-LR. The LBTR-SVM method uses the support vector machine (SVM) and the LBTR-LR method uses the logistic regression (LR) to classify the vertices of the network. While retaining the quality of identified network community structures, the LBTR approach requires notably low computational time. This method has a unique perspective of the incremental method which is stable on the dynamic network that evolves over time.

Singh et al. [38] presented an analysis, in details, of the dynamic methods of detecting communities in terms of the computation time and the accuracy. They have also provided few important guidelines to choose the best method for the identification of dynamic communities for given evolving networks.

5. Evaluation metrics

The detection of community structures in real network is very crucial to understand the underlying network structures. Once the community structures from the networks are identified using the community identification methods, the next step is to assess the quality of identified community structures. Evaluation of identified communities becomes easier, when the network's "ground-truth" communities (i.e., original or actual communities) are known a priori. The validation metric (e.g., NMI, ARI, etc.) measures similarity between "ground-truth" communities and identified communities of large-scale real networks. Scoring metric (e.g., Modularity) assigns a score to the network communities without knowing the underlying ground truth communities of the network. We have briefly discussed several evaluation metrics below.

5.1 Modularity

Modularity is a scoring metric that measures *"how densely the network structure is partitioned into community structures (or modules)"* [34]. The modularity, represented as 'Q', is formally defined as,

$$Q = \sum_{S=1}^{N_m} \left[\frac{l_S}{L} - \left(\frac{d_S}{2L} \right)^2 \right]$$

where, N_m represents number of communities, l_S represents number of links between vertices within community S, d_S represents total summation of degrees of vertices inside community S, and L denotes total number of links present in network.

5.2 Normalized mutual information

NMI (Normalized mutual information) is the information-theory-based metric which is used to assess similarity score between identified community structures and actual community structures (i.e., "ground-truth" communities) [39]. Formally, $NMI(X, Y)$ is defined as follows:

$$NMI(X, Y) = \frac{2 \times I(X, Y)}{H(X) + H(Y)}$$

where, $I(X, Y)$ represents the mutual information transferred between structures X and Y; $H(X)$ and $H(Y)$ represents entropy of structure X and Y respectively.

NMI is always a number between 0 (no mutual information) and 1 (perfect correlation).

5.3 Adjusted rand index

ARI (Adjusted Rand Index) [40] is often used in cluster validation. The ARI computes similarity between discovered communities and "ground-truth" communities. Adjusted Rand Index, represented as $ARI(X,Y)$, is defined as follows:

$$ARI(X, Y) = \frac{\sum_{ij} \binom{n_{x_i y_j}}{2} - \sum_i \binom{n_{x_i}}{2} \sum_j \binom{n_{y_j}}{2} / \binom{n}{2}}{\frac{1}{2} \left(\sum_i \binom{n_{x_i}}{2} + \sum_j \binom{n_{y_j}}{2} \right) - \sum_i \binom{n_{x_i}}{2} \sum_j \binom{n_{y_j}}{2} / \binom{n}{2}}$$

where, $X = \{x_1, x_2, \ldots, x_i\}$ represents set of detected communities, $Y = \{y_1, y_2, \ldots, y_j\}$ represents set of "ground-truth" communities, n represents total number of nodes, $n_{x_i} = |x_i|$, and $n_{x_iyj} = |x_i \cap y_i|$.

ARI is a symmetric measure that ranges from -1 to $+1$. When both the communities are totally different then the value of ARI is -1 and when both the communities are completely similar then the value of ARI is $+1$.

5.4 F-score

The F-score, also called the F_1-score, or F-measure is a widely used metric for assessing performance of community identification methods. F-score is harmonic mean of precision and recall. Precision is number of correctly classified vertices divided by number of discovered community vertices. Recall is number of correctly classified vertices divided by number of "ground-truth" community vertices. F-score is defined as follows:

$$F - score = 2 \times \frac{\text{Precision} \times \text{Recall}}{\text{Precision} + \text{Recall}}$$

The value of F-score ranges from 0 (if the detected community is invalid) to 1 (if the algorithm classifies correctly).

6. Applications

There are many potential applications of community detection such as identifying likeminded users for recommendations, finding common research area in collaboration networks, discovering a set of functionally associated proteins in protein–protein interaction networks, uncovering hidden relations among the nodes in the network, etc. Basically, predominated uses of community detection are those areas which require the identification of highly connected groups of individuals or objects having similar characteristics. Below, some of prominent applications of community detection are briefly discussed.

6.1 Epidemic spreading

Community detection plays significant roles in epidemic spreading [41]. The strong community structures in the networks are useful in minimizing the danger that brought by the epidemic prevalence. It is also observed that if the networks having strong community structures, then the special attention should be paid to the vertices that are close to the initial infected seeds.

6.2 Link prediction

One of the significant applications of community identification in the network theory is the missing link prediction. For example, detecting hidden associations between members of a criminal network, discovering potential collaborators in collaboration networks, etc. Some of the links may not get observed during measurement process for a number of reasons. The concept of community structures handled these issues because it allows one to assign probability of existence of a link between the given pair of vertices.

6.3 Information diffusion

Communities in networks play a significant role in diffusion of novel information. Community identification helps in the designing of an effective approximation method for influence maximization problem. In [42], the authors observed that the effects of community can help to solve most important problems (e.g., influence maximization problem) in information diffusion process. They also proposed a model, which is based on community intensifying effects, that takes advantage of community structure effects on the dissemination of information.

6.4 Clustering web clients

In a Web application, identifying groups of web clients that are geographically close to each other and have similar interests can improve the service performance provided on the WWW (World Wide Web). It is advantageous to move the content near to the clusters of web clients that are responsible for a large subset of requests to an origin server, that is, each group of clients can be served by a dedicated mirror server [43].

6.5 Recommendation systems

A recommendation system (RS) predicts a rating or preference a user would like to give an item. RSs usually make use of either (or both) content-based filtering (CB) and collaborative filtering (CF). CF relies on user's preferences or taste information, as well as similar decisions made by other users. CF further classified into model and neighborhood-based CF. The neighborhood-based CF predicts the user preference based on the preferences given by neighbor users. In neighborhood based CF, community detection is used to generate user neighborhoods. Identifying clusters of customers having similar interests can help in establishing an efficient recommendation system.

6.6 Forwarding strategies in communication networks

The overlapping communities play a significant role in message forwarding strategies in communication networks by significant reduction of duplicate messages. A good forwarding strategy can proactively forward messages to destination devices that share a greater amount of common community labels with destination devices. A community-based forwarding strategy will considerably bring down the redundant messages while maintaining the better delivery ratios.

6.7 Software package refactoring

At a higher level, complex networks are used to represent software structures. In Ref. [44], the authors used community detection in dependency network for class level software refactoring of object oriented software. Pan, Jiang and Li [45] use the concept of complex networks to represent classes, where nodes denote the classes, and class dependencies, where an edge represents the dependency between the classes; and they also proposed a method (constrained community identification) to get optimized communities in networks.

6.8 Suspicious events detection

Community detection helps in the identification of suspicious events in telecommunication networks. In Ref. [46], the author has discussed "*how community detection helps to understand the customer's behaviour on the basis of their text messages and phone calls over communication networks?*". Community structure evaluation might be helpful in highlighting the outlier's groups of customers (i.e., suspicious communities) in terms of usage.

6.9 Terrorist group detection

Community identification techniques are used to detect terrorist/criminal groups in the criminal networks (e.g., phone call networks). Using graph analysis techniques, such as centrality measures, the identification of important key members of the criminal communities is possible and that members might be targeted to disrupt the criminal communities.

7. Conclusions

In this chapter, we presented an overview of community structures and their detection in large networks (or big graphs). We discussed several methods/approaches for identifying communities in real networks.

Once communities from the real-world networks are identified using any community identification methods, the next step is to evaluate the identified community structures. Therefore, we highlighted some of the important evaluation metrics that is used to assess detected communities. Issues and challenges associated with community detection in the networks are assessed thoroughly. Further, we have also highlighted the important application areas of community detection algorithms in various domains.

References

[1] G. Caldarelli, Scale-Free Networks: Complex Webs in Nature and Technology, Oxford University Press, 2007.

[2] R. Cohen, S. Havlin, Complex Networks: Structure, Robustness and Function, Cambridge University Press, 2010.

[3] M.E.J. Newman, Networks: An Introduction, Oxford University Press, 2010.

[4] D.K. Singh, R. Patgiri, Big graph: tools, techniques, issues, challenges and future directions, in: Sixth International Conference on Advances in Computing and Information Technology (ACITY 2016), 2016, pp. 119–128.

[5] E. Estrada, The Structure of Complex Networks: Theory and Applications, Oxford University Press, 2011.

[6] S.N. Dorogovtsev, J.F.F. Mendes, Evolution of Networks: From Biological Nets to the Internet and WWW, Oxford University Press, 2013.

[7] D.K. Singh, P.K.D. Pramanik, P. Choudhury, Big graph analytics: techniques, tools, challenges, and applications, in: M. Ahmed, A.-S.K. Pathan (Eds.), Data Analytics: Concepts, Techniques, and Applications, CRC Press, Boca Raton, 2018, pp. 195–222.

[8] M. Girvan, M.E.J. Newman, Community structure in social and biological networks, Proc. Natl. Acad. Sci. U. S. A. 99 (12) (2002) 7821–7826.

[9] M.A. Porter, J.-P. Onnela, P.J. Mucha, Communities in networks, Not. Am. Math. Soc. 56 (9) (2009) 1082–1097.

[10] W. Liu, M. Pellegrini, X. Wang, Detecting communities based on network topology, Sci. Rep. 4 (2014) 5739.

[11] M. Rosvall, C.T. Bergstrom, Maps of random walks on complex networks reveal community structure, Proc. Natl. Acad. Sci. 105 (4) (2008) 1118–1123.

[12] V.D. Blondel, J.L. Guillaume, R. Lambiotte, E. Lefebvre, Fast unfolding of communities in large networks, J. Stat. Mech. Theory Exp. 10 (2008) P10008.

[13] U.N. Raghavan, R. Albert, S. Kumara, Near linear time algorithm to detect community structures in large-scale networks, Phys. Rev. E 76 (3) (2007) 036106.

[14] G. Palla, I. Dernyi, I. Farkas, T. Vicsek, Uncovering the overlapping community structure of complex networks in nature and society, Nature 435 (7043) (2005) 814–818.

[15] M.E.J. Newman, Fast algorithm for detecting community structure in networks, Phys. Rev. E. 69 (6) (2004) 066133.

[16] A. Clauset, Finding local community structure in networks, Phys. Rev. E Statist. Nonlinear Soft Matter Phys. 72 (2) (2005) 026132.

[17] F. Luo, J.Z. Wang, E. Promislow, Exploring local community structures in large networks, in: IEEE/WIC/ACM International Conference on Web Intelligence (WI'06), 2006, pp. 233–239.

[18] M. Fanrong, Z. Mu, Z. Yong, Z. Ranran, Local community detection in complex networks based on maximum cliques extension, Math. Probl. Eng. 2014 (4) (2014) 653670.

[19] X. Ding, J. Zhang, Y. Jing, A robust two-stage algorithm for local community detection, Knowl. Based Syst. 152 (2018) S0950705118301771.

[20] J. Xie, B.K. Szymanski, X. Liu, SLPA: uncovering overlapping communities in social networks via a speaker-listener interaction dynamic process, in: 2011 IEEE 11th International Conference on Data Mining Workshops, 2011, pp. 344–349.

[21] A. Lancichinetti, F. Radicchi, J.J. Ramasco, S. Fortunato, Finding statistically significant communities in networks, PLoS One 6 (4) (2011) e18961.

[22] S. Gregory, Finding overlapping communities in networks by label propagation, New J. Phys. 12 (10) (2010) 103018.

[23] P. Holme, J. Saramki, Temporal networks, Phys. Rep. 519 (3) (2012) 97–125.

[24] N.P. Nguyen, T.N. Dinh, Y. Xuan, M.T. Thai, Adaptive algorithms for detecting community structure in dynamic social networks, in: 2011 Proceedings of IEEE INFOCOM, 2011, pp. 2282–2290.

[25] W.H. Chong, L.N. Teow, An incremental batch technique for community detection, in: Proceedings of the 16th International Conference on Information Fusion, 2013, pp. 750–757.

[26] J. Shang, L. Liu, F. Xie, Z. Chen, J. Miao, X. Fang, C. Wu, A real-time detecting algorithm for tracking community structure of dynamic networks, CoRR (2014) abs/1407.2683.

[27] J. Shang, L. Liu, X. Li, F. Xie, C. Wu, Targeted revision: a learning-based approach for incremental community detection in dynamic networks, Phys. A Statist. Mech. Appl. 443 (2016) 70–85.

[28] F. Radicchi, C. Castellano, F. Cecconi, V. Loreto, D. Parisi, Defining and identifying communities in networks, Proc. Natl. Acad. Sci. U. S. A. 101 (9) (2004) 2658–2663.

[29] M.C.V. Nascimento, Community detection in networks via a spectral heuristic based on the clustering coefficient, Discret. Appl. Math. 176 (2014) 89–99.

[30] Y.J. Wu, H. Huang, Z.F. Hao, F. Chen, Local community detection using link similarity, J. Comput. Sci. Tech. 27 (6) (2012) 1261–1268.

[31] A. Clauset, M.E.J. Newman, C. Moore, Finding community structure in very large networks, Phys. Rev. E 70 (6) (2004) 066111.

[32] M.E.J. Newman, Finding community structure in networks using the eigenvectors of matrices, Phys. Rev. E 74 (3) (2006) 036104.

[33] P. Pons, M. Latapy, Computing communities in large networks using random walks, in: International Symposium on Computer and Information Sciences (ISCIS 2005), 2005, pp. 284–293.

[34] M.E. Newman, M. Girvan, Finding and evaluating community structure in networks, Phys. Rev. E. 69 (2) (2004) 026113.

[35] M.E.J. Newman, Modularity and community structure in networks, Proc. Natl. Acad. Sci. U. S. A. 103 (23) (2006) 8577–8582.

[36] A. Lancichinetti, S. Fortunato, Community detection algorithms: a comparative analysis, Phys. Rev. E 80 (5) (2009) 056117.

[37] B. Adamcsek, G. Palla, I.J. Farkas, I. Derényi, T. Vicsek, CFinder: locating cliques and overlapping modules in biological networks, Bioinformatics 22 (8) (2006) 1021–1023.

[38] D.K. Singh, R.A. Haraty, N.C. Debnath, P. Choudhury, An analysis of the dynamic community detection algorithms in complex networks, in: 2020 IEEE International Conference on Industrial Technology (ICIT), 2020, pp. 989–994.

[39] D. Leon, A. Díaz-Guilera, J. Duch, A. Arenas, Comparing community structure identification, J. Stat. Mech. Theory Exp. (9) (2005) P09008.

[40] L. Hubert, P. Arabie, Comparing partitions, J. Classif. 2 (1) (1985) 193–218.

[41] H. Wei, C. Li, Epidemic spreading in scale-free networks with community structure, J. Stat. Mech. Theory Exp. 2007 (1) (2007) P01014.

[42] S. Lin, Q. Hu, G. Wang, S.Y. Philip, Understanding community effects on information diffusion, in: Pacific-Asia Conference on Knowledge Discovery and Data Mining, 2015, pp. 82–95.

[43] B. Krishnamurthy, J. Wang, On network-aware clustering of web clients, in: Proceedings of the conference on Applications, Technologies, Architectures, and Protocols for Computer Communication, 2000, pp. 97–110.

[44] W.F. Pan, B. Li, Y.T. Ma, J. Liu, Y.Y. Qin, Class structure refactoring of object-oriented softwares using community detection in dependency networks, Front. Comput. Sci. China 3 (3) (2009) 396–404.

[45] W.-F. Pan, B. Jiang, B. Li, Refactoring software packages via community detection in complex software networks, Int. J. Autom. Comput. 10 (2) (2013) 157–166.

[46] C.A.R. Pinheiro, Community detection to identify fraud events in telecommunications networks, in: SAS SUGI Proceedings: Customer Intelligence, 2012.

About the authors

Dhananjay Kumar Singh received the M.Tech. degree in Computer Science and Engineering from the Department of Computer Science and Engineering, National Institute of Technology, Silchar, India, in May 2016. He is currently a Ph.D. Research Scholar in the Department of Computer Science and Engineering, National Institute of Technology, Durgapur, India. His primary research interests include community structure analysis, complex networks, big graph mining, and recommendation systems.

Prasenjit Choudhury received the Ph.D. degree in Computer Science and Engineering from the National Institute of Technology, Durgapur, India. He is currently an Assistant Professor in the Department of Computer Science and Engineering, National Institute of Technology, Durgapur, India. He has published more than 60 research articles in International Journals and Conferences. His research interests include wireless network, complex networks, information retrieval, data science, mobile and crowd computing, and recommendation systems.

CHAPTER FOURTEEN

Power rank: An interactive web page ranking algorithm

Ankit Vidyarthi[a] and Pawan Singh[b]

[a]Department of CSE&IT, Jaypee Institute of Information Technology, Noida, Uttar Pradesh, India
[b]Department of CSE, Amity University, Lucknow, Uttar Pradesh, India

Contents

Abstract

In the present era, one of the most important assets, information may be availed through different medium out of which the web has the status of most popular medium. Nowadays, several businesses, public domain enterprises and government organizations uses web to float and fetch information, supervision, and communications with external bodies. The importance of web pages thus becomes an inherently subjective matter whose dependency is on the reader's interests and understanding. But there exists much more than this that can be said impartially about the solitary web pages importance. In this book chapter, a new iterative web page ranking algorithm is proposed named as Power Rank. It uses the web graph where every outgoing edge of a page is weight on the basis of in-links and out-links as a power of a number to calculate the rank of web page in website. The objective of the proposed algorithm is to maximize the power of a path to select the page, indicating the page with high power. The experimental result gains the significant difference in terms of performance measures and proves that proposed algorithm outperforms over state-of-art algorithms like Page Rank, HITS and Weighted Page Rank in comparative analysis.

Advances in Computers, Volume 128
ISSN 0065-2458
https://doi.org/10.1016/bs.adcom.2021.10.008

1. Introduction

The World Wide Web (WWW) generates several challenges in the field of information retrieval where the target is to fetch the relevant information from the millions of web resources. These web resources are in millions in number with extremely diverse content and heterogeneous in nature. In addition to these challenges, the experience and information searching text by user also manipulates the page ranks on the web by search engines.

In the present time the wide use of WWW in different field of science, commerce, learning, several businesses, public domain enterprises, government organizations, news, etc. makes its use important. This leads the fast growth of the web where users get mystified due to more than one options. Therefore searching the relevant options for the user and cater them, has become the herculean task for website owners. Now days in teaching and learning process the use of multimedia is encountered which is using web based applications [1,2].

The vast expansion of web can be visualized from the statistics that within a short span of time a few dozens of web servers have increased up to more than 400 million today [3]. The estimations provided by the Google, they have indexed more than 3 billion documents up to the year 2018 and more than 8 billion pages are indexed currently. According to the analytical research on data, 8% is the rate of creation of new pages per week [4]. Observation says that only about 20% of the pages will be accessed by the users after 1 year [5]. Due to the enormous size of the web the finding of good quality page of required information is still a big challenge.

Moreover, the quality analysis of the searched web page is the reflection of user interest. It is not possible to collect all the related pages from the web on user request to get the exact required information [6]. Even then it is mandatory to rank the web pages and provide the best high ranked page to the user on his request which can also be done by graph model [7–9]. In literature there exists some page ranking mechanism working on different principle even then there is a great scope to empower the ranking mechanism by devising a better algorithm, which considers the better criteria of ranking with improved efficiency.

Unlike flat documentation, the WWW is considered as a hypertext that provides auxiliary information of the web pages with the help of link structure and link text. In this book chapter a web page ranking mechanism is proposed named as Power-Rank. Power Rank algorithm considers the in-link as well as out-links of a web page to calculate the ranking of a page in a same fashion like weighted page rank but the way of calculation using consideration of the combination of in-links and out-links is extremely different. In proposed algorithm the power of every outgoing link of a node is considered as $2^{I_i/O_i}$, where I_i is number of input links and O_i is the number of output links on page i, if the output link considered is from same i to j. These weights are used to calculate importance of any page in the form of power. Through this mechanism the advantage of the link structure is used to find the importance factor of web pages in terms of ranking, called Power Rank.

Rest of the book chapter is formatted as follows: the detailed description of the work already done in this field is presented in Section 2. Later, in Section 3, proposed methodology is presented followed by experimental setup results in Section 4. The final conclusion of the proposed work is presented in Section 5.

2. Literature study

In the past different mechanism has been developed which exists in literature that were used for finding relevant pages in hyper link environment [10]. Web mining methodologies are one of them which were deployed by the search engines to find required information page from the web [11]. Now days the trend is to mix the web mining with artificial intelligence for smart mining [12]. The most successful search engine Google is using Page Ranking mechanism to search the page as per user interest. Few of the algorithms works on the actual contents on the page while some algorithms uses user behavior and web structure made up of hyper-links to find the relevant pages from the large corpus of web [13]. In literature, one of the most widely used and considered as most popular algorithm is the Page Ranking algorithms [14].

The general page rank algorithm uses the web structure (link structure) to find the importance of the page. In page rank algorithm the back links are considered to calculate the importance as well as the rank score of a page is equally divided into the outgoing links of the page [15]. The importance of any page is represented by summing the page ranks forwarded by back links. A page will be more importance if it has an in-links from an importance pages. Literature also holds method that was considered as the expansion of standard page rank into the Weighted Page Rank [16]. In this mechanism the outgoing links will not get rank distribution equally among them instead more vital pages will get larger rank values. Every outgoing links gets a value proportional to its popularity which is calculated on the basis of in-links and out-links.

A new mechanism of a page ranking algorithm named PRLV is proposed in Ref. [17] which is based on link visits. It consider the structure of web as well as the usage mining, which relates the number of times user visits on respective pages/links, to find the popularity of the web page in the form of page rank. In this the rank is also distributed in the outgoing links in the proportion of popularity but the popularity is calculated using the visit count of that link saved in the sever log file [18,19].

Another algorithm named as Distance rank algorithm is proposed in literature which is based on reinforcement learning. In this algorithm the distance between pages is considered as punishment [20]. The distance is the number of average clicks between two pages. The sum of received punishments (distance) is to be minimized by the agent and the web page having low distance value will receive high rank. In the mechanism it is considered that user will surf the web randomly and initially the user have no background about the web pages so he/she clicks on the basis of current page's content but as the time passes the user select the pages on the basis of current content of each page as well as background. As user gives more time they gain more knowledge from the environment and pages.

In the late 90s, i.e., from the year 1995 to 1998, there exist several literature studies in the form of research articles and thesis that focuses on the link structure of the web. These studies put their emphasis on how to make the most of the link structure of outsized hypertext systems such as the web.

A work published in Refs. [21,22] uses a broad range of link based analysis for the ranking of the web pages. Moreover, a clustering based method that takes the link structure into account for finding the relevant page is discussed in Ref. [23]. The author discussed the information that can be obtained from the link structure for a variety of applications. Good visualization demands added structure on the hypertext and is discussed in Ref. [24,25]. Later, one author had developed an interesting model of the web as Husband Authorities, based on an Eigen vector calculation on the co-citation matrix of the web [26].

Later in the present research arena there are several papers which were focused on the various features and factors involving in the page ranking of the web. The survey of various factors on search engine result pages is presented in Ref. [27] where relevant factors were computed that affect search results on the web. In addition, the author also used the manual phrasing for getting relevant web links. Some of the other works based on the modifications in the strategy of the page rank algorithms also exist in literature, i.e., Genetic Algorithm (GA) based page rank algorithm [28] and Swarm optimization for page ranking [29].

The involvement of the soft computing approaches and nature inspired computing algorithms in the state of art algorithms for page rank on web has gained the significant interest of the research community. The use of the GA [28] in page rank algorithm is found significant where the hybrid methodology is used for optimization of the web search while keeping the page ranking advantages as usual. On the other hand, swarm optimization is used in Ref. [29] to boost the overall performance of the search result over web. The search result is boosted using the fitness function that actually maximizes the page ranking scores over the web. One of the limitations of this approach is that it is used for the local model while for global model the performance has to be still tested.

Recently the page ranking algorithm performance is also boosted using the vector normalization in page ranking named as sNorm (p) algorithm [30]. The key concept of the algorithm is that it is designed for the efficient ranking of web resources. It is the extension of the state of art algorithm named SALSA where vector norm is used for ranking computations.

In the next section, the detailed description of the proposed methodology is given that is used for the ranking of the web pages in solitary web based environment.

3. Proposed methodology

3.1 Description of proposed power rank algorithm

This book chapter proposed an intelligent page ranking algorithm named as Power Rank which is based on reinforcement learning. The proposed algorithm overcomes the drawbacks of other traditional page ranking algorithms. It is considered generally that related pages are linked to each other. It propagates a power to the outgoing links on the basis of number of incoming and outgoing links. The pages having high average power will be the one with good rank. Here power is measured as a reward. The popularity provides power hence the page with more power will be the more popular and full of relevancy for user. The working of our algorithm can be understood by the following propositions:

Preposition 1. A web is made up of in-links and out-links so the rank must be calculated using both types of links. For a web page every outgoing link must be used for power distribution. The power of outgoing link P_{ij} that is from node "i" to node "j" is shown in Fig. 1.

Mathematically, the power computation between the nodes is defined as:

$$P_{ij} = n^{I_i/O_i} \tag{1}$$

where, n represents the number of possible actions can be taken on a link. As the link can be selected or may not be selected so the value of "n" is considered "2," I_i represents the number of in-links on page I, O_i represents the number of out-links on page i.

According to popularity as many numbers of links will be coming to a page they will bring the power to it hence it is found that power is directly proportional to node in-links, i.e.,

$$P_{ij} \propto I_i \tag{2}$$

Moreover, according to popularity as less number of links will be going out the power will be distributed in less number of links and every link will get big share, i.e.,

$$P_{ij} \propto \frac{I}{O_i} \tag{3}$$

Finally considering the value of n and other facts the equation to compute the power of link will be given as,

$$P_{ij} = 2^{\frac{I_i}{O_i}} \tag{4}$$

Preposition 2. A good path from some initiator node i to destination node j will be the path through high power links. For example, if we consider the sample graph in Fig. 2, which is supposed as a sub graph of web then the path from node i to node j have two possible paths, i.e., path "$i \rightarrow p \rightarrow q \rightarrow j$" and path "$i \rightarrow r \rightarrow s \rightarrow j$."

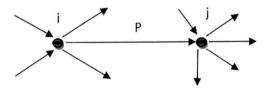

Fig. 1 Power distribution between two nodes, i.e., node "i" to node "j."

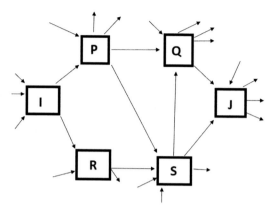

Fig. 2 Sample graph of web scenario representing pages with in-links and out-links.

Table 1 Out-link power of nodes of sub graph in Fig. 2.

Nodes/pages	In-links	Out-links	Out-links power
i	3	2	$2^{3/2}$
p	2	3	$2^{2/3}$
q	2	4	$2^{2/4}$
r	2	2	$2^{2/2}$
s	3	2	$2^{3/2}$
j	2	3	$2^{2/3}$

According to the number of out-links and in-links based on Fig. 2, the power of out-links of different (pages) nodes using Eq. (4) is given in Table 1.

Next, the power of the path from node i to node j can be calculated as the summation of the power of individual nodes lying intermediate between nodes i to node j. Mathematically it is computed as, assuming that there is a path $i \rightarrow a \rightarrow b \rightarrow c \rightarrow d \rightarrow - - - x \rightarrow y \rightarrow z \rightarrow j$ the power of the path is calculated as,

$$P_{ij} = P_{ia} + P_{ab} + P_{bc} + - - + P_{xy} + P_{yz} + P_{zj} \qquad (5)$$

Using Eq. (4), the individual power between links is calculated as,

$$P_{ij} = 2^{Ii/Oi} + 2^{Ia/Oa} + 2^{Ib/Ob} + - - - - - + 2^{Ix/Ox} + 2^{Iy/Oy} + 2^{Iz/Oz} \qquad (6)$$

The general representation of Eq. (6) is given as,

$$P_{ij} = \sum_{k=i}^{j-1} 2^{I_k/O_k} \tag{7}$$

Based on Eq. (7), the power path value of the path "$i \to p \to q \to j$" and path "$i \to r \to s \to j$" using Fig. 2 is computed as,

$$\text{Path}"i \to p \to q \to j" = \left(2^{3/2} + 2^{2/3} + 2^{2/4}\right) = 5.83$$

$$\text{Path}"i \to r \to s \to j" = \left(2^{3/2} + 2^{2/2} + 2^{3/2}\right) = 7.6568$$

Hence the preferred path which will be selected is the second path that has more power value, i.e., 7.6568.

Preposition 3. If there is no link between two nodes, i.e., nodes i to node j then the power between them will be considered as zero i.e. the power exists only if the link exists.

$$P_{ij} = 0 \ \{\text{iff no path exist in between nodes i to node j}\} \tag{8}$$

Preposition 4. For two different outgoing links of two different pages, if they have different power so will be their rank proposition.

Let the two pages are p and q and their power calculated using Eq. (4), have a relation such as

$$2^{I_p/O_p} > 2^{I_q/O_q} \tag{9}$$

By taking $Log2$ on both sides of inequalities

$$\log_2 2^{I_p/O_p} > \log_2 2^{I_q/O_q}$$
$$\Rightarrow I_p/O_p > I_q/O_q \tag{10}$$

As it is already been discussed that the rank propagation $(R^p) \propto$ Power (P) in preposition 2 and power is directly proportional to in-links and inversely with out-links in Eqs. (2) and (3), thus it is called that

$P \propto 1/O$

$\Rightarrow (R^p) \propto P \propto 1/O$
$\Rightarrow (R^p) \propto 1/O$ $\Rightarrow (R^p) > (R^q)$ {Using Eq. (10)}
$\Rightarrow (R^p) > (R^q)$

Preposition 5. Let the P_{ij} indicates the power between page i and page j as a power of link $i \to j$ which is one of outgoing link of page i then P_j will denote the average power of page j. It can be represented as,

$$P_j = \frac{1}{N} \sum_{k=1}^{N} P_{kj} \qquad (11)$$

where, N is the total number of web pages.

To explore the computation of average page power consider that the page j is having only one incoming link and it is from page i than it can be such as:

$$P_j = \frac{1}{N} \sum_{k=1}^{N} P_{kj}$$

$$P_j = \frac{1}{N} \left[\sum_{k=1 \& k \neq i}^{N} P_{kj} + P_{ij} \right]$$

As there is only one way to reach j, that is through i, thus

$$P_j = \frac{1}{N} \left[\sum_{k=1 \& k \neq i}^{N} \left(P_{ki} + P_{ij} \right) + P_{ij} \right] \qquad (12)$$

Now it is being observed and computed mathematically that the total count of number of paths for $\sum_{k=1 \& k \neq i}^{N} \left(P_{ki} + P_{ij} \right)$ is $(N-1)$. Thus Eq. (12) is simplified as,

$$P_j = \frac{1}{N} \left[\sum_{k=1 \& k \neq i}^{N} \left(P_{ki} \right) + N.P_{ij} \right]$$

$$= P_{ij} + \frac{1}{N} \left[\sum_{k=1 \& k \neq i}^{N} \left(P_{ki} \right) \right]$$

$$= P_{ij} + \frac{1}{N} \left[\sum_{k=1 \& k \neq i}^{N} \left(P_{ki} \right) + P_{ii} - P_{ii} \right]$$

$$= P_{ij} + \frac{1}{N} \left[\sum_{k=1}^{N} \left(P_{ki} \right) - P_{ii} \right]$$

$$= P_{ij} + \frac{1}{N} \left[\sum_{k=1}^{N} \left(P_{ki} \right) \right] - \frac{P_{ii}}{N} \quad \left\{ \because P_i = \frac{1}{N} \sum_{k=1}^{N} P_{ki} \right\}$$

$$= P_{ij} + P_i - \frac{P_{ii}}{N}$$

$$\left\{ \because N \, is \, very \, very \, big \, so \lim N \to \infty \, \& \lim \frac{P_{ii}}{N} \to 0 \right\}$$

$$P_j \approx P_i + P_{ij} = P_i + 2^{I_i / O_i}$$

Hence the Power Rank of page j will be equal to

$$P_j = \max_i \left[P_i + 2^{\ I_i/O_i} \right] \tag{13}$$

For example, as in Fig. 2 the Power Rank of page j can be calculated as follows:

$$P_j = \max \left[P_q + 2^{2/4}, \ P_s + 2^{3/2} \right]$$

$$= \max \left[P_p + 2^{2/3} + 2^{2/4}, \ P_r + 2^{2/2} + 2^{3/2} \right]$$

$$= \max \left[P_i + 2^{3/2} + 2^{2/3} + 2^{2/4}, \ P_i + 2^{3/2} + 2^{2/2} + 2^{3/2} \right]$$

$$P_j = P_i + 2^{3/2} + 2^{2/2} + 2^{3/2}$$

As per the Sutton and Barto [31], by using Q learning off policy temporal difference reinforcement learning in one step can be represented as:

$$Q(s_t, a_t) = Q(s_t, a_t) + \alpha \left[r_{t+1} + \gamma \max_a Q(s_{t+1}, a) - Q(s_t, a_t) \right] \tag{14}$$

where, s_t represents state at time t, a_t is action at t, r_t is reward at t, α is the step size parameter, γ is the discount rate parameter.

Further Eq. (14) can be simplified as:

$$Q(s_t, a_t) = Q(s_t, a_t) + \alpha[r_{t+1} + \gamma \max_a Q(s_{t+1}, a) - Q(s_t, a_t)]$$

$$= Q(s_t, a_t) + \alpha[r_{t+1} + \gamma \max_a Q(s_{t+1}, a)] - \alpha Q(s_t, a_t) \tag{15}$$

$$= (1 - \alpha) Q(s_t, a_t) + \alpha[r_{t+1} + \gamma \max_a Q(s_{t+1}, a)]$$

Since the parameter γ and r_{t+1} are constant thus,

$$\implies \gamma \max_a Q(s_{t+1}, a) \equiv \max_a \gamma \, Q(s_{t+1}, a)$$

$$\implies r_{t+1} + \gamma \max_a Q(s_{t+1}, a) \equiv r_{t+1} + \max_a \gamma \, Q(s_{t+1}, a)$$

$$\implies r_{t+1} + \gamma \max_a Q(s_{t+1}, a) \equiv \max_a \left[r_{t+1} + \gamma \, Q(s_{t+1}, a) \right]$$

Using the above inequality the resultant of Eq. (15) will be

$$Q(s_t, a_t) = (1 - \alpha) Q(s_t, a_t) + \alpha \max_a \left[r_{t+1} + \gamma \, Q(s_{t+1}, a) \right] \tag{16}$$

On the basis of Eq. (16) we generate an equation for the calculation of Power Rank, i.e.,

$$P_{j_{t+1}} = (1 - \alpha) * P_{j_t} + \alpha * \max_{\frac{I_i}{O_i}} \left[2^{\frac{I_i}{O_i}} + \gamma * P_{i_t} \right] \qquad (17)$$

This equation is working on the reinforcement learning algorithm, it will get converge and then reach to global optimal state. Where $i \in Bc(j)$ back links on page j, $0 < \alpha \leq 1$, $0 \leq \gamma \leq 1$.

As per the Zareh Bidoki [20] the power at the initial point is not known so we take its initial value as 1 and then get it experimentally decreased up to zero value exponentially. In addition γ, the discount factor is utilized to regulate the power as passing to final node and α is the rate of learning while walking in the net which is computed as,

$$\alpha = \frac{1}{e^{\beta * t}} \qquad (18)$$

where β is a constant value used to control the regularity of learning rate, it can be calculated with the help of average human brain capability of remembering (0.85 is in our case).

3.2 Power rank algorithm

The Power Rank of a page will be calculated iteratively and easily with the help of following algorithm. The algorithm of the power rank is taking vector of all the web pages as an input and returns the vector of power rank of respective web pages. It is given as:

Algorithm 1. Power rank algorithm
Input: Web pages $W = \{W1, W2, ..., Wn\}$
Output: Power Rank of Web pages $P = \{P1, P2, ..., Pn\}$
Initialization:
Step1: $\forall W_i \in W$ do,
    set $P_i = 0$; i. e. $\{P = [0, 0, ..., 0]\}$
    Set the iteration counter as $T = 0$
    Set the mpe (minimum permitted error)
Main Block:
Step 2: Compute the parameter α as
    $\alpha = \frac{1}{e^{\beta * t}}$ {Using Equation 18}
Iterative loop:
Step 3: for each web page $W_i \in W$ do,
    Loop

$$P_T[Wi] = (1 - \alpha)*P_{T-1}[Wi] + \alpha * \max_{\frac{I_u}{O_u}} \left(2^{\frac{I_u}{O_u}} + \gamma *P_{T-1[U]} \right)$$

{Using equation 17}

   Where $0 < \alpha \leq 1$, $0 \leq \gamma \leq 1$, $U \in Bc(W)$
   //back links on page W
Step 4: Compute temoral difference error δ
   $\delta = |P_T - P_{T-1}|$
   $T++$
while($\delta > mpe$)   //mpe is a minimum permitted error
END Loop
Return vector P

3.3 Relevancy rule

The relevancy of a web page relies on its kind and its position in the page-list. The larger the relevancy value better is the result. The relevancy "κ" of a page-list is a function of its category and position which is given as,

$$\kappa = \sum_{i \in R(p)} (n - i + 1) \times W_i \tag{19}$$

This is the modified formula of actual relevancy rule [16]. At the end result web page-list $R(p)$, i represents the ith web page and the first n web pages selected from the end result web page list are represented by n. Wi represents the weight of page i such that $Wi = \{v1, v2, v3, v4\}$, here if page is Very relevant (VR), Relevant (R), Weak Relevant (WR) and Irrelevant (IR) respectively v1, v2, v3 and v4 are the values assigned to a page. Also the values are selected in such manner so that $v1 > v2 > v3 > v4$. Through experimental studies the value of Wi for an experiment could be decided. Page Ranking can be compared by categorizing the resultant pages of a query into four classes on the basis of their relevancy calculated.

3.4 Comparative analysis

There exist several state-of-art algorithms that are used in this work in form of comparative analysis of the experimentation results. The results of the proposed Power Rank algorithm are compared with three well known algorithms used in past for various research studies known as Page Rank algorithm [15], Weighted Page Rank algorithm [16] and HITS [32].

Page Rank: S. Brin and L. Page [15] developed a novel ranking procedure which is now utilized by Google, named as Page Rank. It uses the link structure of the WWW to demonstrate the importance of pages of any web sites. The calculation of the Page Rank (PR) of any linked web graph is calculated by using the given equation:

$$PR(u) = (1 - d) + d \sum_{v \in B(u)} \frac{PR(v)}{N_v} \tag{20}$$

where, d is damping factor and its value considered is 0.85, $B(u)$ is set of back link of page u, $PR(v)$ is the page rank of incoming links on page u and N_v is number of out-links on page u. It is noted that the above equation uses the concept of probability distribution over web pages whose final summation will be 1.

Weighted Page Rank: The extended version of the page rank algorithm is proposed later which is named as weighted page rank. The algorithm provides larger rank values to more crucial pages instead of distributing the rank value of a page evenly among its outgoing linked pages. All out-link web page receives a value according to its popularity. The computation of the popularity of a web page is done through the counting of in-links and out-links. In a similar manner like Page Rank, the Weighted Page Rank is calculated for the linked graph by using the given equation:

$$PR(u) = (1 - d) + d \sum_{v \in B(u)} PR(v) W^{in}_{(v,u)} W^{out}_{(v,u)} \tag{21}$$

where the parameter $W^{in}_{(v,u)}$ and $W^{out}_{(v,u)}$ is given as,

$$W^{in}_{(v,u)} = \frac{I_u}{\sum_{p \in R(v)} I_p}, \ W^{out}_{(v,u)} = \frac{O_u}{\sum_{p \in R(v)} O_p}$$

where, I_u and I_p represents the count of in-links of web page u and web page p, correspondingly O_u and O_p provides the count of out-links of web page u and web page p.

Hyperlink Topic Induced Search (HITS): It is a link analysis algorithm that ranks the web pages by processing its entire in links and out links. Thus, ranking of the web pages is decided by analyzing its textual contents against a given query. The algorithm associates two non-negative weights to each of the page in website called as authority weight (AW_p) and Hub weight (HW_p). The algorithm maintains the invariant normalization such that the $\sum_{p \in P} A W_p^2 = 1$ and $\sum_{p \in P} H W_p^2 = 1$. The pages which will get the higher value of authorities will be considered as better. The detail description on the computation of the authorities and values are presented in [33].

4. Experimental setup

To test the performance of the proposed algorithm, two experiments were setup. Both the experiments use the demo website created for reference purpose. The first experiment is performed using the website data created for one of the reputed university while the second experiment is performed by creating the sample website for motivational thoughts of various categories.

4.1 Experiment 1

The first experimentation is performed by using the website data for one of the reputed university in India named as "IIMT University". For experimentation perspective with proper consent with university management a sample demo website is created including 13 web pages. The sample linking of the web pages of the demo website is shown in Fig. 3. These web pages are interlinked with each other in some fashion most preferably in form of links.

Each of the web pages of the demo website has some information contents related to students interests on various domains. The website was

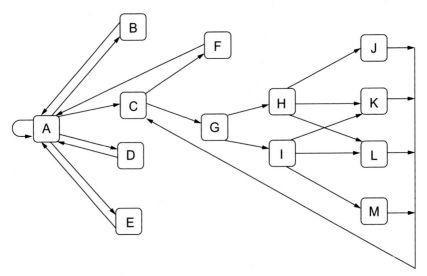

Fig. 3 Web page linking of the sample demo website of IIMT.

hosted and made available to access by the students of all branches and every year for more than 3 months. During this duration the student's feedback towards the content of the web pages were collected.

Evaluation of the experiment 1: To evaluate the proposed algorithm and the comparison with other methods, using the web linking graph as shown in Fig. 3, a sub graphs is considered which is in Fig. 4 where the comparison of the Power Rank is done with the Page Rank and Weighted Page Rank algorithms.

The rank of each page is generated as per the graph link between the pages. The relevancy is been calculated for the two queries as topic "IIMT" and "Student" then the evaluation of the algorithms is done on the basis of number of relevant pages for these queries as shown in Tables 2–4. The relevancy value indicates the importance of order of web pages generated by different ranking algorithms. The values selected for the relevance indicator {VR, R, WR, IR} are {1, 0.5, 0.1, 0} respectively.

As the final result of queries for "IIMT" and "Student" the counting of number relevant pages is done from the sets of pages. We have constructed the page sets in the count of 2, 4, 6 and 9. Later on the value of relevancy is computed using Eq. (19) as shown in Tables 5 and 6.

The comparative experimentation results of the proposed algorithm with the other two state-of-art methods is shown in Figs. 5 and 6 on the basis of the total number of relevant pages to page set for query IIMT and Student.

As per the calculated values of number of relevant pages for the query IIMT and Student, it is found that Weighted Page Rank is better in giving

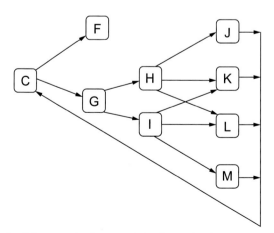

Fig. 4 Sub graph of the sample demo website for evaluation.

Table 2 Relevancy status of different pages for query IIMT and student using page rank algorithm.

Web page	Rank	Query IIMT	Query student
C	1.311	IR	WR
F	0.70716	IR	IR
G	0.70716	R	R
H	0.4506	IR	WR
I	0.4506	IR	WR
K	0.40542	WR	IR
L	0.40542	R	VR
J	0.27771	R	R
M	0.27771	IR	IR

Table 3 Relevancy status of different pages for query IIMT and student using weighted page rank algorithm.

Web page	Rank	Query IIMT	Query student
C	0.76999	IR	WR
F	0.36824	R	R
G	0.25912	IR	IR
H	0.22837	IR	WR
I	0.22837	IR	WR
K	0.20174	WR	IR
L	0.20174	R	VR
J	0.16293	R	R
M	0.16293	IR	IR

the good rank to the relevant pages than Page Rank. While Power Rank is best among these three in providing the high rank to more relevant pages. The calculated relevance values also suggest that the working of Power Rank is better than WPR and PR. It also shows that as the number of pages crawled is increased then the performance of the Power Rank also increased simultaneously as shown in Fig. 7. The value associated in Fig. 7 reflects a pair $<x, y>$ indicating number of iterations required "x" to find the number of relevant pages "y."

Table 4 Relevancy status of different pages for query IIMT and student using power rank algorithm.

Web page	Rank	Query IIMT	Query student
C	4.90484	IR	IR
F	4.90484	R	R
G	4.28501	IR	WR
H	2.31905	IR	WR
I	2.31905	IR	WR
K	1.57983	R	R
L	1.57983	R	VR
J	1.57983	WR	IR
M	1.57983	IR	IR

Table 5 The number of relevant pages and relevance value for query IIMT.

Query IIMT	Number of relevant pages			Relevancy value		
Size of page set	PR	WPR	Power rank	PR	WPR	Power rank
2	0	1	1	0	0.5	0.5
4	1	1	1	1	1.5	1.5
6	2	2	2	2.1	2.6	3
9	4	4	4	6.4	6.9	7.7

Table 6 The number of relevant pages and relevance value for query student.

Query student	Number of relevant pages			Relevancy value		
Size of page set	PR	WPR	Power rank	PR	WPR	Power rank
2	1	2	1	0.2	0.7	0.5
4	3	3	3	1.5	2	1.8
6	4	4	5	3.1	3.6	3.9
9	6	6	6	9.5	10	10.8

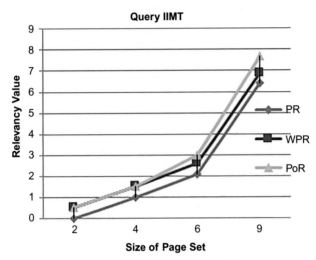

Fig. 5 Number of relevant pages vs page set for query IIMT.

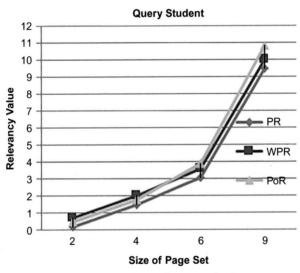

Fig. 6 Number of relevant pages vs page set for query Student.

Discussion on the results of experiment 1: In this experiment the value of β is set to 0.1 as discussed in state-of-art [20] and the value of γ is taken 0.25. With the help of crawling the relevant pages can be discovered with high ranking of page rank. The algorithm which is capable to find the more relevant pages is considered as the better algorithm. In this experiment the

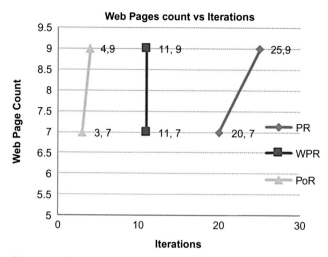

Fig. 7 Simulation results in between number of iterations to web page count.

Power Rank outperforms in terms of finding relevant pages with respect to other two algorithms. In addition, with Power Rank these relevant pages are been found in lesser time. The main advantage of Power Rank is to model areal user browsing on the web. Initially the user browses the web to find the desired page by clicking and visiting new pages slowly user gains more knowledge and then the selection is improved. The learning rate α plays a vital role in the convergence of the system in the reinforcement learning algorithms, but as the time passes in the searching the required pages the rate of learning decreases slowly. It shows that the best learning rate function is with soft ramp [20]. In the calculation the number of iteration required for the Power Rank is least with respected to Page Rank and Weighted Page Rank, this shows that the convergence quality of the Power Rank is also better than other two algorithms.

4.2 Experiment 2

In the second experiment, a sample website is created which includes motivational thoughts on various topics. This website includes 13 different categories of thoughts with an overall of 74 web pages in total. These web pages are interlinked dynamically to other page based on the relevant terms found in page and the ranking provided by the user. The sample website is hosted online for the duration of 6 months and during this period the website is made available to approximately 2000 students of Jaypee

Institute of Information Technology Noida. These students visited the website and gave their ranking for each page based on their interest.

Proposed methodology architecture: The architecture of the proposed adopted methodology for the experimentation is given in the following Fig. 8. The block architecture of the proposed demo website experimentation uses the feedback of the various visitors on each page to find the relevancy order of the page dynamically.

General description of the experimentation 2: To test the proposed Power Rank algorithm performance, a bigger web scenario is developed where the relevancy order of the web page is computed dynamically using user feedback. In total 12 categories of quotes is selected based on which 74 web pages created with one motivational code on each page with respect to domain. The home page of the website with quote category labels is shown in Fig. 9.

Initially the end user will select any of the quote categories from the home page as per choice. After clicking on the result, the website navigates the user to related quote page. The user responds in terms of the feedback ranging in between 0 and 5 such that 0 relates to irrelevant while 5 refers to strongly relevant. The overall ratings are divided as {Irrelevant, Weakly Irrelevant, Nominal, weakly Relevant, Relevant, and Strongly Relevant}. Based on the feedback rating, the system identifies the relevant terms among the quote and prepare indexing. The index is matched with the relevant terms and respective web page will be displayed to the user. The sample quote web page is shown in Fig. 10.

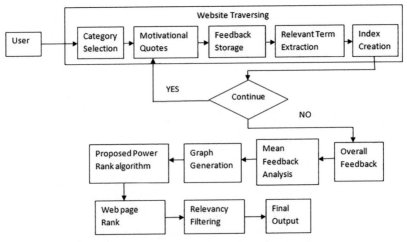

Fig. 8 Architecture of the proposed methodology for Quote website.

Fig. 9 Home page of website with category label.

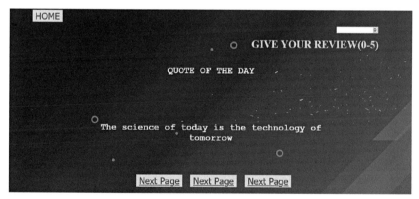

Fig. 10 Sample quote web page of the website with feedback ranking.

As the user traverse the web pages iteratively, the feedback is stored for all the pages and the process continues for all users who ever is accessing the website. Thus for every page the list of 'n' feedbacks will be stored where "n" is number of users visited specific web page, result in the formation of the overall feedback of the page in website. Later, the mean of the average feedback is computed of every page. In addition, the system also generated the dynamic graph among the web pages as shown in Fig. 3, based on the relevant terms found in page, indexing of documents and feedback of the page. This graph is further used to test the proposed Power Rank algorithm and other state-of-art algorithms.

At the end, the fetched web page rank will be mapped to test the relevancy of the page, i.e., how better is the algorithm to rank the page as compared to the feedback given by the user while accessing the web pages. Based on the average user ranking for the whole website, the distribution of the web pages on basis of relevancy is shown in Table 7.

Evaluation of the experiment 2: Using the similar setting and parameters value as set in experiment 1, the dynamic interlinked graph of the web page is generated. This graph is used to find the page rank of individual pages using state-of-art algorithms and proposed Power Rank algorithm. The experimentation setup also made an assumption that based on the page ranking of all the web pages of a website, here the adopted methodology will display the top 10 pages among the overall pages using each algorithm. The experimental results of experiment 2 are given in Table 8.

Table 7 The number of relevant pages and irrelevant pages based on user feedback.

Relevancy	Count of pages	Page numbers
Relevant page (RP)	52	{1, 2, 4, 5, 6, 7, 9, 10, 11,12, 14, 15, 16, 17, 19, 21, 22, 24, 25, 28, 31, 32, 33, 34, 36, 37, 38, 39, 41, 42, 43, 45, 47, 48, 49, 50, 51, 52, 54, 56, 57, 58, 59, 61, 62, 65, 67, 68, 70, 71, 73, 74}
Irrelevant page (IRP)	22	{3, 8, 13, 18, 20, 23, 26, 27, 29, 30, 35, 40, 44, 46, 53, 55, 60, 63, 64, 66, 69, 72}

Table 8 Order of the top 10 retrieved pages using different algorithms.

Page rank	Weighted page rank	HITS	Power rank
48	48	72	69
69	21	69	51
51	47	70	48
70	27	53	70
72	69	48	47
21	70	21	21
27	51	44	44
47	20	47	3
44	26	20	16
3	14	9	5

The values indicated in bold in Table 8 reflects the relevant pages while the values in normal are for irrelevant pages. In addition, the face value highlighted in Table 8 suggests the respective web page number which the algorithm returns in specific order based on relevancy graph and user rating.

Statistical validation of experimentation 2: To validate or check the performance of the proposed algorithm, few statistical measurements are computed which are given below.

Precision: It is defined as the ratio of the number of documents retrieved which are relevant out of total retrieved documents as per the user query or choice. Mathematically it is derived as,

$$P = \frac{|\{Relevant\ Documents\} \cap \{Retrieved\ Documents\}|}{|Retrieved\ Documents|}$$

It is noted that for any specific user query the precision of the system must be high.

Recall: Recall is defined as the ratio of the number of documents retrieved which are relevant out of total relevant documents exist in total as per the user query or choice. Mathematically it is derived as,

$$R = \frac{|\{Relevant\ Documents\} \cap \{Retrieved\ Documents\}|}{|Relevant\ Documents|}$$

It is noted that for any specific user the recall of the system must be high.

F-Measure: It is defined as the harmonic mean of precision and recall which is given as,

$$F = \frac{2*P*R}{P+R}$$

The comparative result of the proposed algorithm with other state-of-art algorithms based on the above said parameters are given in Table 9.

Table 9 Comparative analysis of the algorithms based on top 10 retrieved documents.

	Page rank	Weighted page rank	HITS	Power rank
Precision	50%	60%	50%	**70%**
Recall	9.61%	11.53%	9.61%	**13.46%**
F-Measure	16.12%	19.34%	16.12%	**22.57%**

The values indicated in bold suggests that the proposed Power Rank algorithm is giving good precision, recall and F-measure values as compared with other state-of-art algorithms and found suitable for application.

5. Conclusion

This book chapter proposed a new iterative web page ranking algorithm, based on reinforcement learning named as Power Rank. This algorithm considers the input and output links count on power of two to calculate the rank of pages. The maximum power rank provides its rank to the respective web page. The main advantage of Power Rank is to model user's real time browsing on the web. Initially the user browses the web to find the desired page by clicking on the available links. As the user visits the new pages slowly user gains more knowledge and then the selection is improved. The learning rate α plays a vital role in the convergence of the system in the reinforcement learning algorithms, but as the time passes in searching the required pages the rate of learning decreases slowly. The number of iterations required to calculate the rank is comparatively small enough so that it is having a fast convergence speed.

Acknowledgments

The authors would like to thanks every individual who ever helped them either directly are indirectly to complete the work. The authors would like to give special gratitude toward U.G. students Astitva, Shagun and Rachit for providing their help in implementation of the some modules.

References

[1] Y. Lee, Developing an efficient computational method that estimates the ability of students in a web based learning environment, Comput. Educ. 58 (1) (2012) 579–589.
[2] M. Wang, et al., A web based learning system for software test professionals, IEEE Trans. Educ. 54 (2) (2011) 263–272.
[3] ISC, 2007. http://www.isc.org/index.pl?/ops/ds/host-count-history.php.
[4] A. Gulli, A. Signorini, The indexable web is more than 11.5 billion pages, in: Special Interest Tracks and the Posters of the 14th International Conference on World Wide Web Chiba, Japan, 2004.
[5] A. Arasu, J. Cho, C. Olston, What's new on the web? The evolution of the web from a search engine perspective, in: Proceeding of the 13th Conference on World Wide Web, 2004, pp. 1–12.
[6] T.B. Lee, J. Hendler, O. Lassila, The semantic web, Sci. Am. 284 (5) (2001) 34–43.
[7] B. Li, L. Zhou, S. Feng, K. Wong, A unified graph model for sentence-based opinion retrieval, in: Proceedings of the 48th Annual Meeting of the Association for Computational Linguistics, Uppsala, Sweden, 2010, pp. 1367–1375.

[8] M. Hu, B. Liu, Mining and summarizing customer reviews, in: Proceedings of the 10th ACM SIGKDD International Conference on Knowledge Discovery and Data Mining, KDD' 04, ACM, 2004, pp. 168–177.

[9] A. Broder, R. Kumar, F. Maghoul, P. Raghavan, S. Rajagopalan, R. Stata, A. Tomkins, J. Wiener, Graph structure in the web, Comput. Netw. 33 (1–6) (2000) 309–320.

[10] N. Duhan, A.K. Sharma, K.K. Bhatia, PageRanking algorithms: a survey, in: Proceedings of the IEEE International Conference on Advance Computing, 2009.

[11] K.D. Satokar, S.Z. Gawali, Web search result personalization using web mining, Int. J. Comput. Appl. 2 (5) (2010) 29–32.

[12] C.F. Lin, Y.C. Yeh, Y.H. Hung, R.I. Chang, Data mining for providing a personalized learning path in creativity: an application of decision tree, Comput. Educ. 68 (2013) 199–210.

[13] R.B. Yates, E. Davis, Web page ranking using link attributes, in: Proceedings of the 13th International World Wide Web Conference on Alternate Track Papers & Posters, 2004, pp. 328–329.

[14] P. Kolari, A. Joshi, Web mining: research and practices, Comput. Sci. Eng. 6 (4) (2004) 49–53.

[15] S. Brin, L. Page, The anatomy of a large-scale hypertextual web search engine, Comput. Netw. ISDN Syst. 30 (1–7) (1998) 107–117.

[16] X. Wenpu, G. Ali, Weighted PageRank algorithm, in: Proceedings of the Second Annual Conference on Communication Networks and Services Research, 2004, pp. 19–21.

[17] A.K. Sharma, N. Duhan, G. Kumar, Page ranking based on number of visits of links of web page, in: 2nd International Conference on Computer and Communication Technology, 2011, pp. 11–14.

[18] L.K.J. Grace, Web log data analysis and mining, in: International Conference on Computer Science and Information Technology: Advanced Computing, 2011, pp. 459–469.

[19] A.K. Sharma, P.C. Gupta, Analysis of web server log files to increase the effectiveness of the website using web mining tool, Int. J. Adv. Comput. Math. Sci. 4 (1) (2013) 1–8.

[20] B. Zareh, et al., Distance rank: an intelligent ranking algorithm for web pages, Inform. Process. Manag. 44 (2) (2008) 877–892.

[21] P. Pirolli, J. Pitkow, R. Rao, Silk from a sow's ear: extracting usable structures from the Web, in: Proceedings of the SIGCHI Conference on Human Factors in Computing Systems, Association for Computing Machinery, 1996, pp. 118–125.

[22] J. Pitkow, Characterizing World Wide Web Ecologies, Ph.D. Thesis, Georgia Institute of Technology, 1997.

[23] R. Weiss, et al., HyPursuit: a hierarchical network search engine that exploits content-link hypertext clustering, in: Proceedings of the Seventh ACM Conference on Hypertext, Association for Computing Machinery, 1996, pp. 180–193.

[24] M. Sougata, D.F. James, H. Scott, Visualizing complex hypermedia networks through multiple hierarchical views, in: Proceedings of the SIGCHI Conference on Human Factors in Computing Systems, ACM Press/Addison-Wesley Publishing Co, USA, 1995, pp. 331–337.

[25] M. Sougata, D.F. James, Showing the context of nodes in the World-Wide Web, in: Conference Companion on Human Factors in Computing Systems, Association for Computing Machinery, 1995, pp. 326–327.

[26] M.K. Jon, Authoritative sources in a hyperlinked environment, J. ACM 46 (5) (1999) 604–632.

[27] Z. Christos, et al., Important factors for improving google search rank, J. Fut. Internet 11 (2019) 1–12. article 32.

[28] Y. Lili, et al., An improved Page rank method based on genetic algorithm for web search, Procedia Eng. 15 (2011) 2983–2987.
[29] M. Coppola, et al., The Page Rank algorithm as a method to optimize swarm behavior through local analysis, J. Swarm Intell. 13 (3–4) (2019) 277–319.
[30] G. Shubham, et al., An efficient page ranking approach based on vector norms using sNorm(p) algorithm, J. Inf. Process. Manag. 56 (3) (2019) 1053–1066.
[31] S.S. Richard, G.B. Andrew, Reinforcement Learning: An Introduction, second ed., The MIT Press, Cambridge, MA; London, UK, 2014.
[32] J.M. Kleinberg, Authoritative sources in hyperlink environment, J. ACM 46 (1998) 668–677.
[33] HTS, Lecture #4: HITS Algorithm—Hubs and Authorities on the Internet, 2009. http://pi.math.cornell.edu/~mec/Winter2009/RalucaRemus/Lecture4/lecture4.html. Accessed on December 2019.

About the authors

Ankit Vidyarthi presently holds a post of Assistant Professor (Sr. Grade) in the Department of Computer Science Engineering & Information Technology, Jaypee Institute of Information Technology Noida. He joined the Institute in June 2018 and from then onwards he is associated with academics and research activities at the university level. He had completed his Postdoc (Machine Learning) from Bennett University in 2018. He obtained his Ph.D. from the Department of CSE, Malaviya National Institute of Technology Jaipur in 2017. He had several research papers in reputed SCI/(E) indexed journals with good impact factors. He had also published 20 research articles in various peer-reviewed conferences of IEEE/Springer/ACM/Elsevier which were indexed in Scopus. He is a member of ACM with professional members and SIGACT membership. He is also associated with various journals as a reviewer which are of high standards and indexed in SCI/(E). He is also a member of the Technical Program Committee in various conferences and workshops. He is associated with one journal (JIEEE) as an associate Editor-in-Chief, Senior Editor with AI Foundation Trust, India, and Associate Editor (GE) with IEEE Transactions on Industrial Informatics, and Interdisciplinary Sciences: Computational Life Science Springer Journal.

Pawan Singh received the B.E. (Computer Science and Engineering) from CCS University, Meerut, India, M.Tech. (Information Technology) from GGSIPU, New Delhi, India and Ph.D. (Computer Science) from Magadh University, Bodh Gaya, India in 2013. Currently, he is serving in Department of Computer Science & Engineering, Amity School of Engineering and Technology, Amity University, Lucknow Campus, India. His research interests include software metrics, software cost estimation, web structure mining, energy aware scheduling, cloud computing, medical imaging, nature inspired meta-heuristic optimization techniques and its applications. He has authored and co-authored a number of research papers in the journals of international reputation. Dr. Pawan has served as a reviewer in various SCI and SCIE indexed journal. Dr. Pawan served as a technical committee for many international conferences. He has been the Special Sessions Cochair of the "Emerging Trends towards Communication, Computing and Internet of Things", 2nd International Conference on Communication and Computing Systems (ICCCS - 18) and National Seminar cum Workshop on Data Science and Information Security, 28th Feb-2nd March 2019,. He has been a Guest Editor of Special Issue on Advanced Optimization Techniques for Operation and Control of Intelligent Power Systems, Journal of Control Science and Engineering, Hindawi Publications, London, United Kingdom He is a member of the IEEE and IEEE Computational Intelligence Society (CIS).

CHAPTER FIFTEEN

GA-based energy efficient modeling of a wireless sensor network

Anish Kumar Saha[a], Joseph L. Pachuau[a], Arnab Roy[a], and Chandan Tilak Bhunia[b]
[a]Department of Computer Science and Engineering, National Institute of Technology Silchar, Silchar, India
[b]Durgapur Institute of Advanced Technology & Management, Durgapur, India

Contents

Abstract

A wireless sensor network is a collection of sensing nodes. All nodes necessarily maintain connectivity with the root either through direct or relay communication. Sensors have limited battery capability. Hence, improper coverage causes some sensors to die early, and thus, it could cause other subnetworks to get disconnected from the root. It lets down in maintaining in the sensing area. The power consumption of sensors depends on their coverage area. Proper shrinking or growing in the coverage area of sensors make a longer lifetime of batteries. These types of optimization problems are NP problems that take a long time even with a small increase in network size. In this article, an optimization model is proposed for the energy saving of the network. Evolutionary computation is one of the techniques for solving NP problems. A genetic algorithm is proposed here to get an optimal/suboptimal result under certain constraints. Tree-based crossover and mutation are proposed for the generation of better offspring.

Advances in Computers, Volume 128
ISSN 0065-2458
https://doi.org/10.1016/bs.adcom.2021.10.009

381

1. Introduction

A wireless sensor network (WSN) is a distributed collection of sensors working together to sense certain areas. Sensors have the capabilities to sense and forward data to their neighbor sensors. The goal of all sensors is to reach a special node, called the root node, either through direct or relay communication. WSN maintains a destination-oriented directed acyclic graph (DODAG) for making a connectivity with the roots [1, 2]. Sensors have a limited battery capacity which causes network failure due to fast power dissipation. With the growth in coverage area of a sensor, drastic power dissipation occurs. Relay communication is one of the solutions in which the distances from sensors to the root are divided into smaller distances via intermediate nodes. Different approaches are already proposed for efficient energy harvesting. Wang et al. [3] analysed the effect of the cluster shape on the power consumption of WSN. They found that hexagonal cluster is more efficient and ideal for clustering large areas. Jemal et al. [4] discussed a probabilistic approach to estimate quality of service (QoS) monitoring thereby minimizing the transmission energy usage of WSN. They proposed the use of a Hidden Markov chain and fuzzy logic to predict and control the service. Saha et al. [5] proposed energy-efficient modelling in which power consumption of each port and data rate are considered as a parameter for the modelling of the network. They optimized link data rate and port up/down approach under traffic load. They added multipath for maximum utilization of resources to minimize time delay for data transfer. It is observed that an inefficient routing algorithm leads to unnecessary wastage of battery life. Rani et al. [6] investigated the reduction in total transmission time and the energy consumption of WSN. They proposed the use of a routing algorithm that increases network lifetime as well as speed up data communication. Their algorithm, minimized transmission distance by clustering and intracluster relay communication. Physical changes in the location of nodes can lead to failure in communication. Zhou et al. [7] proposed to deal with the failures that happen due to changes in the physical location. Their approach is flexible enough to rearrange cluster and cluster heads after failure. Sivakumar et al. [8] carried out the analysis of the network lifetime of WSN using a LEACH protocol. They have combined LEACH with a genetic algorithm (GA) and compared it with other known methods. They observed that LEACH-GA performs better compared to the other known methods.

2. Motivation

Sensors of WSN suffer from improper management of power dissipation. These harvesting optimizations are generally NP problems and require a longer time for computation. Mixed-integer or integer programming with a heuristic approach are used to cut-off the required time for computation [9].

Some articles already proposed evolutionary-based algorithms. Elhoseny *et al.* [10] proposed a GA to optimize coverage and resources for a longer sensor lifetime. Yadav *et al.* [11] proposed the use of particle swarm optimization (PSO) for finding the optimum cluster head. Their objective function considered the energy dissipation of the cluster head and sensors power consumption due to data aggregation and communication to neighbors. Praveena *et al.* [12] discussed the performances of the PSO algorithm in WSN. They considered re-path calculation for nonfunctional sensors. Guo *et al.* [13] proposed a hybrid genetic algorithm to improve routing protocol in WSN. The initial population of the gene pool formed from a single-stage parent evolution algorithm. They gave higher priority for the high-quality gene pool. A high-quality gene pool improved the performance of the system. Dhami *et al.* [14] proposed the use of GA with Virtual Grid-based Dynamic routes Adjustment (VGDRA). They claimed better results in terms of the number of fewer loops in GA for energy-efficient computation. Bojan *et al.* [15] proposed a method of minimization of energy consumption by careful observation of optimization space and customization of GA's to suit their specific energy-saving function. Customization in their GA lowered the CPU consumption. Gupta *et al.* [16] pointed out the problem of faster energy depletion of nodes closer to the base station in most of the clustering techniques. They resolved the problem by changing the role of cluster head (CH) with rotation. They used GA based energy distribution for the selection of CH. Clustering is a common technique for the fragmentation of sensors into groups. Sujee *et al.* [17] proposed a GA-based algorithm for the selection of CH. In their algorithm, the selection of cluster head, energy distribution with optimal probability is used. Yuan *et al.* [18] also proposed GA-based self-organizing network clustering. Their algorithm determined CH and then dynamic clusters are formed. In their approach, no node was allowed to deplete its energy before any other to maximize the network life. Nodes with a lower amount of residual energy were less likely to be selected as CH. The residual energy of all the nodes was mostly the same throughout the network life.

Based on the above motivation of energy harvesting, this article proposed energy-efficient modelling in WSN. GA evolutionary tool is used to solve the optimization problem. Tree-based population initialization, crossover, mutation etc. are proposed. A network of 16 number of nodes with arbitrary distances is taken here for computational observation and analysis.

3. Proposed algorithm and structure

The objective of the article is to calculate the radius of the coverage area of sensors so that the overall power consumption of a network is minimum. Sensors maintain direct/indirect communication to the root. A sensor is capable of sensing, forwarding and aggregating data from neighbors. The forwarding of aggregated data continues until reaching the root node. The power consumption of a sensor is proportional to the square of the radius of its coverage area. Hence, lowering the coverage gives better battery life. Such coverage optimization takes a longer computational time even for a small increase in size. Evolutionary algorithms show good direction for solving these problems. One of the optimization algorithms is GA which mimics natural reproduction.

Before going to the framework of the problem, the common steps in GA is shown in Fig. 1. GA maintains a population which are a collection of solutions. The population are initialized with arbitrary solutions, these solutions are called chromosomes. Parents chromosome from population generate offspring chromosome. Weak chromosomes are removed and the processes continue until a certain termination criterion is reached. The best chromosome from the population is a solution for the problem. The best chromosome is determined from its fitness value, known as an objective function. Selection of parents, generation of offspring and sudden change in genes are called selection, crossover and mutation, respectively. One of the major steps in GA is mutation, which explores the uncover solutions in the solution space. The probability of mutation is generally less. Parent selection, crossover point, and mutation location are carried out randomly for better results.

3.1 Structure of chromosome

Chromosome represents a solution for a problem and the structure of the chromosome depends on the type of problem. Here a one-dimensional array of size $\{1 \times n\}$ represents a chromosome, each value for different indices is called a gene. The index represents child nodes and the value of different

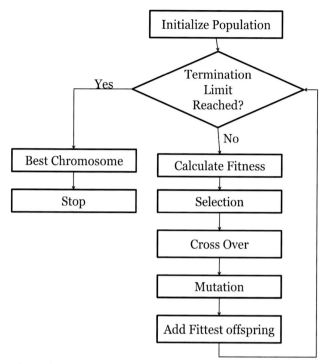

Fig. 1 Flowchart of GA.

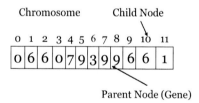

Fig. 2 Structure of the chromosome.

indices represent the parent node. A chromosome and its tree structure for our optimization problem are shown in Figs. 2 and 3, respectively. Node 0 is considered here as root, as root has no parents hence value at index 0 is fixed to 0. The structure says it is up to the child node to cover its parent node for relay communication as shown in Fig. 4.

3.2 Fitness function

Sensors follow the equation $P_i \propto r_i^2$, where P_i represents power dissipation for covering of radius r_i. Here the sum of the square of the radius of coverage

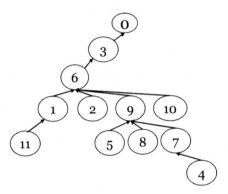

Fig. 3 Tree structure of the chromosome.

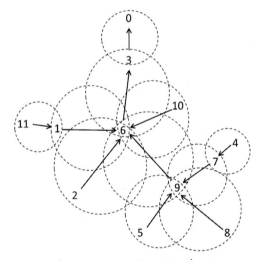

Fig. 4 Variation in radius of sensors connecting toward root.

is used as a fitness function, as shown below (1). The objective is to minimize the fitness and determine r_i, $\forall$ $i \in \{0 \ldots n\}$.

$$Fitness\ function = \sum_{i=0}^{n} r_i^2 \qquad (1)$$

Note:
- n is the number of sensors in integer, label 0 is considered as root and other labels are sensors in the network.

- *Distance*[][] is a two dimensional matrix of size {*nxn*}, it contains the distance from a node *i* to another node *j*.
- *Gene*[] is a chromosome of size {1x*n*}, contain list of nodes acting as parent node for that index value child node.

3.3 Generation of initial population

The first step of GA is the initialization of the population. The population is a collection of chromosomes and a subset of all possible solutions are randomly placed in the population. Algorithm 1 shows the steps of initialization. Initialization needs to be careful done to avoid loop, that's why *Element*{} ∩ *Index*{} = ∅ is maintained for all loop. Initialization of population is required only for a one-time generation.

ALGORITHM 1 Fitness function.
procedure FITNESS(Float *D*[][],Integer *n*,Chromosome *Gene*[])
    FITNESS ← 0
    for *i* ← 1 to *n* **do**
        FITNESS ← *FITNESS* + (*D*[*i*][*Gene*[*i*]])^2
    end for
    return *FITNESS*
end procedure

ALGORITHM 2 Initialization a chromosome.
procedure INITIALIZATION(Integer *n*)
    Gene[0] ← 0
    Element{}← {0}
    Index{}←{1, 2, ..., *n*}
    while *Index*{}! = *Empty* **do**
        INDEX ← *Random_Selection*(*Index*{})
        ELEMENT ← *Random_Selection*(*Element*{})
        Gene[*INDEX*] ← *ELEMENT*
        Index{}← *Index*{}− *INDEX*
        Element{}← *Element*{} + *INDEX*
    end while
    return *Gene*[]
end procedure

Note:

Function *Random_Selection(Set)* returns a random element from *Set*. Algorithm maintain two sets named *Index{}* and *Element{}*. Initially, *Index{}* contain with all nodes except 0 and *Element{}* is set with root 0. *Gene*[0] is initialized to 0 to make root with no parent. An *INDEX* and an *ELEMENT* is randomly selected from sets *Index{}* and *Element{}*, respectively. *ELEMENT* is set for the gene position *Gene[INDEX]*. After, *INDEX* is removed from *Index{}* to avoid loop. *INDEX* is then added in *Element[]* to make a random chance for the index to become a parent.

Suppose for a WSN of 07 numbers of nodes is taken an example as below,

$Index\{\} = \{1, 2, 3, 4, 5, 6\}$
$Element\{\} = \{0\}$
$Gene[0] = 0$

Loop 1:

$INDEX \leftarrow Random\_Selction(Index\{\}) = 3$
$ELEMENT \leftarrow Random\_Selection(Element\{\}) = 0$
$Index\{\} \leftarrow Index\{\} - \{3\} = \{1,2,4,5,6\}$
$Element\{\} \leftarrow Element\{\} + \{3\} = \{0, 3\}$

Chromosome:	0			**0**			

Loop 2:

$INDEX \leftarrow Random\_Selection(Index\{\}) = 5$
$ELEMENT \leftarrow Random\_Selection(Element\{\}) = 3$
$Index\{\} \leftarrow Index\{\} - \{5\} = \{1, 2, 4, 6\}$
$Element\{\} \leftarrow Element\{\} + \{5\} = \{0, 3, 5\}$

Chromosome:	0			0		**3**	

Loop 3:

$INDEX \leftarrow Random\_Selection(Index\{\}) = 6$
$ELEMENT \leftarrow Random\_Selection(Element\{\}) = 3$
$Index\{\} \leftarrow Index\{\} - \{6\} = \{1, 2, 4\}$
$Element\{\} \leftarrow Element\{\} + \{6\} = \{0, 3, 5, 6\}$

Chromosome:	0			0		3	**3**

Loop 4:

$INDEX \leftarrow Random\_Selection(Index\{\}) = 2$

$ELEMENT \leftarrow Random\_Selection(Element\{\}) = 6$

$Index\{\} \leftarrow Index\{\} - \{2\} = \{1, 4\}$

$Element\{\} \leftarrow Element\{\} + \{2\} = \{0, 2, 3, 5, 6\}$

Chromosome:

0		6	0		3	3

Loop 5:

$INDEX \leftarrow Random\_Selection(Index\{\}) = 1$

$ELEMENT \leftarrow Random\_Selection(Element\{\}) = 6$

$Index\{\} \leftarrow Index\{\} - \{1\} = \{4\}$

$Element\{\} \leftarrow Element\{\} + \{1\} = \{0, 1, 2, 3, 5, 6\}$

Chromosome:

0	6	6	0		3	3

Loop 6:

$INDEX \leftarrow Random\_Selection(Index\{\}) = 4$

$ELEMENT \leftarrow Random\_Selection(Element\{\}) = 1$

$Index\{\} \leftarrow Index\{\} - \{4\} = \{\}$

$Element\{\} \leftarrow Element\{\} + \{4\} = \{0, 1, 2, 3, 4, 5, 6\}$

Chromosome:

0	6	6	0	1	3	3

3.4 Selection

Randomness in different steps gives a better result. Random selection of parents helps in diversifying offspring. Paired parents are randomly selected for each generation. Selected parents then perform crossover in every generation. The selection process is a repetitional step followed by crossover and mutation in every generation.

3.5 Crossover

Crossover is the process in which parents exchange their information for offspring. In this article, a random node is selected and then a random path containing that random node is chosen for each parent. Two paths from two

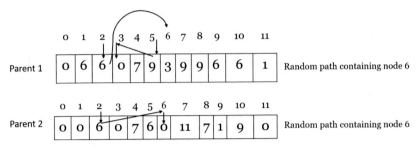

After crossover, two paths are exchanged containing node 6

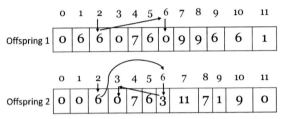

Fig. 5 Structure of chromosome before and after crossover.

parents containing that random node are swapped. Other nodes present in the chromosome are adjusted accordingly with the changes after swapping. The same is shown in Fig. 5. In the figure, a random node 6 gets selected, and a random path from two different parents, $\{2 - 6 - 3 - 0\}$ and $\{2 - 6 - 0\}$ are chosen for swapping. The crossover happening in tree form is also shown in Fig. 6.

ALGORITHM 3 Crossover.

procedure CHROMOSOME(Chromosome *PARENT1*,Chromosome *PARENT2*)
    Node{}← {0, 1, 2, ..., n}
    NODE ← *Random_Selection(Nodes{})*
    PATH1 ← *Random_path_Selection(NODE, PARENT1)*
    PATH2 ← *Random_path_Selection(NODE, PARENT2)*
    OFFSPRING1 ← *Generate_offspring(PATH2, PARENT1)*
    OFFSPRING2 ← *Generate_offspring(PATH1, PARENT2)*
    return *OFFSPRING1, OFFSPRING2*
end procedure

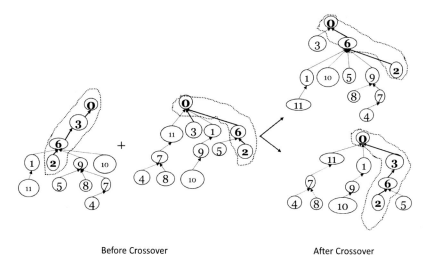

Before Crossover   After Crossover

Fig. 6 Crossover operation for tree structure.

Note:

Function *Random_path_Selection(NODE, chromosome)* returns a random path which contains the random node *NODE* in the tree. Function *Generate_offspring(PATH2, PARENT1)* returns an offspring after insertion of a path PATH2 into PARENT1 chromosome. The same function is called for second time *Generate_offspring(PATH1, PARENT2)* for generating second offspring.

3.6 Mutation

Mutation causes sudden changes in the gene. The probability of occurring of mutation generally varies from 1% to 10%. Mutation happens individually in an offspring. A random internal node and a random leaf node is selected here from an offspring and then leaf and internal node position are swapped. If the chromosome structure is observed, leaf nodes are identified by all missing nodes and all internal nodes are identified by finding the existing node from the chromosome. Mutation makes a leaf node become an internal node and vice versa. This swapping of internal and leaf nodes could not be possible during Crossover. Mutation makes leaf nodes to be visible inside the gene of the chromosome. The presence of leaf nodes helps in chromosome for finding versatile solutions, which could be far-away without mutation for the next generation. The mutation process is shown in Figs. 7 and 8.

ALGORITHM 4 Mutation.

procedure CHROMOSOME(Chromosome Gene[])
    LEAF ← Random_Leaf(Gene[])
    INTERNAL_NODE ← Random_Internal_Node(Gene[])
    OFFSPRING ← Swap_Position(Gene[], INTERNAL_NODE,
        LEAF)
    return OFFSPRING
end procedure

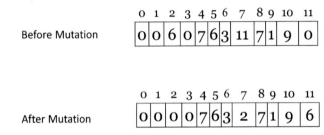

Fig. 7 Mutation operation in the chromosome.

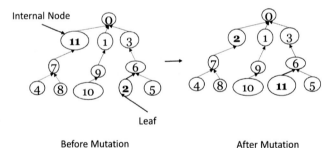

Before Mutation   After Mutation

Fig. 8 Mutation operation in the tree structure.

3.7 Removal of population and exit criteria

After crossover and mutation, the size of the population gets double. Old populations with offspring are sorted in ascending order and half of the population is selected for the next generation. The remaining half chromosomes are discarded, makeing the population size fixed. The process of selection, crossover, mutation, and removal will continue until the termination criterion is satisfied. Termination occurs when all chromosomes contain the same fitness value in the population. It happens when all chromosomes reach the same tree structure. Another termination can be used with a fixed number of generations.

4. Result analysis

We have used two different types of termination criteria in GA. One of them is when all chromosome reaches to same fitness value. Generally, it happens when all chromosome has the same tree structure with the same gene value. Standard deviation (SD), Mean, RMS, and coefficient of variation (CV) are calculated from a set of 20 numbers of fitness values under the same GA environment. Fig. 9 shows the optimal fitness value in different statistical terms with respect to variation in population size. It also shows the maximum and minimum obtained value of optimum fitness in 20 different samples. The plot shows almost the same optimal results for population size around 50 or above. Fig. 10 shows the changes in SD for fitness, low value of SD indicates low variation with respect to mean. Fig. 11 shows CV, which is SD/mean. CV shows extend of variation of the values to the mean. From 50 or more in population size, CV is 5%–10%, SD below 600–700, RMS around 9000. It is concluded that population size between 50 and 80 is sufficient for the GA. Fig. 12 shows the number of required

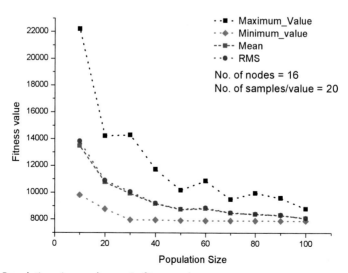

Fig. 9 Population size vs change in fitness values.

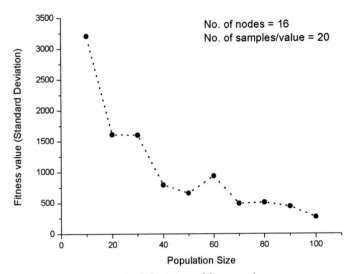

Fig. 10 Population size vs standard deviation of fitness values.

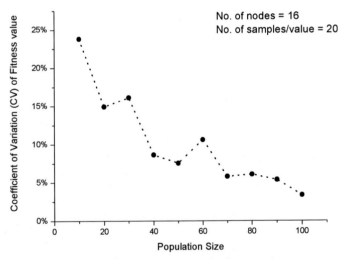

Fig. 11 Population size vs coefficient of variation for fitness values.

generations before termination and around 50–65 generations is sufficient to have the optimum result in the range of population size around 50–95. The algorithm is checked for a large value in generation count for the same population size 50, it is concluded that from Table 1 around 50 generations is sufficient for the GA.

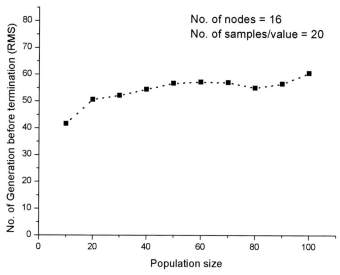

Fig. 12 Population size vs no. of generation before reaching termination.

Table 1 RMS of fitness value for different number of generation.
No of nodes: 16

No. of samples/values = 20

Population size: 50

No of generation	Fitness value (RMS)
50	8760.00
1000	8699.30
2000	8867.89
3000	8931.05
4000	8830.88

5. Conclusion

WSN suffers from battery power limitations. Proper harvesting of energy through multihop communication gives a longer lifetime. A GA-based optimization algorithm is proposed here for power saving. The variation in the coverage area of sensors is considered. All nodes keep minimum coverage and maintain a constraint of connecting to the root node. The algorithm explained the steps in initialization of population, selection, crossover, and Mutation. Tree-based crossover and mutation are explained. In a crossover, two random paths from parents are swapped. In mutation, leaf and internal node exchange their position in the offspring. The result shows that for the network structure of 16 nodes, around 50 numbers of population size with 50–60 generation is sufficient to obtain optimal results.

References

[1] J. Yick, B. Mukherjee, D. Ghosal, Wireless sensor network survey, Comput. Netw. 52 (12) (2008) 2292–2330.

[2] T. Watteyne, A. Molinaro, M.G. Richichi, M. Dohler, From manet to ietf roll standardization: a paradigm shift in wsn routing protocols, IEEE Commun. Surv. Tutor. 13 (4) (2011) 688–707.

[3] D. Wang, L. Lin, L. Xu, A study of subdividing hexagon-clustered WSN for power saving: analysis and simulation, Ad Hoc Netw. 9 (7) (2011) 1302–1311.

[4] A. Jemal, M. Hachicha, R.B. Halima, A.H. Kacem, K. Drira, M. Jmaiel, Energy saving in WSN using monitoring values prediction, Procedia Comput. Sci. 32 (2014) 1154–1159.

[5] A.K. Saha, K. Sambyo, C.T. Bhunia, Energy efficient modelling of a network, China Commun. 15 (1) (2018) 107–117.

[6] S. Rani, J. Malhotra, R. Talwar, Energy efficient chain based cooperative routing protocol for WSN, Appl. Soft Comput. 35 (2015) 386–397.

[7] R. Zhou, M. Chen, G. Feng, H. Liu, S. He, Genetic clustering route algorithm in WSN, in: 2010 Sixth International Conference on Natural Computation, vol. 8, IEEE, 2010, pp. 4023–4026.

[8] P. Sivakumar, M. Radhika, Performance analysis of LEACH-GA over leach and LEACH-C in WSN, Procedia Comput. Sci. 125 (2018) 248–256.

[9] S. Basagni, A. Carosi, E. Melachrinoudis, C. Petrioli, Z.M. Wang, A new MILP formulation and distributed protocols for wireless sensor networks lifetime maximization, in: 2006 IEEE International Conference on Communications, vol. 8, IEEE, 2006, pp. 3517–3524.

[10] M. Elhoseny, A. Tharwat, X. Yuan, A.E. Hassanien, Optimizing K-coverage of mobile WSNs, Expert Syst. Appl. 92 (2018) 142–153.

[11] A. Yadav, S. Kumar, S. Vijendra, Network life time analysis of WSNs using particle swarm optimization, Procedia Comput. Sci. 132 (2018) 805–815.

[12] K.S. Praveena, K. Bhargavi, K.R. Yogeshwari, Comparision of PSO algorithm and genetic algorithm in WSN using NS-2, in: 2017 International Conference on Current Trends in Computer, Electrical, Electronics and Communication (CTCEEC), IEEE, 2017, pp. 513–516.

[13] L. Guo, Q. Tang, An improved routing protocol in WSN with hybrid genetic algorithm, in: 2010 Second International Conference on Networks Security, Wireless Communications and Trusted Computing, vol. 2, IEEE, 2010, pp. 289–292.

[14] M. Dhami, V. Garg, N.S. Randhawa, Enhanced lifetime with less energy consumption in WSN using genetic algorithm based approach, in: 2018 IEEE Ninth Annual Information Technology, Electronics and Mobile Communication Conference (IEMCON), IEEE, 2018, pp. 865–870.

[15] Š. Bojan, Z. Nikola, Genetic algorithm as energy optimization method in WSN, in: 2013 21st Telecommunications Forum Telfor (TELFOR), IEEE, 2013, pp. 97–100.

[16] S.R. Gupta, N. Bawane, S. Akojwar, A clustering solution for wireless sensor networks based on energy distribution & genetic algorithm, in: 2013 Sixth International Conference on Emerging Trends in Engineering and Technology, IEEE, 2013, pp. 94–95.

[17] R. Sujee, K.E. Kannammal, Energy efficient adaptive clustering protocol based on genetic algorithm and genetic algorithm inter cluster communication for wireless sensor networks, in: 2017 International Conference on Computer Communication and Informatics (ICCCI), IEEE, 2017, pp. 1–6.

[18] X. Yuan, M. Elhoseny, H.K. El-Minir, A.M. Riad, A genetic algorithm-based, dynamic clustering method towards improved WSN longevity, J. Netw. Syst. Manag. 25 (1) (2017) 21–46.

About the authors

Anish Kumar Saha received B.E from National Institute of Technology Agartala, India, M.Tech from Tripura University, India and Ph.D. from National Institute of Technology Arunachal Pradesh, India in Computer Science and Engineering. He is currently working as Assistant Professor in the department of Computer Science & Engineering at National Institute of Technology Silchar, India. His research interests in Optimization, Software Defined Networking, and Quantum Computing.

Joseph L. Pachuau received B.Tech from North Eastern Regional Institute of Science and Technology, Arunachal Pradesh, India, M. Tech from National Institute of Technology, Silchar, India. He is currently pursuing Ph.D in the department of Computer Science and Engineering at National Institute of Technology, Silchar, India.

Arnab Roy received B.Tech from Central Institute of Technology, Kokrajhar, Assam, India, M. Tech from North Eastern Hill University, Shillong, Meghalaya, India. He is currently pursuing Ph.D in the department of Computer Science and Engineering at National Institute of Technology, Silchar, India.

Chandan Tilak Bhunia received B. Tech. in Radiophysics and Electronics, M. Tech. in Radiophysics and Electronics from University of Calcutta, India and Ph.D. in Computer Science & Engineering from Jadavpur University, India. He was the former Director at National Institute of Technology, Arunachal Pradesh, India. He has published many research papers at various national/ international journals/ conferences/seminars/ magazines/technical reports of repute. He is a fellow of the IETE and the IE(I), and a senior member of the IEEE & CSI.

CHAPTER SIXTEEN

The major challenges of big graph and their solutions: A review[☆]

Fitsum Gebreegziabher and Ripon Patgiri
Department of Computer Science and Engineering, National Institute of Technology Silchar, Silchar, Cachar, Assam, India

Contents

[☆]Preprint submitted to Elsevier October 5, 2021

Abstract

Large-scale computing graphs hold a high capacity to manage the flow of big data circulated throughout the complex systems of the virtual world. Hence, big-graphs are becoming among the most prevalent research areas of hi-tech industries. Consequently, this chapter focuses on reviewing the recent works on the dominant issues and challenges of big-graph and their solutions. Some of the relevant complex networks that implement large-scale graphs are justified in this chapter. In addition, this chapter describes the significant attributes of efficient and effective graphs. The core part of this review illustrates the issues and challenges related to graph visualization, partition and data allocation, benchmarking and evaluation, dynamic graph analysis, data exchange and integration, and to knowledge graph problems. Finally, the chapter invites all concerned researchers to contribute their efforts to find a solution for the justified open challenges of big-graph.

1. Introduction

Big-graph is a large-scale dataset with complex structures to show the complex interaction of actors in the digital system. Big-graph representation includes social networks, Wireless networks, biological information networks, and the World Wide Web [1]. Big-graph processing systems usually use various distributed models to manage the shared resources that they depict. The shared resources include storage, computing power, and management services [2].

The demand to manage enormous and complex data using powerful mathematical graphs is increasing. So, several implementation issues need further detailed studies to satisfy the demand. Large number of studies have continued delivering new frameworks for decades to enhance the capacity of big graphs. Nevertheless, several new issues are becoming imperative topics of the future. So, this chapter critically appraises different existing studies on the issues and challenges of big graphs.

The next part of this chapter consists of five sections. Section 2 explains the complex networks that widely apply big graphs to manage the big data they own. Section 3 holds a brief discussion about the basic requirements of efficient and effective graph data. In Section 4, issues and challenges of using big graphs are critically reviewed based on the recent works' results. Section 5 presents the summary of big graph issues addressed as well as the open challenges. Finally, Section 6 contains the conclusive statement of this review.

2. Big graph applications

Big graph data is generated from the globally known application systems that dominantly circulate exponentially increasing digital data. Such application systems include social networks, biological networks, communication networks, and semantic networks.

2.1 Biological network

A biological network is a relevant molecular computation model that deals with biology problems representing cells, genes, or proteins using a vertex and the relationships between these parts using edges. Biological networks are modeled inside the basic biological unit-cell (gene regulation networks (GRNs)) or outside it in a higher macroscopic level (brain networks, neural networks, phylogenetic networks). The fundamental unit of heredity with a sequence of nucleotides forming deoxyribonucleic acid (DNA) is known as a gene. Genes model to make amino acids chained in making the large molecules proteins outside the cell. This model contains all relevant functions related to the life of the organism. The biological network uses to model the interaction between proteins as well as the metabolic networks mathematically. The topological index serves as a computational function associating the structure with real numbers to measure the physical, chemical, and biological characters. For instance, assume a graph $G = (V, E)$ is an undirected biological graph with a set of vertices V and set of edges E. We can mathematically define the topological index as a function f: $G \rightarrow R^+$ that maps every graph to positive real numbers [3,4].

2.2 Social networks

The technology through which billions of users communicate with one another by exchanging their concerns in different media types such as text, audio, and video is known as a social network. Graphs can model social networks as nonlinear structures in which nodes depict the users, and the edges represent the interaction between users [5]. According to Ref. [6], Facebook, YouTube, WhatsApp, Facebook Messenger, and Instagram are the most popular social network platforms where billions of users interact. Within these dominant platforms, frequent interaction between users takes place based on mutual and often cyclic relationships, which affects the size and complexity of the system [7]. For example, When any user (vertex) of a Facebook gets a like, comment, tag, or any other interaction from another user, a new association (edge) is created with its attributes. So, the billions of interactions performed within few hours or even within minutes escalate the size of data complexity demanding powerful graph processing.

The following terms show the characteristics of the relationship (edge) between users.

Triadic closure is the chance of creating a link with friends of a friend in the social network. The second property that the probability a user can join in a relationship with users having a common interest and a similar range of age is known as *homophily*. The type of relationship defined as positive or negative is another property used to label the edge of the graph based on the like (+) or unlike (−) interaction of the nodes (users). The aforementioned three and many more dependency properties exponentially increase the size and complexity of the graph data. So, the need for efficient and effective graph models is incrementing from time to time.

Ríos *et al.* [8] state three strategies of creating edges in modeling graphs for social networks. The first strategy is creating an edge for every reply when a user creates a thread of posted documents. The edge is defined between the user that creates the original post in the thread and the user that replies to the created thread. The second strategy defines an arc between the user who creates a previous post in the thread and the user generates a reply. The last and most complicated strategy considers that every reply of a thread as a response to all posts that exist in a particular thread. Accordingly, the edges are created between users who created all last posts in the thread and the user that generates a reply.

2.3 Wireless *ad hoc* networks

A collection of decentralized mobile nodes that dynamically form a temporary network without preexisting infrastructure is known as wireless *ad hoc* network [9]. The two commonly used wireless *ad hoc* network types are wireless sensor networks (WSNs) and mobile *ad hoc* networks (MANETs). WSNs consist of small sensing nodes used for monitoring the health of individuals remotely, environmental monitoring, home automation. MANETs are the dynamic and movable nodes interconnected to serve in a vehicular network so that vehicles can communicate and interlink on expressways to avoid accidents by exchanging information. The connectivity between the devices that exchange information is illustrated as graph $G = (V, E)$, referring to a path e^1 between two vertices v_1 and v_2 in the graph G. Unlike static computers fixed in place, the moving devices in a MANET probably interrupt the connectivity. The disruptive property of wireless *ad hoc* networks makes graph management more complex and unstable when compared with the issues caused due to the failure of a fixed core device in a wired network. Nodes within similar radio transmission ranges are clustered together. This classification assigns a particular node from the group as cluster head (CH) with the responsibility of managing the cluster [3].

2.4 The Internet

The Internet is a thoroughly engineered communication infrastructure of the world. It consists of billions of devices connected and tremendously growing without limit. The Internet comprises several autonomous systems (ASs), which individually own routing policies to administer their respected networks. Internet uses routing as a means to find out the efficient path from origin to multiple destinations. Routers within the same autonomous systems run similar protocols, whereas routers in separate autonomous systems have the probability of using different routing protocols. Autonomous systems use an inter-ASs standard protocol known as Border Gateway Protocol (BGP) to exchange information throughout the global routing system of the Internet. Thus, in graph modeling, the vertices represent the autonomous systems, and the edges represent the BGP peering relationship among the autonomous systems. Distance vector and link-state algorithms are the most fundamental routing algorithms. Every node uses the first algorithm to send its shortest distance to neighboring nodes. Then, they update their routing table and can use the efficient path for exchanging information with it. The link-state algorithm uses a globally distributed algorithm in the way that every router can understand the entire topology and calculate the shortest path to every other node using the Dijkstra's single-source shortest path algorithm. Despite plenty of problems of Internet routing, different graph algorithms are on hand to solve the challenges sufficiently [3,10].

2.5 The World Wide Web networks

World Wide Web (WWW) is the information system over the Internet that contains billions of interlinked web pages. A graph that models the web pages as nodes and the hyperlinks as edges is called Web graph [11,12]. A web graph is a complex network that consists of millions of nodes having common complex network properties. These nodes are dynamically added to and removed from the web graph. This dynamic change process increases the difficulty of graph management. The two basic algorithms of a web graph are Hyperlink-Induced Topic Selection (HITS) and Page Rank.

HITS is applied to find the significant authorities based on the user's query. These relevant authorities are the web pages that contain the relevant information for the web query. Web pages that point to the authorities are known as hubs. Page rank algorithm uses a procedure to select the most relevant page of the graph depending on the number of pages that point to it. The rank of a page increases when a large number of pages in the graph votes for it. The page with the highest rank is the most important and popular in the graph [3].

2.6 Semantic networks

A semantic network is the logical collection of objects in the natural world with understandable meanings [13]. Semantic networks use graphs created from a set of directed and labeled edges and from a set of vertices to represent logic-based knowledge. The vertices and edges represent the concepts and semantic relations between the concepts respectively [14]. A knowledge graph is a preeminent type of semantic network to depict highly complex and big data. Initially, Google used knowledge graphs on an industry level. However, in the subsequent years, Amazon, Microsoft, Wikidata, DBpedia, and many other companies extend implementing more powerful knowledge graphs. Due to the increase of hype in knowledge graphs and the invention of advanced artificial intelligence-based services, most companies and service organizations assert adapting knowledge graphs. According to the layered perspective of Bellomarini *et al.* [15], the Knowledge graph has three views discussed below.

2.6.1 Knowledge representation tools

This view focuses on how the graph help to depict a particular type of knowledge. In knowledge graphs, the issue of representing the knowledge entirely or partially is raised in addition to the usual technique of modeling the entities and their relationship as a graph. Modeling knowledge graph starts with the acquisition of knowledge from the natural world as a teaching-leaning process. So teaching the entity with the necessary knowledge is taken as a precondition of building the graph. Since knowledge graphs are the base of intelligence, the degree of abstraction is integrated with the mathematical representation as knowledge units in the nodes and as connectors in the edges. In knowledge graphs, threefold of Object, subject, and Predicate constitute the knowledge. While the object and subject represent the entities, the predicate points to association. An attribute depicts the metadata of every entity.

2.6.2 Knowledge graph management systems (KGMS)

The focus of this view is on how a system manages the knowledge graph. The management includes- incorporating new knowledge into the knowledge graph, deriving new knowledge based on the previous understanding, and retrieving data using a query of a different format. The reasoning engine is the core component of a KGMS. This engine accesses the rule repository to draw inferences. Bellomarini *et al.* [15] assert that one fully developed KGMS should consist of three interrelated requirements. The first requirement is the comprising of language and system for reasoning. This requirement has to incorporate a logical formalism for expressing facts and rules and a reasoning engine that uses this language to deliver different powerful reasoning features. The second required feature of a matured KGMS is efficient and effective accessing, exchanging, cleaning of data, and handling queries. The last requirement is the ability of the system to embed procedural and third-party code intending to access the features of existing powerful software. This capability will scale up the capacity of the KGMS.

2.6.3 Knowledge application services

This view refers to the ability of a graph to support certain applications that define the knowledge graph. From the perspective of libraries and humanities, knowledge graphs are considered as knowledge organization systems. They are taken as rule mining and reasoners when used in database management systems.

Knowledge graph's end goal is delivering effective intelligence services. The services include smart analytics, recommendation and forecasting services, and expediting effective user query processing. To provide these services learning and reasoning are considered as essential components of most knowledge management tasks. Different systems can define a variety of models and methods based on the target they aim to achieve [16].

3. Features of powerful big graph

Big-graph needs to effectively and efficiently manage the exponentially growing data of complex systems. To solve the complicated problems of the systems the large-scale graph is required to have the following features [17].

3.1 Capability of handling flexible data model

An efficient and effective graph should be capable enough to incorporate the graph with heterogeneous edges and vertices. The flexible handling of edges and vertices with different types and attributes allows the graph to extend its model in parallel to the growth of the complex network systems to which it represents. Besides, a big-graph model has to depict and handle a graph or an aggregation of graphs. Such aggregation is known as a community within a system. Conclusively, graph data needs to support a data model that can yield robust operators to analyze and process the graph data.

3.2 Efficaciousness in managing complex query and analysis

It is known that any system aims to satisfy its users with a high rank of functionality and usability. While relational databases use structural query language (SQL), graph databases need to have a more powerful declarative query language to handle the database. Some of the query languages implemented to manipulate specific graph databases are Cipher on Neo4J or AQL on ArangoDB [18]. So a graph is said to be robust if it is capable enough to effectively and efficiently serve the users who retrieve or analyze data using declarative query languages. In addition, the system ought to support the iterative processing of the big ratio or the entire complex graph analysis task of the graph. The complex graph analysis tasks include detection of patterns (graph mining) [19], evaluation of application-specific or universal metrics. Constituting a robust model in the graph empowers the system to implement the analytical graph algorithms easily. And to minimize the effort of developing all analysis work-flows. So, efficient and effective management of complex queries and analysis is a necessary feature of robust graph data.

3.3 Scalability

AS the technology (processing units, storage units, and the network infrastructure) is rapidly excelling to higher capacity innovations, the ability of the graph to scale up with the fast-growing capacity of the technology to handle the large size of processing is required [18]. Nowadays, the most challenging feature of a graph is the scalability to efficiently handle the highly growing size of data [20]. A robust graph data is flexible to extend to a large scale handling billions of vertices and edges. However, effective scalability requires implementing distributed classes and in-memory graph processing supported by efficient application operators. Furthermore, the partitioning of the graph data among the nodes minimizes communication cost and redistribution of dynamic data and balances a load of computation evenly [17].

3.4 Permanent storage and transaction support

Graph databases apply the adjacency list format and store adjacency information related to the neighbor entity. This process helps the graph to provide efficient operations for retrieving the neighbors of a vertex [21]. All processed data in a graph must be stored persistently so that other queries can use it at any time. Furthermore, the dynamically growing complex and concurrent transaction processing demands big-graph to support online transaction processing. Online transaction processing is an essential function of a large number of complex systems.

3.5 Simplicity of use and visualization

Simplifying the usability and analysis of graph data is a very crucial issue. So that users can easily understand and use it. The use of relevant interaction controls (keyboard, mouse, visual or virtual reality related, and other advanced tools) simplifies the usability challenges. Ease of use and interaction is not limited to visualization of the information in a dataset but includes the weight, multitude, or direction of relations of the structure. These relations can have further attributes when the datasets support multiple variables. The vertices of the graph can also hold additional properties. The dynamic behavior of graphs demands features to handle the relational data over time change [22]. Additionally, the definition of graph workflows needs to be assisted by graphical editors. And the graph data and analysis results are expected to be delivered in a powerful visualization and customizable form.

4. Big graph issues and challenges

As a result of the fast-growing diversity and volume of graph data originated from different convoluted systems, the challenges of computational complexity and processing time cost are increasing enormously [23]. The traditional centralized graph computing models and systems are incapable of processing the rapidly growing massive graph data easily and effectively. Thus, efficient handling of big graphs is becoming a substantial problem of the current and the coming cyber-world [23–25].

Although constructing embeddings into a homogeneous graph is effective, it is rarely successful with heterogeneous graphs. But most big graphs are heterogeneous in design. So, building up a model for various types of connection edges and nodes of a complex graph hardens the challenge of constructing embeddings [26]. Besides, Erciyes [3] states that connectivity, backbone construction, clustering, and routing as the essential graph-related complications confronted in MANETs and WSNs. According to current researches, the main challenges faced in big graphs are discussed in the following subsections.

4.1 Graph visualization

Graph visualization has a significant impact on the usability of several application areas. Such applications include fraud detection reports in the financial systems analysis, information propagation exploration in social media graphs and demonstration of the interaction among proteins in biological graphs. Implementing successful visualizations for the various application areas is an issue that has become an important topic for researchers and developers to address it [22,27]. So far, several studies have taken place to improve the interactivity and efficiency of graphs for decades showing significant progress. Due to the increase in size and complexity of data and the dynamic graphs become more dependent on time, significant challenges are demanding further studies [22,28]. Imrea *et al.* [29] introduced new spectrum-preserving sparsification (SPS) framework to enhance the visualization of graphs. In addition, the team produced a scheme of node reduction depending on the fundamental spectral graph properties allowing a significant degree of detail simplification. SPS integrates spectral graph sparsification and clusters analysis to attain scalable visualization and efficient management of large graphs at a time. Recently, Han *et al.* [27] delivered an open-source JavaScript library named as NetV.js. This tool intends to complement the rapid visualization of big-graph data that can model up to 1 million edges and 50,000 vertices at a high level of interactivity using a personal computer. Net.js is a novel that incorporates three components known as *Rendering Engine, Graph Model Manager,* and *Interaction Manager.*

These three components help to implement the algorithms formulated to satisfy the requirement of authoring usability and accessibility. However, extending to support heterogeneous graphs and incorporating more parts and analysis algorithms has left as future works.

4.2 Graph partition and data allocation

A big graph holds enormous size data that a memory of a single microcomputer can't process. Therefore, a big-graph data system has to cluster its model into efficient and effective subgraphs dedicated to specific system communities. Appropriate data has to be allocated to the distributed parts of the graph. Partitioning of big-graph data is a highly challenging task because of computational complexity and sluggish performing time. Most distributed graph data systems apply a simple hash partition to randomly partition the graph. This way supports an equal allocation of vertices and subsequently, it assures load balancing. On the contrary, this leads to the repetitious assignment of neighboring vertices to various clusters resulting in the poor locality of processing and communication overheads. Other data systems support an edge-based partitioning instead of a vertex-driven approach. This approach balances the number of edges across the partitions, but the nodes are duplicated across their related edges. This approach is necessary for graphs that contain unevenly distributed vertex degrees, but it has the drawback of creating imbalanced loads on the partition. [17,23,30] Hence, a balanced graph partitioning (BGP) approach helps the partitioning of a graph into a required quantity of balanced partitions. Every partition will have a nearly similar amount of vertices or links implementing the vertex-cut and edge-cut partitioning methods. Moreover, computational resources in the client's machines have a limited capacity of managing the enormous size of big graph data entirely. Thus, the balanced data allocation over the clustered subgraphs with strong integrity is essential so that users can easily access and manage their data [23,30].

Accordingly, Yang et al. [23] suggested a new parallel balanced graph partitioning framework known as JA-BE-JA-RIK to improve the efficiency of partitioning to have better edge-cut value and a higher degree of knowledge coherence. In parallel BGP, a large-scale graph is partitioned into smaller subgraphs in the way a memory of a microcomputer can process each subgraph. JA-BE-JA-RIK is an extension of JA-BE-JA [31] but adds a concept of the richness of implicit knowledge (RIK) along with a quantitative measurement of dynamic evaluation of RIK values amid partitioning. The three stages this algorithm uses are data preprocessing, node color exchange, and RIK evaluation.

In preprocessing stage, the original graph is classified in the form of node-set and edge-set files. Node-set file holds a set of all vertices as a column and an edge-set file holds every vertex affixing all its direct neighbors storing in a row.

Candidate pair selection and color exchange are performed in the second stage. This stage delivers all the node colors in the updated color table through iterative energy computation and color exchange.

RIK evaluation is the third stage that calculates the average path lengths of total partitions and total edge-cuts resulted from partitioning.

RIK evaluation passes through an iterative reading of node ids and the color of these nodes retrieved from table color. After getting a maximum RIK value, all node colors will become outcomes of the final graph partitioning. Yet, this approach requires more features to comply with the demand of the user's graph query.

Another research [32] presents a distributed k-partition algorithm named DPHV (Distributed Placement of Hub-Vertices) for large-scale graph partitioning. The target DPHV is delivering a fast and efficient partitioning of a big graph by optimizing the number of cut-edges and ensuring load balancing and network bandwidth of the cluster nodes. Unlike the fast and load balancing achievements, DPHV is limited in optimizing the cut-edges to their highest degree. In addition, big size memory is required to handle the process of this algorithm.

4.3 Benchmarking and evaluation

A benchmark [33] is a framework expected to satisfy the predefined require-ments of graph data systems. The absence of a common base benchmark for the architectural design of different systems creates challenges for the end-users to select the best system [34].

The demand of analyzing the exponentially increasing size and diversity of graphs is motivating researchers and industries to develop better graph eval-uation software platforms. Different researchers try to define various graph analysis benchmarks. LinkedBench [35] is an open-source database bench-mark that assists to predict the performance of a database in the persistent storage of Facebook's production data.

Another benchmark, LDBC Graphalytics [33] tries to fulfill the standards of Linked Data Benchmark Council (LDBC). LDBC a council that focuses on developing standards benchmark specifications, practices, and results for graph data processing tools. Accordingly, LDBC presented Graphalytics to assess the graph analysis platforms that include stress-testing and performance variability. The team defined five main parameters that have to be addressed by LDBC Graphalytics. This work introduced a large number of new fea-tures. Nonetheless, the rapid growth in scope and parameters of large-scale graphs is demanding more powerful universal benchmarks.

gMark [36] is the other benchmark that holds the ability to target and control the heterogeneous properties of generated instances and their work-load. It aims to satisfy the requirements such as domain independence, extensibility, schemaderivation, and configuration. gMark is good at query processing, but it is limited to small graph instances. In quadratic queries the time of processing is slow.

Graph-Based Benchmark Suite (GBBS) [37] a recent work that extends the features of previous benchmarks to solve broad problems efficiently. This benchmark is applicable on 20 fundamental graph problems categorized under clusters of; shortest path problem, connectivity problem, covering problem, substructure problem, and Eigenvector problem. GBBS is designed to be implemented on big graphs that hold up to 200 billion edges using a single multi-core machine. Despite its efficiency to handle huge data and address a relatively large number of problems, GBBS should have to upgrade its functionality to satisfy the requirements of the latest systems.

To sum up, most of the aforementioned benchmarks define evaluation guidelines to assess the cost of the systems under evaluation. Besides, there is rapid growth and change in features and performance of graph data management systems. So, this rapid change requires a powerful and universal benchmark that can scales up its functionality accordingly.

4.4 Dynamic graph analysis

Static graphs are limited to process the highly evolving complex networks. Thus, dynamic graphs which can vary and attune their capacity with the fastgrowing complexity of systems are strongly demanded. As a result analysis of the dynamic graphs is among the preeminent topics of current researches. For instance, in the highly dynamic social networks (such as Twitter and Facebook) community detection, PageRank metrics, interaction notifications, and identification of new topics as well as popular articles require fast graph analysis [17].

The analysis of dynamic graphs that depend on community detection is slow, so it is an important topic of researches. The target of community detection [38] is determining the firmly integrated collections of nodes coming in through an input network, where the components of the communities share concentrated edges. Hence, Zarayeneh and Kalyanaraman [38] demonstrated an efficient incremental technique to solve the challenges of dynamic community detection. The technique tries to evaluate the newly introduced collection of changes to an input graph and chooses a subset of nodes to reexamine for potential community assignment. This technique is also capable to generate the required performance preserving the output quality. However, this study is still limited in thoroughly detecting the edges.

Another study on community detection [39] delivers two algorithms to determine when to update the structure of dynamic community structures to manage the frequent change of nodes (vertices). In complex social networks, a large number of users can join or be removed frequently. So, Puranik and Narayanan [39] categorize the edges and define the minimum threshold to the number of edge types that must be allowed to join the network before the community structure is updated. The two algorithms provided to solve the challenge of snapshot identifications are known as the Edge Distribution Analysis algorithm (EDA), and the Modularity Change Rate algorithm (MCR).

Despite the improvements in analyzing the structure change of the network in terms of adding or deleting nodes, the frequent change in content (attribute) demands more studies to accommodate the change in users' interest for different contents (topics).

4.5 Data exchange and integration

Data exchange expresses the process of transforming data situated under the structure of source schema into target schema compatible data [40]. Besides the process of aggregating data stored in various origins in the way of providing a unified view to the user via global schema is known as data integration. Screening and transforming graph data extracted from original data sources to the targeted schema is a necessary task. Additionally, it is essential to merge and interlink the data extracted from different sources into the graph. These requirements are prerequisites for successful graph analysis and processing. Hence it is possible to consider data exchange and data integration are among the necessary interoperability activities of graph databases [17]. Although different studies have taken place on the two activities, some limitations need to be solved. Some of the current challenges of graph data integration and exchange are *complexity of computation* and *Query answering*. The complexity of computation is derived from the data and the schema of the source and the complication in the combined target. Similarly, the query answering problem comes out due to loosed integration of data combined in the global schema [40].

Francis and Libkin [41] discussed the impact of schema mappings on query answering problems while avoiding iteratively defined navigation over target data. The authors conclude that adding data to graph structure increases the complexity of answering queries. However, the paper suggests more study on the use of SQL nulls to allow efficient answering of a big class of query in integration and exchange scenarios. Arshad and Anjum [42] provided a new graph-based method to accommodate data using an in-memory environment. This study de- livered an approach of representing the data sources as graphs for integration. It aims to ensure fast querying and to guarantee consistency of sources while evolving and provide scalable storage. Whence, the effective way of aggregating coherent data accommodation and performance in the globally distributed system is delivered without loading data at the time of analysis. However, the approach has been implemented in a limited domain. So, graph data integration and exchange is continuing as an open challenge of future studies.

4.6 Challenges on knowledge graph implementation

Many companies are using a knowledge graph as their fundamental tool of facilitating services. Hence, the technology is demanding more improvements. A team of authors from Microsoft, Google, Facebook, eBay, and IBM made a joint work [43] in defining the challenges of large-scale industrial knowledge graphs. According to the joint study, the following challenges are the consistently occurring issues across most implementations.

4.6.1 Management of entity disambiguation

The process of linking the ambiguous terms in surface text with its semantic forms (entities) in a knowledge base is called entity disambiguation [44]. Disambiguation of entities is a significant challenge in the semantic systems such as in natural language processing, knowledge base, question answering systems [45]. Despite many studies continuing on entity disambiguation solutions, the challenge is remaining as a topic of future researches [44,45]. The challenge is faced while assigning a unique normalized identity and a type to the label of an entity. Automatically retrieving entities is mostly faces the problem of getting a large number of entities with a similar form. Inadequate disambiguation and linking lead to the wrong association of entities and fact which ends up with incorrect inference. Ma et al. [45] proposed a Markov Logic Network (MLN) Knowledge graph model to disambiguate the entities by inferring the inconsistent relationship within the knowledge base. However, the issue is demanding more studies to have better performance in resolving the ambiguity of entities.

4.6.2 Resolution of membership type

Many entities of a knowledge graph can have more than one type of membership as per different circumstances. Ren et al. [46] defined a new task to reduce the noise of label in entity typing and proposed a heterogeneous partial-label embedding framework to resolve the noise problem of candidate types. Ma *et al.* [47] also delivered a label embedding method with prototypical and hierarchical information for zero-shot fine-grained named entity typing. The team used the zero-shot framework to enhance the prediction of both seen and unseen entity types. However, while the domain of entities grows more, applying criteria to preserve semantic stability becomes more challenging.

4.6.3 Knowledge change handling

Managing the rapidly evolving demand of a dynamic graph with changing the existing knowledge is becoming a challenge in large-scale industries. Hence, more studies are expected to come with better frameworks for handling dynamic knowledge change.

4.6.4 Knowledge extraction from multiple sources

Knowledge extraction from structured and unstructured sources continues as a challenge regardless of many advancements made in the previous studies. To solve the challenges, Wu *et al.* [48] presented a new framework for developing a knowledge graph from multiple online encyclopedias. The team applies knowledge extraction and knowledge linking modules. The first module extracts different types of articles and the second module utilizes entity matching and learning methods. However, the framework doesn't support the extraction of unstructured data. Thus, novel solutions are expected in the future to alleviate or eliminate the gaps.

4.6.5 Managing operations at scale

Managing scale is a very basic challenge that alters the operations related to workload and performance. It can also affect the management of quick upgrades to large-scale graphs. The trade-off between computational cost and model scalability remains a significant challenge despite various researches made on the issue [49]. Therefore, large-scale knowledge graphs will efficiently support the industries if and only if the current challenges are resolved using new all-inclusive powerful frameworks.

5. Summary

Due to the fast invention of new and sophisticated technologies in the digital world, all the systems where the large-scale graphs are applied, are demanding more improvements on their existing features. So, large graph data systems are on the way to frequent change to keep up the rate of the expeditious growth of technologies. Accordingly, various studies are taking place to solve the bottlenecks of the graph systems and some challenges are still waiting for more improvements. Hence, the issues addressed and the challenges open for future works discussed in the previous section, are summarized in Table 1.

Table 1 Summary of issues and challenges.

Paper	Issues solution	Open	Challenges
Imrea *et al.* [29]	Graph Complexity	Sparsification of big graph for efficient visual exploration	Ease of interactivity
Han *et al.* [27]	High-efficiency visualization of big graph data	A powerful visualization tool(NetV.js)	Supporting Heterogeneous Graphs
Yang *et al.* [23]	Graph partitioning and load balancing	JA-BE-JA-RIK	Querying usability
Adoni *et al.* [32]	Graph partitioning and load balancing	Distributed k-partition algorithm named DPHV	Cut-edge optimization and memory cost minimization
Armstrong *et al.* [35]	Benchmarking	Open-source database benchmark for performance prediction	Limitation to evaluate large number of parameters
Iosup *et al.* [33]	Benchmarking and evaluation	Graphalytics, for evaluating graph analysis platforms	Cost Evaluation
Bagan *et al.* [36]	Benchmarking	gMark to control the heterogeneous properties of instances and their workload	Efficient processing of quadratic queries
Dhulipala *et al.* [37]	Exhaustive benchmarking	GBBS, to implement up to 200 billion edges using a single multi-core machine	implementing on streaming graphs

Continued

Table 1 Summary of issues and challenges.—cont'd

Paper	Issues solution	Open	Challenges
Zarayeneh and Kalyanaraman [38]	community detection	Incremental technique for dynamic community detection	Thorough detection of the Edges
Puranik and Narayanan [39]	Dynamic graph analysis	EDA and MCR algorithms for managing the frequent change of nodes	Attribute(content) change Management
Francis and Libkin [41]	exchange and integration of graph data	Showing the impact of Schema mapping on query answering	Applying the powerful approach of data exchange and integration
Arshad and Anjum [42]	Graph data integration	Graph-based model to integrate data using in-memory environment	Applicability in various Domains
Ma et al. [45]	Entity disambiguation in KG	Markov Logic Network(MLN) Knowledge graph model	Adjustment with new Datasets
Xiang et al. [46]	Resolution of membership type in KG	Framework, to reduce the noise of label in entity typing	Improvement to a higher degree of accuracy in an archaic approach
Yukun et al. [47]	Resolution of membership type in KG	Label embedding method with zeroshot framework	Preserving semantic stability in large varieties of entities
Wu et al. [48]	Knowledge extraction from multiple sources	Framework, for developing a knowledge graph from multiple online encyclopedias	Knowledge extraction from unstructured source

6. Conclusion

This review briefly discusses the relevant applications and requirements. And it critically prevails the significant issues and challenges of big-graph. We explained some relevant examples of the most widely complex computing systems that implement big graphs intending efficient use of big data. Simultaneously, the way how the systems implement the graph is described. The second point raised in this chapter is about the prominent attributes of an efficient and effective graph. Finally, we review several related studies on the issues and challenges of big-graph. This work discussed some of the solutions presented by recent studies. Despite multiple solutions provided for the previous problems, this chapter shows that numerous challenges are open for future studies. So, we recommend all interested researchers focus on the identified challenges of big-graph to support the all-inclusive digital world.

References

[1] D.K. Singh, R. Patgiri, Big graph: tools, techniques, issues, challenges and future directions, in: Sixth International Conference on Advances in Computing and Information Technology (ACITY 2016), 2016, pp. 119–128, https://doi.org/10.5121/csit.2016. 60911.

[2] F. Hu, Y. Lu, Secure dynamic big graph data: scalable, low-cost remote data integrity checking, IEEE Open Access J. 7 (2019) 12888–12900, https://doi.org/10.1109/ ACCESS.2019.2892442.

[3] K. Erciyes, Guide to Graph Algorithms: Sequential, Parallel and Distributed, first ed., Springer International Publishing, 2018, https://doi.org/10.1007/978-3-319-73235-0.

[4] W. Gao, H. Wu, M.K. Siddiqui, A.Q. Baig, Study of biological networks using graph theory, Saudi J. Biol. Sci. 25 (6) (2018) 1212–1219, https://doi.org/10.1016/j.sjbs.2017. 11.022. https://www.sciencedirect.com/science/article/pii/S1319562X17302966.

[5] R.K. Behera, M. Jena, D. Naik, B. Sahoo, S.K. Rath, Linkage-based social network analysis in distributed platform, Big Data Anal. A Soc. Netw. Approach (2018) 1–25.

[6] Statista, 2021. https://www.statista.com/statistics/272014/global-social-networksranked-by-number-of-users/, accessed on March 27, 2021.

[7] T. Jan, The Ins and Outs of Network-Oriented Modeling: From Biological Networks and Mental Networks to Social Networks and Beyond, Springer, Berlin Heidelberg, Berlin, Heidelberg, 2019, pp. 120–139, https://doi.org/10.1007/978-3-662-58611-2_2. https://doi.org/10.1007/978-3-662-58611-2_2.

[8] S.A. Ríos, F. Aguilera, J.D. Nuñez-Gonzalez, M. Graña, Semantically enhanced network analysis for influencer identification in online social networks, Neurocomputing 326–327 (2019) 71–81, https://doi.org/10.1016/j.neucom.2017.01.123. https://www. sciencedirect.com/science/article/pii/S0925231217315242.

[9] A. Kush, C.J. Hwang, V. Dattana, Big data analytics on manet routing standardization using quality assurance metrics, in: 2016 Future Technologies Conference (FTC), 2016, https://doi.org/10.1109/FTC.2016.7821610.

[10] F. Agostino, The graph structure of the internet at the autonomous systems level during ten years, J. Comput. Commun. 7 (2019) 17–32, https://doi.org/10.4236/jcc.2019. 78003.

[11] S. Sa'adah, R.S.W. Kemas, S.A. Hartomo, Implementation of vocabulary based summarization of graph (vog) on the web graph, in: 2019 IEEE International Conference on Signals and Systems (ICSigSys), 2019, pp. 156–159, https://doi.org/10.1109/ ICSIGSYS.2019.8811091.

[12] G. Lee, S. Kang, J.J. Whang, Hyperlink classification via structured graph embedding, in: Proceedings of the 42nd International ACM SIGIR Conference on Research and Development in Information Retrieval, SIGIR'19, Association for Computing Machinery, New York, NY, USA, 2019, pp. 1017–1020, https://doi.org/10.1145/ 3331184.3331325. https://doi.org/10.1145/3331184.3331325.

[13] L. Yong, J.L. Lu, Y. Qing, Z. Bo, Robot task planning based on state semantic network, in: 2017 10th International Conference on Intelligent Computation Technology and Automation, 2017, pp. 420–424, https://doi.org/10.1109/ICICTA.2017.101.

[14] P. Pirnay-Dummer, D. Ifenthaler, N.M. Seel, Semantic Networks, Springer US, Boston, MA, 2012, pp. 3025–3029, https://doi.org/10.1007/978-1-4419-1428-6_1933. https://doi.org/10.1007/978-1-4419-1428-6_1933.

[15] L. Bellomarini, E. Sallinger, S. Vahdati, Chapter 2 Knowledge Graphs: The Layered Perspective, Springer International Publishing, Cham, 2020, pp. 20–34, https://doi.org/10.1007/978-3-030-53199-7_2. https://doi.org/10.1007/978-3-030-53199-7_2.

[16] L. Bellomarini, E. Sallinger, S. Vahdati, Chapter 6 Reasoning in Knowledge Graphs: An Embeddings Spotlight, Springer International Publishing, Cham, 2020, pp. 87–101, https://doi.org/10.1007/978-3-030-53199-7_6. https://doi.org/10.1007/978-3-030-53199-7_6.

[17] M. Junghanns, A. Petermann, M. Neumann, E. Rahm, Management and analysis of big graph data: current systems and open challenges, in: Handbook of Big Data Technologies, Springer, 2017, pp. 457–505, https://doi.org/10.1007/978-3-319-49340-4_14.

[18] D. Fernandes, J. Bernardino, Graph databases comparison: allegrograph, arangodb, infinitegraph, neo4j, and orientdb, in: Proceedings of the 7th International Conference on Data Science, Technology and Applications - DATA, SCITEPRESS - Science and Technology Publications, Lda, 2018, pp. 373–380. https://www.scitepress.org/papers/2018/69102/69102.pdf.

[19] D. Robinson, C. Scogings, The detection of criminal groups in real-world fused data: using the graph-mining algorithm "graphextract", Secur. Inform. 7 (2018) 2, https://doi.org/10.1186/s13388-018-0031-9.

[20] S. Sahu, A. Mhedhbi, S. Salihoglu, J. Lin, M.T. Özsu, The ubiquity of large graphs and surprising challenges of graph processing, Proc. VLDB Endow. 11 (4) (2017) 420–431, https://doi.org/10.1145/3164135.3164139. https://doi.org/10.1145/3164135.3164139.

[21] A. Pacaci, A. Zhou, J. Lin, M.T. Özsu, Do we need specialized graph databases? Benchmarking real-time social networking applications, in: Proceedings of the Fifth International Workshop on Graph Data-Management Experiences and Systems, GRADES'17, Association for Computing Machinery, New York, NY, USA, 2017, https://doi.org/10.1145/3078447.3078459. URL. https://doi.org/10.1145/3078447.3078459.

[22] M. Burch, W. Huang, M. Wakefield, H.C. Purchase, D. Weiskopf, J. Hua, The state of the art in empirical user evaluation of graph visualizations, IEEE Access 9 (2021) 4173–4198, https://doi.org/10.1109/ACCESS.2020.3047616.

[23] Z. Yang, R. Zheng, Y. Ma, Parallel heuristics for balanced graph partitioning based on richness of implicit knowledge, IEEE Access 7 (2019) 96444–96454, https://doi.org/10.1109/ACCESS.2019.2926753.

[24] X. Liu, Y. Zhou, X. Guan, C. Shen, A feasible graph partition framework for parallel computing of big graph, Knowl.-Based Syst. 134 (2017) 228–239, https://doi.org/10.1016/j.knosys.2017.08.001. https://www.sciencedirect.com/science/article/pii/S095070511730357X.

[25] H.S. Chemseddine Nabti, Querying massive graph data: a compress and search approach, Futur. Gener. Comput. Syst. 74 (2017) 63–75.

[26] M. Korenevaa, A.A. Visheratina, D. Nasonova, Decoupling graph convolutional networks for large-scale supervised classification, Proc. Comput. Sci. 178 (2020) 337–344, https://doi.org/10.1016/j.procs.2020.11.035.

[27] D. Han, J. Pan, X. Zhao, W. Chen, Netv.js: A web-based library for high-efficiency visualization of large-scale graphs and networks, Visual Inform. 5 (1) (2021) 61–66, https://doi.org/10.1016/j.visinf.2021.01.002. https://www.sciencedirect.com/science/article/pii/S2468502X21000048.

[28] Y. Hu, M. Nöllenburg, Graph Visualization, Springer International Publishing, Cham, 2019, pp. 904–912, https://doi.org/10.1007/978-3-319-77525-8_324. https://doi.org/10.1007/978-3-319-77525-8_324.

[29] M. Imrea, J. Taob, Y. Wangc, Z. Zhaoc, Z. Fengc, C. Wang, Spectrumpreserving sparsification for visualization of big graphs, Comput. Graph. 87 (2020) 89–102, https://doi.org/10.1016/j.cag.2020.02.004.

[30] Z. Shi, J. Li, P. Guo, S. Li, D. Feng, Y. Su, Partitioning dynamic graph asynchronously with distributed fennel, Futur. Gener. Comput. Syst. 71 (2017) 32–42, https://doi.org/10.1016/j.future.2017.01.014. https://www.sciencedirect.com/science/article/pii/S0167739X1730033X.

[31] F. Rahimian, A.H. Payberah, S. Girdzijauskas, M. Jelasity, S. Haridi, A distributed algorithm for balanced graph partitioning, in: Jabeja: 2013 IEEE 7th International Conference on Self-Adaptive and Self-Organizing Systems, IEEE, 2013, pp. 51–60.

[32] W. Y. H. Adoni, T. Nahhal, M. Krichen, A. El Byed, I. Assayad, Dhpv: a distributed algorithm for large-scale graph partitioning, Big Data 7. https://doi.org/10.1186/s40537-020-00357-y.

[33] A. Iosup, T. Hegeman, W.L. Ngai, S. Heldens, A. Prat-Perez, T. Manhardt, H. Chafi, M. Capota, N. Sundaram, M. Anderson, I.G. Tanase, Y. Xia, L. Nai, P. Boncz, Ldbc graphalytics: a benchmark for large-scale graph analysis on parallel and distributed platforms, VLDB Endow. 9 (2016) 1317–1328, https://doi.org/10.14778/3007263.3007270.

[34] H.W.Y. Adoni, T. Nahhal, M. Krichen, B. Aghezzaf, A. Elbyed, A survey of current challenges in partitioning and processing of graphstructured data in parallel and distributed systems, Distrib. Parallel Databases (2019) 1–36.

[35] T.G. Armstrong, V. Ponnekanti, D. Borthakur, M. Callaghan, Linkbench: a database benchmark based on the Facebook social graph, in: 2013 ACM SIGMOD International Conference on Management of Data, 2013, pp. 1185–1196, https://doi.org/10.1145/2463676.2465296.

[36] G. Bagan, A. Bonifati, R. Ciucanu, G.H.L. Fletcher, A. Lemay, N. Advokaat, Gmark: schema-driven generation of graphs and queries, IEEE Trans. Knowl. Data Eng. 29 (2016) 856–869, https://doi.org/10.1109/TKDE.2016.2633993.

[37] L. Dhulipala, J. Shi, T. Tseng, G.E. Blelloch, J. Shun, The Graph Based Benchmark Suite (GBBS), in: Proceedings of the 3rd Joint International Workshop on Graph Data Management Experiences & Systems (GRADES) and Network Data Analytics (NDA), GRADES-NDA'20, Association for Computing Machinery, New York, NY, USA, 2020, https://doi.org/10.1145/3398682.3399168. https://doi.org/10.1145/3398682.3399168.

[38] N. Zarayeneh, A. Kalyanaraman, A fast and efficient incremental approach toward dynamic community detection, in: 2019 IEEE/ACM International Conference on Advances in Social Networks Analysis and Mining, 2019, pp. 9–16, https://doi.org/10.1145/3341161.3342877.

[39] T. Puranik, L. Narayanan, Community detection in evolving networks, in: 2017 IEEE/ACM International Conference on Advances in Social Networks Analysis and Mining, 2017, pp. 385–390, https://doi.org/10.1145/3110025.3110067.

[40] A. Bonifati, I. Ileana, Graph Data Integration and Exchange, Springer International Publishing, Cham, 2019, pp. 815–822, https://doi.org/10.1007/978-3-319-77525-8_209. https://doi.org/10.1007/978-3-319-77525-8_209.

[41] N. Francis, L. Libkin, Schema mappings for data graphs, in: Proceedings of the 36th ACM SIGMOD-SIGACT-SIGAI Symposium on Principles of Database, PODS'17, 2017, pp. 389–401, https://doi.org/10.1145/3034786.3056113.

[42] B. Arshad, A. Anjum, High performance dynamic graph model for consistent data integration, in: Proceedings of the 12th IEEE/ACM International Conference on Utility and Cloud Computing UCC'19, 2019, pp. 263–272, https://doi.org/10.1145/3344341.3368806.

[43] N. Noy, Y. Gao, A. Jain, A. Narayanan, A. Patterson, J. Taylor, Industry scale knowledge graphs: lessons and challenges, Commun. ACM 62 (8) (2019) 36–43, https://doi.org/10.1145/3331166.

[44] S. Zwicklbauer, C. Seifert, M. Granitzer, in: Doser—a knowledge-base-agnostic framework for entity disambiguation using semantic embeddings, Proceedings of the 13th International Conference on The Semantic Web. Latest Advances and New Domains, 9678, Springer-Verlag, Berlin, Heidelberg, 2016, pp. 182–198, https://doi.org/10.1007/978-3-319-34129-3_12. https://doi.org/10.1007/978-3-319-34129-3_12.

[45] J. Ma, T. Wei, Y. Qiao, Y. Huang, W. Xie, C. Zhang, Y. Wang, R. Zhang, Entity disambiguation with Markov logic network knowledge graphs, Int. J. Perform. Eng. 13 (8) (2017) 1293, https://doi.org/10.23940/ijpe.17.08.p11.12931303.

[46] X. Ren, W. He, M. Qu, C.R. Voss, H. Ji, J. Han, Label noise reduction in entity typing by heterogeneous partial-label embedding, in: Proceedings of the 22nd ACM SIGKDD International Conference on Knowledge Discovery and Data Mining, 2016, pp. 1825–1834.

[47] Y. Ma, E. Cambria, S. Gao, Label embedding for zero-shot fine-grained named entity typing, in: Proceedings of COLING 2016, the 26th International Conference on Computational Linguistics: Technical Papers, 2016, pp. 171–180. https://www.aclweb.org/anthology/C16-1017.pdf.

[48] T. Wu, H. Wang, C. Li, G. Qi, X. Niu, M. Wang, L. Li, C. Shi, Knowledge graph construction from multiple online encyclopedias, World Wide Web (2019) 1–28, https://doi.org/10.1007/s11280-019-00719-4.

[49] S. Ji, S. Pan, E. Cambria, P. Marttinen, S. Y. Philip, A survey on knowledge graphs: representation, acquisition, and applications, IEEE Trans. Neural Networks Learn. Syst. https://doi.org/10.1109/TNNLS.2021.3070843.

About the authors

Fitsum Gebreegziabher received the BSc. degree in Information Systems from Addis Ababa University, Ethiopia. And, he got MSc degree from Madras University, India. Currently, he is pursuing PhD in Computer Science and Engineering at National Institute of Technology Silchar, India. His research interests are cloud security, big data, and information security.

Dr. Ripon Patgiri has received his bachelor's degree from Institution of Electronics and Telecommunication Engineers, New Delhi in 2009. He has received his M.Tech. degree from Indian Institute of Technology Guwahati in 2012. He has received his Doctor of Philosophy from National Institute of Technology Silchar in 2019. After M.Tech. degree, he has joined as an assistant professor at the Department of Computer Science & Engineering, National Institute of Technology Silchar in 2013. He has published numerous papers in reputed journals, conferences, and books. His research interests include distributed systems, file systems, Hadoop and MapReduce, big data, bloom filter, storage systems, and data-intensive computing. He is a senior member of IEEE. He is a member of ACM and EAI. He is a lifetime member of ACCS, India. Also, he is an associate member of IETE. He was General Chair of 6th International Conference on Advanced Computing, Networking, and Informatics (ICACNI 2018, http://www.icacni.com) and International Conference on Big Data, Machine Learning and Applications (BigDML 2019, http://bigdml.nits.ac.in). He is Organizing Chair of 25th International Symposium on Frontiers of Research in Speech and Music (*FRSM 2020*, http://frsm2020.nits.ac.in) and International Conference on Modeling, Simulations and Applications (CoMSO 2020, http://comso.nits.ac.in). He is convenor, Organizing Chair and Program Chair of 26th annual International Conference on Advanced Computing and Communications (ADCOM 2020). He is guest editor in special issue "Big Data: Exascale computation and beyond" of EAI Endorsed Transactions on Scalable Information Systems. He is also an editor in a multiauthored book, title "*Health Informatics: A Computational Perspective in Healthcare*," in the book series of *Studies in Computational Intelligence*, Springer. Also, he is writing a monograph book, titled "*Bloom Filter: A Data Structure for Computer Networking, Big Data, Cloud Computing, Internet of Things, Bioinformatics and Beyond*," Elsevier.

CHAPTER SEVENTEEN

An investigation on socio-cyber crime graph

V.S. Nageswara Rao Kadiyala and Ripon Patgiri
Department of Computer Science and Engineering, National Institute of Technology Silchar, Silchar, Cachar, Assam, India

Contents

Advances in Computers, Volume 128
ISSN 0065-2458
https://doi.org/10.1016/bs.adcom.2021.10.011

423

Abstract

Today, cyber crime has done much to harm individuals, organizations and public authorities. Cyber crime detection and classification methods have succeeded in protecting and preventing data against these attacks by varying degrees. Many rules and procedures to combat cyber crimes were put in place and offenders were subjected to severe sanctions. However, the study reveals that many countries are still dealing with this problem, with the United States of America suffering the most harm as a result of cyber crime over the years. A new report shows that the financial losses in 2018 were almost 781.84 million dollars. This chapter discusses the different forms of cyber crime today and the popular environments in which cyber crime occurs.

1. Introduction

Crime is known an illegal act which is punishable by a government. No mandated interpretation is however given for those kind of reasons. Not only some people, but the community or the government are harmful. The law enforced is not related to cyber crime. In cyber, the authors say a label is being used in the technology for data era, to depict an individual, something or concept. Basically, a computer network collects communicating nodes that help transmit data across. The nodes could be a computer at any given time, the laptop, smart phones etc. Cyber crime covers any crime involving computers and networks. This includes Internet-based crime [1]. The Web is essentially a channel of communication and data-sharing networks across countries. The need for a tool for illicit use, including forgery, child exploitation trafficking, proprietary information, identification robbery and infringement of data, also called cyber crime. Table 1 demonstrates six crime categories through Internet. A country-wise distribution of crime are given in the Table 1. There are six factors for crime communities, namely, Computer malicious activity share, malware code, spam zombies, phishing website, bot, and attack origin. Crimes are higher in the developed countries due to availability of the resources and advancement of the technologies. However, the Internet play vital roles in promoting the crimes.

The international village efficiently shares and communicates essential data on progress in Internet technologies such as 2G and 3G throughout the network. However, people are trying to monitor and unlawfully retrieve crucial and sensitive information for their individual usages, financial gains as well as other purposes. Cyber crime comprises a wide variety of crimes, along with theft, forgery and economic crime, pornographic material, smuggling, downloading of unlawful files, and so on. Cyber crime refers to every crime involving the Internet and a computer [2].

Table 1 List of cyber crime community action of six factors.

Country	Computer malicious activity share	Rank of malware code	Ranked spam zombies	Phishing website hosts rank	Bot rank	Attack origin rank
USA	23%	1	3	1	2	1
China	9%	2	4	6	1	2
Germany	6%	12	2	2	4	4
Britain	5%	4	10	5	9	3
Brazil	4%	16	1	16	5	9
Spain	4%	10	8	13	3	6
Italy	3%	11	6	14	6	8
France	3%	8	14	9	10	5
Turkey	3%	15	5	24	8	12
Poland	3%	23	9	8	7	17
India	3%	3	11	22	20	19
Russia	2%	18	7	7	17	14
Canada	2%	5	40	3	14	10
South Korea	2%	21	19	4	15	7
Taiwan	2%	11	21	12	11	15
Japan	2%	7	29	11	22	11
Mexico	2%	6	18	31	21	16
Argentina	1%	44	12	20	12	18
Australia	1%	14	37	17	27	13
Israel	1%	40	16	15	16	22

Source: W. Patterson, C.E. Winston-Proctor. An international extension of Sweeney's data privacy research, In: T. Ahram, W. Karwowski (Eds.), Advances in Human Factors in Cybersecurity, Springer International Publishing, Cham, 2020, pp. 28–37.

As the Internet is becoming an increasingly important part of the world's lives, online strategies such as stock spam email campaigns have been used increasingly by criminals to reach numerous potential victims cost-effectively and efficiently. The results show that 85% of respondents are of the opinion that spam email is another causal factor in the effects shown in Fig. 1. As the discussion has shown, we may wonder why

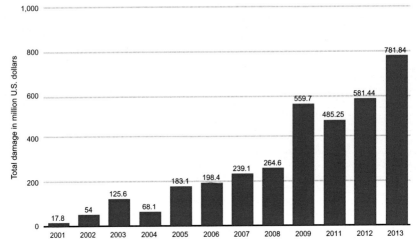

Fig. 1 Currency cyber crime harm damage between 2001 and 2013 (US dollars in million).

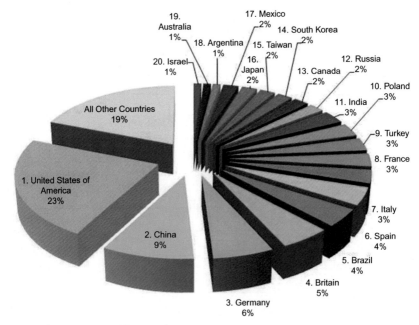

Fig. 2 Cyber crime top 20 countries.

computers or networking are used as instruments for such cyber crimes. Fig. 2 says that 90% of respondents said that access is easy, 85% said that access is easy, and 80% said that the computers are affordable, for almost three reasons.

Cyber crime has greatly contributed to the availability and ease of access to computers and the Internet. The fact that a computer is bought and connected to the Internet is now more affordable, as many people incorporate technology, whether at home or at work, into their lives. Thanks to technology advancements, innovation became an instrument, aim as well as site for illegal activities. People have significantly increased possibilities for electronic abuse or internet fraud as Internet and computational communication technologies become more affordable. Advancements in technology unintentionally served to boost it by attempting to make it easier and faster to download music. A considerable amount of music was illegally transferred as a result of these improvements. Third of the CDs are pirated by phonographic industries [3]. Internet criminals know these circumstances and use their own ever growing ability to achieve success in their criminal activities, as well as computer security and internet knowledge.

2. Online fraud

2.1 Financial crimes

The demand for online banking has grown, and financial crimes have become extremely worrisome. The exemplars are financial crimes such as credit card fraud, money stolen from online banks and other financial crimes. Fraudsters of credit cards often receive information from their victims by impersonating officials or persons from financial institutions to ask for information from their credit card. Without proper due diligence, victims fall victim to this, and provide the criminals with credit card information. This allows criminals to Rock your identity and most of the implications are financial. Fraud and financial crimes represent a form of robbery that takes when someone or organization takes or uses money or property unlawfully in order to gain a profit from it. All such crimes usually undergo some type of lies and deception, subterfuge or abuse of a trusting position that distinguishes between robbery and robbery. Fraud and financial crimes can take various forms in today's complex economy. The below resources introduce you to more common means of economic crimes like falsification, fraud on credit card payments, malpractice and money laundering.

2.2 Cyber terrorism

Cyber terrorism refers to cyber-space terrorist acts. Cyber terrorism can be as easy as transmitting information on the Internet at a particular moment about potential bomb attacks. Cyber terrorism is a personal, political or social gain by using computers and networks to threaten and compel a person, organization or even a government. Cyber terrorism could even impact or target almost everybody, so preparing is always a good idea. If you want to ensure that your organization's safety role is adequate to oppose ongoing and/or advanced attacks, Testing, detecting and mitigating vulnerabilities your security measures regularly. In addition, make sure which you regularly upgrade your tools, software and firewalls. On either hand, a few experts think that the term cyber attacks is not appropriate at global level, since this does not cause harm and follows a system similar to a traditional 'war,' violent cyber attack.

2.3 E-mail spoofing and phishing scams

Spoofing emails from both known and unknown persons is a common criminal tactic. The practice of e-mail spoofing is to send an e-mail from an apparently sent source. The common source of monetary loss is e-mail spoofing. Phishing is the attempt, as a trusted entity in an electronic company, to obtains essential information like passwords and credit card details. Phishing mails may contain links to websites which are infected with malware. The primary objective of spoofing is to make people think that the email belongs to somebody they can or know regarding – a coworker, a supplier or brand in most cases. When the attacker exploits this trust, he asks the receiver to divulge or take other measures. For instance, an attacker could create a PayPal-style email. Your message tells the user that if you don't visit a website, log in to the website or change ones login credentials, your account will be halted. When the user is deceived successfully and credentials are typed, the attacker now has credentials to authenticate into his PayPal account, which could rob the user of money.

2.4 Cyber pornography

In this category, you can download pornographic website, pornographic movies, videos, photos and magazines for pornography (photos, writings, and so on). "Computer pornography is a fresh horror," according to the report Porno gram Report Carnegie Mellon University in the US has also conducted extensive research on child and computer pornography and gathered evidence [4]. The results showed gender gap in cyber pornography exposed for both initiatives. The extent of their exposure was underestimated by women. Men have shown the opposite trend, interestingly. The relationships between gender and specific study bias were also mediated by perceived social acceptance. In comparison to past findings, we start debating the results of this study and the importance of subjective factors to study communally controversies including cyber pornography. The use of cyber pornography has become a common practice and has the capability to promote various results for the user. Studies indicate repetitively that, when questioned clearly, men are more revealed to cyber-pornography unlike women. Even so, such explicit reports can dramatically bias societal acquiescence of cyber media usage. The study was carried out to increase differences between men and women in cyber-pornography publicity through using explicit personality initiatives and a new quantify formed to adversely examine visibility to cyber-pornography.

3. Types of cyber crimes

3.1 Cyber terrorism

Cyber-terrorism refers to cyberspace terrorist acts. Cyber terrorism can be so simple as broadcasting information on the Internet at a specific point in the future concerning potential bomb attacks. Cyber terrorists are people who use their PCs and systems that threaten and force a person, organization or even a government to act for individual, social and political gains. It is not the first time in which cyber terrorism is used as a strategic susceptibility to a modern innovation. Although the match among cyber warfare concepts and air support is not clear, it is relevant to evaluate the four. In response to the First World War, the ability of enemies to wage war would be disrupted and vital infrastructure, far behind the front segments.

3.2 E-money laundry and tax evasion

Transfers of electronic funds have been helping to disguise and transport criminals for a long time. The origin of bad profits is greatly hidden by the new technologies of today. Tax bodies can disguise income derived legitimately. Supervision by central banks will circumvent the unofficial banking or concurrent payment systems development. A PC and a system committing this type of crime, there is no separate law but rather it is subject to the laws which apply in general to these crimes. In the occlusion and movements of the criminal proceeds, online fund transfers helped. Evolving techniques will enable greatly to hide the origin of discourse profits. Genuinely derived earnings may also be masked from tax authorities more easily. Investment banks will no sooner did be the only ones with the order to convey electronic funds at supersonic speeds through many jurisdictions. The development of informal financial firms and parallel central banks may enable central bank supervision to be circumvented, but may also enable the evasion of disclosure rules for money transfers in those countries which have it.

3.3 Theft of telecommunications services

Crimes as well as individuals may have full rights and access to a company's switchboard plug or switch circuits. Free phone calls to a local and long distance number are permitted. Telecommunications robbery is now a crime and one of the first kinds of cyber crime. The criminal has to pay a fine and has to be sent to gaol for a brief period. The extent of complexity and frequency of data thefts is increasing by just day with advancement of technology. Because of the large amount of data on the web, compliance and real-time audits are very difficult to track. Additionally, an automated analysis process will indeed large change packet safety checks for all internet data packets that could mean invasion on their information customers' privacy.

3.3.1 Telecommunications piracy

Digital technologies of the present day ensure that prints are perfectly replicated as well as that graphics, Graphical, audio and many other possibilities of digital media are distributed. For copyrighted material owners, this was a major concern. In cases where the creators of a specific work cannot take advantage of their own creations, this results in serious financial loss and generally has a major impact on creative work. Latest expected technological developments that emerge from the consolidation of computing and communication have already had an effects on many areas of life. Financial sector, stock markets, aviation, telephones, electric power, and a wide range of health, education and welfare institutions are highly dependent on IT and telecommunication services. We are travelling quickly to the limited extend that" all depends on software" can be claimed.

3.4 Fraud in transfer of electronic funds

There is a proliferation of electronic transmission systems as is the risk of intercepting or diversion of such transactions. There is no question of the rapid and widespread worldwide acceptance of the e-funds transfer system. The use of an electronic money transmission system will undoubtedly increase the risk. Hackers are capable of diverting funds that have been transferred over the Internet. Electronic transfer is very vulnerable to crimes like robbery and fraud. There is no doubt that technological progress in enterprise services has benefited to a greater degree both companies and consumers. There is a darker side to this technological development, however, because of cyber crime. Although organizations are aware of different cyber crime types, the capabilities of cyber criminals are extremely difficult to grasp [5]. In fact, organizations have difficulty grasping the next goal and the value of a cyber criminal. When the organizations realise that their targets are being targeted, the damage is too late and already done. They fall into the hands of cyber criminals despite efforts made by organizations to prevent such cyber crimes. The chart in Fig. 1 shows the overall sum of money damage from 2001 to 2013 caused by cyber crime. These data were sent to IC3. In 2001, the total losses totalled 17.8 million dollars. In 2009, the amount was some 559.7 thousand dollars, with almost certainly significant financial losses in 2013, amounting to 781.84 thousand dollars.

A well-known security firm, Symantec has performed a comprehensive and were able to learn identify the top high 20 countries that faced most cyber crime activities or that caused them. Symantec did find and identified the number phishing site when this list was compiled. These phishing sites were designed to prevent the end user from distinguishing the actual from the fake sites. The design has been created to mislead users in the disclosure of personal or bank account details [6].

During the subsequent Symantec has been able to examine obtain information including the Bot-infected devices number. Those Devices can be able to thoroughly monitored by cyber criminals. In US, the highest rate of cyber crime has been found. This could be due to the fact that the country is well-equipped with widespread internet connections. Fig. 1 demonstrates that countries have were sufferers of the cyber crimes, like malware sharing, spam messages, phishing, etc. The six factors are indicated in Fig. 1.

- Share of malicious computer activity,
- Malicious code rank,
- Rank of spam zombies,
- Hosts rank on the Phishing website,
- Rank of bot and
- Origin of attack

The United States of America has experienced significant cyber crime suffering according to the pie chart in Fig. 2. There have also been substantial losses in other countries. Today, China is in second place. The countries that were to suffer the losses caused by cyber crime came like Germany and the UK. Cyber crime is almost the same in countries such as Brazil and Spain. Likewise, in their respective countries, Italy, France, Turkey, and Poland share cyber crime. India represents 3% of all computer malicious activity. Argentina, Australia and Israel account for less than 1% of all malicious computing activities. Mexico is the world's first host to websites for phishing. The bot-activities in Australia are ranked highest.

4. Crimes on the Internet

Internet-based crimes are commonly referred to as crimes committed through Internet use. The word "cyber crime" refers to the occurrence of harmful activities on digital devices, mainly over the internet, according to David Wall. Cyber crime does not practically refer to the law, and is more widely established by the media. Computer crime usually refers to Offences like phishing, theft of banks, credit and debit card fraud, child abduction through Chat rooms, virus formation or transmission, etc. These are all made easier computer crimes. Web crimes are committed made known, while others remain hidden until someone or a firm is committed. Fake accounts can also be created or pattern sent in social networks to the other customers. If a Facebook user gives a false profile to embody a different person, that action may be unlawful even without permission of the other individual. Even then, a crime has been committed only if the false user had the objective to harm, fraud or intimidate somebody with the fake Facebook account or notifications.

4.1 Crimes associated with e-mail

An e-mail is becoming the universe popular mode of communication. Every day lots of e-mails are flowing around the world. Criminals also use e-mail for abuse, like any other kind of communication. Thanks to its ease, speed of transmission and relative anonymity, it has become a powerful criminal tool. E-mail also provides both conventional lawbreakers with a device. Although liberals praise use of such cryptography algorithms in communication systems, terrorists and criminals could use encryption methods to hide their plans. Authorities at law report that some terror organizations are incorporating pictures with commands and data via a procedure known stenography, an advanced technique of concealing information in the clear view. Even realising that something is so obscured almost always demands a lot of computational power; it is extremely difficult to decode information if someone doesn't have the key to disconnect the confidential message.

4.2 Spoofing of electronic mail

Electronic mail spoofing is defined as an e-mail which always appears to have come through one origin when sent from a different one. Spoofing of email usually takes place through the email address and/or name forging of the sender. The following information should typically be entered to send an e-mail. While most fake emails could indeed conveniently be identified and re mediated by actually delete the message, a few kinds could lead to serious issues and safety risks. For instance, a spoofed email can pretend to be from a popular online shop and request that the beneficiary offer confidential material such as a login credentials.

- The recipient's e-mail address.
- The recipient's email accounts (CC means Carbon copy).
- The e - mail addresses of recipients a copy.
- Subject of a text which is a brief description or message summary.

4.2.1 Bombing of e-mail

The result of a significant number of e-mails a perpetrator receives is known as an e-mail attack. It could be quickly increased by utilizing a variety of mailing lists to the victim's email address, groups of special interest formed to email each others and share information and information on a common subject. Mailing lists can generate enough e-mail traffic every day. The list varies depending. When an individual unwittingly adheres to many email list, his e-mail transportation has become too much and the service provider could indeed delete his account. E-mail bombing frequently occurs from an unified system where a user sends millions or thousands of emails to some other user. The email bomber can use a script to immediately send the messages rapidly. You can send several thousand messages a minute by e-mailing with a script. This results in a painful situation wherein the user would have to remove messages manual process. If the user is unable to select all the unsolicited messages at once by the receiver's e-mail client or web - based email system, it might be a lengthy process to complete.

4.2.2 E-mail defamation

For anyone with basic computer knowledge, cyber-distraction or cyber-slander can be extremely dangerous and fatal because blackmailers often threaten their victim via email. If an individual publishes a defamation declaration against any other person on a web page or sends a email to a person to whom the declaration is made, it is a cyber defamation. Defamation protects the reputation of a person from an unwarranted attack. Defamation generally involves one party making a false allegation or statement to another party that damages the reputation of another party. Cyber defamation is a new phenomenon; however, the actual definition of information in relation constitutes injury to the credibility of a 3rd party and can be completed via words or pictures or signs and clear interpretations. The declaration should relate to the claimant and aim at reducing the credibility of the individual the declaration was made to. Cyber defamation, on either side, includes defaming an individual by an innovative and much more efficient way, such as the use of modern electronic instruments.

4.2.3 Malicious code spreading
E-mails are often the most frequent and quickest way of spreading malicious codes. A love bug virus spread to thousands in the Philippines 30 months after it was published via e-mail. Trojans, malware, worms, and others are often bundled with e-cards and sent to suspects by e-mail. Backdoor attacks, scripting attacks, worms, Trojan horse and spyware are malicious examples. Every type of code attack can very quickly cause problems on a helpless IT facilities or wait for a default time on servers or trigger to activate the attack. Industry studies have shown that it often takes weeks or months before losses are noticed and attacks are defeated to detect malicious code. Malicious code can expose the systems, sensitivity and valuable information of an institution by altering, damaging or committing fraud, obtaining or permitting unauthorized access to data, and project requires a customer never expected. The best solution for these kinds of cyber threats is self-protection software.

4.2.4 E-mail-sent threats
Human beings believe e-mails are a useful tool for relatively anonymous technological criminals. Anyone who knows How and when to e-mail anyone without identifying him can easily challenge. Social engineering is a method of investigation and manipulative that is the basis of spam emails, phishing and spear identity theft. The use of malware may also be involved. Spam is one of the threats on our mailing list because everyone with an email account is old acquainted. Spam is a message unwanted that usually promotes products and services to "must view." The remedy to email fraud

detection is based on two factors in combination: human and technical. The human features is the fact to recognise threats and risks for our self, ones coworkers or even your staff. For instance, you must note that the sending address of email is one simple note and must not click on the suspected links.

4.2.5 E-mail frauds

The common method of committing financial offences is e-mail spoofing. It's getting easier to take on the identity of another person and to hide your own identity. It is well known that the criminal is slightly suspected. The recipient is generally aimed at an official site that may look real, but not real. Bank details, user names and login information and login credentials can be included in this information. In the hands of scam artists, everything is dangerous. Phishing emails are best tried to defend by understanding the threat. Financial firms never request this sort of information by e-mail, and many banking institutions have set up safe messaging platforms that will require users to log on. The finest form of e-mail fraud is probably" phishing," a term that refers to messages that request sensitive data or attempt to obtain users to open accesses which configure malicious software.

4.3 Who are cyber criminals?

Cyber criminals have a variety of groups into children, hacker group, employees, and spammer.

4.3.1 Children

While it can be difficult to trust, children could either knowingly or unconsciously be cyber criminals. The most inexperienced hackers are adolescents. It seems as if you're a source of pride for these teens to hack into a electronic device. You can break laws without realising it's illegal what you do. Kids play games, chat, make groups, make video conferences, make tik ticks and do a number of public communication. A few of those, such as kids who willingly or without knowledge misuse technology for unlawful purposes and emotions, misbehaviour, harassment, threats, molesting, damage other people via the internet with mobile devices, tablets, machines, pen drives and other.

4.3.2 Hacker groups

Hacktivists are hacker groups that work together to achieve a certain objective. Mostly political in nature, these organizations. In other cases, social activism, religious activism or anything else could be their motivation. Many of the most difficult crimes to assess is the scale of hacking because sufferers frequently choose not to disclose ones crimes — mostly out of confusion or fear of additional violation of safety. However, researchers say that hacking costs huge sums of money to the global economy. Hacking is not really an external job: it includes activities in companies or government agencies who alter their database documents intentionally for financial gains or political purposes. The biggest loss is due to the stolen information from the owner, and the money extracted by the previous purchaser for the data return is occasionally followed. Hacking by other means is in this sense old-fashioned industrial spy.

4.3.3 Employees

How venomous disgruntled employees can become is hard to imagine. Until now, unhappy employees were able to strike their employers. However, as the reliance on computers and process automation is increased, unfortunate employees can harm their employers much more if they commit offences that can bring down the entire system through computers. Business process outsourcing is the most vulnerable sector of the business process, which creates wealth for a large number of trained Indians and also causes a substantial entry of foreign exchange into the nation. One of it's increasing concern among all these data analysis firms is their vulnerability to cyber threats within the company when employees misuse their personal details at work. Since employees in the organization know the information systems and the safety measures implemented to prevent cyber crime.

4.3.4 Experienced spammers

The electronic storage of information by corporate organizations has resulted in extensive computerisation. Hackers are hired for other industrial information and secrets which could help them by rival organizations. If hacking may obtain the information from rival companies it is considered unnecessary to be physically present to gain access. This also increases the temptation for companies to hire hackers to perform their dirty work [7]. Although spam messages are generally sent out to single email ids, it is also not unusual to send sms emails to cell devices. The Internet is still the favourite mode for hackers, however. Thus, the first process to a hacker is to collect email ids for spamming. No cyber criminal could even have sent spam mails to huge numbers of people without finishing this fundamental task. Such attackers as well start making sure they remain completely undetectable. Different ways to collect these e-mails are done.

5. Case studies

Recently, in India, there have been several instances of cyber crime. Some of the following are described [8].

5.1 Case I

Two students were arrested in April 2019 in Allahabad, India for online fraud, namely Kishan Dubey, Vivek Kumar and Anand Mishra, alias. Kishan Dubey [9]. These two students received a person called the ATM password Mahmood and committed Rs. 1.5 lakhs INR credit card fraud. The students have been apprehended and admitted to several disappointments. According to police, both students does provide Mumbai addresses for both the products they bought online. They would sell the product to unsuspecting consumers after they have been delivered. This petition was brought in accordance with Articles 419, 420 and 66 of the IPC law.

5.2 Case II

In the context of cyber-stalking, Bangalorean youth, Kumar alias Kiran, were detained by the Police in 2020 following the charges Creating a false female Facebook profile against him. The said young woman also harms the female when she rejects her romantic proposal. He created a bogus profile, dejected, and in her name started chatting with others. He called out and wrote threats to the family of the victim. On 20 August 2020, the victim's brother brought a complaint against the Cyber Crime Police.

5.3 Case III

This case is processed by email. A recent spoof was difficult in one of the branches of the World Bank. Many customers have quickly opted for their money and bank accounts to be withdrawn. After an inquiry, It was found that the person decided to send all that and many others manipulated E-mails saying the financial institution would soon be closed because of financial problems [10].

5.4 Case IV

This case involves an e-mail bombing (DoS); an alien who has been living nearly three decades in Simla, India. The Simla House Board was designed to allow him to purchase land at a lower price. His request was refused, however, since the scheme was only open to Indian citizens. This man chose to take revenge. Afterward, he sent a thousand e-mails and e-mails to the Simla Housing Board until the servers collapsed.

6. Experimental evaluation

6.1 Research method

The method of investigating the survey was chosen because the prejudice was decreased due to randomly assigned subjects to treatments and the identity of treatments was ignorant for individuals and researchers. A questionnaire and an interview were used to collect information. Based on literature and related studies, the researcher created the questionnaire. In order to study different types and consequences of cyber crimes, surveys have been used for collecting the information from 2 youth educational institutions, 15 musicians, 10 actors and 5 companies. Data were collected from every school from 100 students. The sample size of the study consists of 250 persons, with 5 managers each. The response was 200 out of 250 people, representing 80%. Due to the device previously examined, the survey participants have been limited but insightful and exact.

6.2 Procedure

After the questionnaire was administered to all five overshadowed, the researcher gathered responses from participants via email and interview and finalised the surveys. The results of all good reports ranging from 5 to 1 for each response category were recorded using a five point rating scale. The four choices are strongly agree (SA), Agree (A), Undecided (U) and Disagreement (DA) (SDA). The data were analysed by percentages.

6.3 Applications of cyber crimes theory

This study uses the theory of routine activity (RAT) because child exploitation is one of the cyber crimes. The theory focused on "criminal opportunities" for the environment. The action takes place essentially between a motivated offender and an appropriate victim of victimization when a potential criminal opportunity arises. Eventually, this criminal activity will occur where a 'suitable target" which is either a vulnerable person or an uncontrolled property is not protected by a suitable guardian [11]. The lack of any such three factors in situations should therefore make it theoretically impossible to commission a crime. This is why the theory of routine activities is seen as a theory of macroeconomics for a wide variety of crimes as it seeks to explain crime victimisation rather than the specific motives behind crime [12]. The crime happens in the nonappearance of an able guardian who may the victim, when the motivated offender interacts with the appropriate target, help stop the theory from committing a crime. The provision of appropriate objectives and worthy guards can understand different criminal rates as well as the hypothesis of the role of motivated perpetrators, as we understand, is rather agnostic [13].

6.4 Findings

The results of the data analysis are shown in Tables 2–4.

Table 2 shows that most respondents (95% age) believe that pornographic (child-exploitation) cyber crime affects children, while just a few (5%) disagree. A large majority (90%) said cyber crime in society is a source

Table 2 Impacts of cyber crime on the society.

S. No.	Impacts	Response	Level of agreement				
			SA	A	U	DA	SDA
1	Child exploitation	N	120	70	0	10	0
		%	60	35	0	5	0
2	Harassment	N	100	80	10	0	10
		%	50	40	5	0	5
3	Digital piracy	N	70	120	10	0	0
		%	35	60	5	0	0
4	Hacking	N	80	80	10	20	10
		%	40	40	5	10	5
5	Intentional damage	N	60	80	40	10	10
		%	30	40	20	5	5
6	Spam	N	110	60	10	20	0
		%	55	30	5	10	0

Table 3 Why computers are tools target for cyber crimes.

S. No.	Impacts	Response	Level of agreement				
			SA	A	U	DA	SDA
1	Availability	N	100	80	10	10	0
		%	50	40	5	5	0
2	Easy access	N	70	100	10	10	10
		%	35	50	5	5	5
3	Affordable	N	60	100	20	10	10
		%	30	50	10	5	5

Table 4 Factors that contributing to cyber crimes.

S. No.	Factors	Response	Level of agreement				
			SA	A	U	DA	SDA
1	Growth of the technology	N	90	70	0	20	20
		%	45	35	0	10	10
2	Economic factor	N	100	60	20	10	10
		%	50	30	10	5	5

of harassment. A individual can misuse others with a computer or a mobile device via a brief message or a social network, like Facebook. This is a problem. A small percentage (5%) denied the social impact of the crime and the rest (5%) did not know about the crime [14]. The results indicate which most (95%), while the remaining (5%) of the participants are not sure, replying positively to the cyber crime. Digital piracy is another effect on social structure. Likewise, an overwhelming majority of respondents (80%) agreed that cyber crime threatens society. Most respondents, either personally or through colleagues, have been affected. 70% of deliberate damage, one of the social effects of cyber crime, were also re-portable. Similarly, a considerable proportion of respondents (85%) agreed that spam mail, with only 10% of disagreement, is another effect of cyber-crime [15]. Why are computers or networks targeting or placing cyber crimes tools?

Table 3 shows that 90% of respondents say that it is because there are now computers everywhere available. Computers connected to the internet are easily accessible according to a significant proportion of respondents (85%). Sometimes, thanks to so-called "bundles" from various mobile telecom companies, it's easy to access the Internet by mobile telephone

[16]. Likewise, an overwhelming majority (80%) agreed that computers and mobile phones now have affordable access to the internet for little money. What are the factors contributing to cyber crimes?

Table 4 shows that, according to the majority of respondents (80%), technological progress is a contributing factor in the existence of cyber crime. Such factors also play a role in the economy, a substantial majority (80%) of respondents agreed.

7. Conclusion

The results of this survey show that a person can commit crimes in cyberspace in numerous ways and means. Cyber crime is a law-punishing crime. The expanding areas of cyber crime were discussed shortly. We have seen the most common cyber crimes and the most common areas in which cyber crimes occur. We have also spoken about the effects of cyber crime, which in many countries, especially sales and investment, lead to massive financial losses. Different fines and sanctions were imposed for this type of crime [17]. These crimes include mail spoofing, email bombing and e-mail distribution. We've also seen many cyber criminals ranging from inexperienced teenagers to hackers often hired by rival organizations to hack other businesses. It is therefore important for everyone to be aware and careful to avoid losses in these crimes. The judiciary has adopted certain laws victims of cyber law to provide access to justice and prosecute lawbreakers. Therefore, Cyber law is important to be aware of these laws for each individual. Moreover, cyber crime is not just a problem of innovation [18]. It is instead a concern of strategy, as machines do not harm as well as attack organizations, but people who use technology to do so. As a result, we must look for the different approaches that can be followed by such criminals. Intellectual thinking is necessary to detect situations which could lead to such damage. It cannot be based on technology alone to resolve such crimes. As a means of tracking and to some extent stopping these activities, the technologies can be utilised [19].

References

[1] K. Kalning, Game piracy runs rampant on the internet, NBC News 25 (2007) 2014. Retrieved July.
[2] E. Ramdinmawii, S. Ghisingh, U.M. Sharma, A study on the cyber-crime and cyber criminals: a global problem, Int. J. Web Technol. 3 (2014) 172–179.
[3] G. P. Black, K. R. Hawk, Computer and internet crimes, San Francisco, California [Online] Available from http://www.fd.org/pdf_lib/WS2010/WS2010_Computer_Crimes.pdf [March 29, 2011].

[4] B.B. Chatterjee, Last of the rainmacs: thinking about pornography in cyberspace, in: D. Wall (Ed.), Crime and the Internet, Routledge, 2001, pp. 74–99. see ncj-213504.

[5] E. Ramdinmawii, S. Ghisingh, U.M. Sharma, A study on the cyber-crime and cyber criminals: a global problem, Int. J. Web Technol. 3 (2014) 172–179.

[6] D.S. Wall, The internet as a conduit for criminal activity, in: A. Pattavina (Ed.), Information Technology and the Criminal Justice System, 2015, pp. 77–98.

[7] N.E. Marion, The Council of Europe's Cyber Crime Treaty: an exercise in symbolic legislation, Int. J. Cyber Criminol. 4 (1/2) (2010) 699.

[8] J.A. Mshana, Cybercrime: an empirical study of its impact in the society—a case study of Tanzania, Huria, J. Open Univ. Tanzania 19 (1) (2015) 72–87.

[9] S. Kharat, Cyber crime—a threat to persons, property, government and societies, Property, Government and Societies (March 1) (2017) 1–14.

[10] S. Srivastava, Pessimistic side of information & communication technology: cyber bullying & legislature laws, Int. J. Adv. Comput. Sci. Technol. 1 (1).

[11] P. Mali, Types of cyber crimes & cyber law in India, CSI Commun. 35 (8) (2011) 33–34.

[12] S. Petre, et al., Cyberwarfare as factor in nation-building and un-building. The case of Assam riots, Rev. Română Stud. Eurasiatice 9 (1 + 2) (2013) 261–288.

[13] A. Nagorski, Global Cyber Deterrence: Views from China, the US, Russia, India, and Norway, EastWest Institute, 2010.

[14] E. Dalla, M. Geeta, Cyber crime a threat to persons, property, government and societies, Int. J. Adv. Res. Comput. Sci. Softw. Eng. 3 (5) (2013) 997–1002.

[15] N. Gopal, V. Maweni, Cybercrime preparedness—a critical snapshot of Brics countries, Africa Insight 49 (2) (2019) 70–88.

[16] P. Grabosky, R. Smith, Telecommunication fraud in the digital age: the convergence of technologies, in: D.S. Wall (Ed.), From Crime and the Internet, Routledge, 2001, pp. 29–43. see ncj-213504.

[17] J. Iqbal, B.M. Beigh, Cybercrime in India: trends and challenges, Int. J. Innov. Advance. Comput. Sci. 6 (12) (2017) 187–196.

[18] B. Hu, T. McInish, L. Zeng, Gambling in penny stocks: the case of stock spam e-mails, Int. J. Cyber Criminol. 4 (1/2) (2010) 610.

[19] M. Gulati, Digital India: challenges and opportunities, Int. J. Manage. Inform. Technol. Eng. 4 (10) (2016) 1–4.

About the authors

Mr. V. S. Nageswara Rao Kadiyala is pursuing his Ph.D. degree at the Department of Computer Science & Engineering, National Institute of Technology Silchar, Assam, India. His research interest spans from Machine Learning, Big Graph, Security, and Big Data.

Dr. Ripon Patgiri has received his Bachelor Degree from Institution of Electronics and Telecommunication Engineers, New Delhi in 2009. He has received his M.Tech. degree from Indian Institute of Technology Guwahati in 2012. He has received his Doctor of Philosophy from National Institute of Technology Silchar in 2019. After M.Tech. degree, he has joined as Assistant Professor at the Department of Computer Science & Engineering, National Institute of Technology Silchar in 2013. He has published numerous papers in reputed journals, conferences, and books. His research interests include distributed systems, file systems, Hadoop and MapReduce, big data, bloom filter, storage systems, and data-intensive computing. He is a senior member of IEEE. He is a member of ACM and EAI. He is a lifetime member of ACCS, India. Also, he is an associate member of IETE. He was General Chair of 6th International Conference on Advanced Computing, Networking, and Informatics (ICACNI 2018, http://www.icacni.com) and International Conference on Big Data, Machine Learning and Applications (BigDML 2019, http://bigdml.nits.ac.in). He is Organizing Chair of 25th International Symposium on Frontiers of Research in Speech and Music (*FRSM 2020, http://frsm2020.nits.ac.in*) and International Conference on Modelling, Simulations and Applications (CoMSO 2020, http://comso.nits. ac.in). He is convenor, Organizing Chair and Program Chair of 26th annual International Conference on Advanced Computing and Communications (ADCOM 2020). He is guest editor in special issue "Big Data: Exascale computation and beyond" of EAI Endorsed Transactions on Scalable Information Systems. He is also an editor in a multi-authored book, title *"Health Informatics: A Computational Perspective in Healthcare"*, in the book series of *Studies in Computational Intelligence*, Springer. Also, he is writing a monograph book, titled *"Bloom Filter: A Data Structure for Computer Networking, Big Data, Cloud Computing, Internet of Things, Bioinformatics and Beyond"*, Elsevier

Printed in the United States
by Baker & Taylor Publisher Services